AF829270

Wideband Continuous-time ΣΔ ADCs, Automotive Electronics, and Power Management

Andrea Baschirotto • Pieter Harpe
Kofi A.A. Makinwa
Editors

Wideband Continuous-time $\Sigma\Delta$ ADCs, Automotive Electronics, and Power Management

Advances in Analog Circuit Design 2016

Springer

Editors
Andrea Baschirotto
Department of Physics "G. Occhialini"
University of Milan
Milano, MI, Italy

Kofi A.A. Makinwa
Electronic Instrumentation Laboratory
Delft University of Technology
Delft, Zuid-Holland
The Netherlands

Pieter Harpe
Department of Electrical Engineering
Eindhoven University of Technology
Eindhoven, Noord-Brabant
The Netherlands

ISBN 978-3-319-41669-4 ISBN 978-3-319-41670-0 (eBook)
DOI 10.1007/978-3-319-41670-0

Library of Congress Control Number: 2016947073

© Springer International Publishing Switzerland 2017
This work is subject to copyright. All rights are reserved by the Publisher, whether the whole or part of the material is concerned, specifically the rights of translation, reprinting, reuse of illustrations, recitation, broadcasting, reproduction on microfilms or in any other physical way, and transmission or information storage and retrieval, electronic adaptation, computer software, or by similar or dissimilar methodology now known or hereafter developed.
The use of general descriptive names, registered names, trademarks, service marks, etc. in this publication does not imply, even in the absence of a specific statement, that such names are exempt from the relevant protective laws and regulations and therefore free for general use.
The publisher, the authors and the editors are safe to assume that the advice and information in this book are believed to be true and accurate at the date of publication. Neither the publisher nor the authors or the editors give a warranty, express or implied, with respect to the material contained herein or for any errors or omissions that may have been made.

Printed on acid-free paper

This Springer imprint is published by Springer Nature
The registered company is Springer International Publishing AG Switzerland

Preface

This book is part of the Analog Circuit Design series and contains the contributions from all 18 speakers of the 25th Workshop on Advances in Analog Circuit Design (AACD). This year, the sponsors of the workshop were Infineon (main sponsor), Joanneum Research Center, INTEL, and IEEE Solid-State Circuits Society Austrian and Italian Chapters. The workshop was held at the Infineon site in Villach (Austria), from April 26 to 28, 2016. The book comprises three parts, covering topics in advanced analog and mixed-signal circuit design that we consider to be of great interest to the circuit design community:

- Continuous-Time Sigma-Delta Modulators for Transceivers
- Automotive Electronics
- Power Management

Each part consists of six chapters written by experts in the field. The aim of the AACD workshop is to bring together a group of expert designers to discuss new developments and future options. Each workshop is then followed by the publication of a book by Springer as part of their successful series on Analog Circuit Design. This book is the 25th in this series (a full list of the previous topics can be found on the following page). The series can be seen as a reference for all people involved in analog and mixed-signal design. We are confident that this book, like its predecessors, will prove to be a valuable contribution to our analog and mixed-signal circuit design community.

Milano, Italy — Andrea Baschirotto
Eindhoven, The Netherlands — Pieter Harpe
Delft, The Netherlands — Kofi A.A. Makinwa

The Topics Covered Before in This Series

2015 Neuchatel (Switzerland)	Efficient Sensor Interfaces Advanced Amplifiers Low Power RF Systems
2014 Lisbon (Portugal)	High-Performance AD and DA Converters IC Design in Scaled Technologies Time-Domain Signal Processing
2013 Grenoble (France)	Frequency References Power Management for SoC Smart Wireless Interfaces
2012 Valkenburg (The Netherlands)	Nyquist A/D Converters Capacitive Sensor Interfaces Beyond Analog Circuit Design
2011 Leuven (Belgium)	Low-Voltage Low-Power Data Converters Short-Range Wireless Front-Ends Power Management and DC-DC
2010 Graz (Austria)	Robust Design Sigma Delta Converters RFID
2009 Lund (Sweden)	Smart Data Converters Filters on Chip Multimode Transmitters
2008 Pavia (Italy)	High-Speed Clock and Data Recovery High-Performance Amplifiers Power Management
2007 Oostende (Belgium)	Sensors, Actuators and Power Drivers for the Automotive and Industrial Environment Integrated PAs from Wireline to RF Very High Frequency Front Ends

2006 Maastricht (The Netherlands)	High-Speed AD Converters Automotive Electronics: EMC Issues Ultra Low Power Wireless
2005 Limerick (Ireland) RF Circuits	Wide Band, Front-Ends, DACs Design Methodology and Verification of RF and Mixed-Signal Systems Low Power and Low Voltage
2004 Montreux (Switzerland)	Sensor and Actuator Interface Electronics Integrated High-Voltage Electronics and Power Management Low-Power and High-Resolution ADCs
2003 Graz (Austria)	Fractional-N Synthesizers Design for Robustness Line and Bus Drivers
2002 Spa (Belgium)	Structured Mixed-Mode Design Multi-bit Sigma-Delta Converters Short-Range RF Circuits
2001 Noordwijk (The Netherlands)	Scalable Analog Circuits High-Speed D/A Converters RF Power Amplifiers
2000 Munich (Germany)	High-Speed A/D Converters Mixed-Signal Design PLLs and Synthesizers
1999 Nice (France)	XDSL and Other Communication Systems RF-MOST Models and Behavioural Modelling Integrated Filters and Oscillators
1998 Copenhagen (Denmark)	1-V Electronics Mixed-Mode Systems LNAs and RF Power Amps for Telecom
1997 Como (Italy)	RF A/D Converters Sensor and Actuator Interfaces Low-Noise Oscillators, PLLs and Synthesizers
1996 Lausanne (Switzerland)	RF CMOS Circuit Design Bandpass Sigma Delta and Other Data Converters Translinear Circuits
1995 Villach (Austria)	Low-Noise/Power/Voltage Mixed-Mode with CAD Tools Voltage, Current and Time References

1994 Eindhoven (The Netherlands)	Low-Power Low-Voltage Integrated Filters Smart Power
1993 Leuven (Belgium)	Mixed-Mode A/D Design Sensor Interfaces Communication Circuits
1992 Scheveningen (The Netherlands)	OpAmps ADCs Analog CAD

Contents

Part I
Continuous-Time $\Sigma\Delta$ Modulators for Transceivers

The first part of this book is dedicated to recent developments in wideband delta-sigma ($\Sigma\Delta$) ADCs for use in transceivers. These have been driven by the spread of mobile devices, most of which must communicate wirelessly with the outside world. To enable wideband and filter-less radio front-ends, continuous-time (CT) $\Sigma\Delta$ modulators with high dynamic and very high linearity are required. The first two chapters begin by discussing various $\Sigma\Delta$ modulator architectures, ranging from the traditional single loop architecture to more recent MASH structures. The next two chapters present various ways of modifying the modulator's signal transfer function (STF) so that it also suppress the out-of-band interferers or blockers that may also be present. The final two chapters discuss the considerations associated with the design of receiver chains for WIFI, and for use in hand-held mobile applications.

The first chapter, by Patrick Torta, Antonio Di Giandomenico, Lukas Dörrer and Jose Luis Ceballos, compares three different WiFi receiver chains based on a high dynamic-range CT $\Sigma\Delta$ ADC. Two implement an active filter stage between the mixer and the ADC: one uses a trans-impedance amplifier and the other one a Gm stage. A third obviates the need for an active filtering stage and replaces it with a passive pole. Measurements show that all three chains achieve roughly the same chain performance, but provide different tradeoffs in terms of blocker rejection, area and power consumption.

The second chapter, by Lucien Breems and Muhammed Bolatkale, provides an introduction to the design of high-resolution wideband $\Sigma\Delta$ modulators, from architectural choices to filter implementation and circuit design. It concludes with a discussion of some recent continuous-time $\Sigma\Delta$ modulator designs that have pushed the envelope in terms of their bandwidth and linearity.

The third chapter, by Rudolf Ritter, Jiazuo Chi and Maurits Ortmanns, also discusses various ways of attenuating interferers within the loop filter of a $\Sigma\Delta$ modulator. It is shown that hybrid combinations of feed-forward and feedback architectures can be made to exhibit enhanced filtering characteristics. Alternatively, a mixed-signal interferer suppression loop can be applied around a modulator. Here, a digital filter extracts out-of-band signals from the modulator's digital output and then cancels them by feeding them back to the modulator's input via a DAC.

The fourth chapter, by Shanthi Pavan and Radha Rajan, explores the possibility of using the built-in filtering of a continuous-time ΣΔ modulator to attenuate out-of-band interferers. A design example is presented that demonstrates that such filtering is indeed quite possible, thus saving the power required by a dedicated filter, while also improving out-of-band linearity and reducing active area.

The fifth paper, by Sebastian Loeda, describes the challenges and design techniques used to exploit the well-known energy-efficiency of FF architectures in the context of a modern consumer radio. Techniques are described to cope with the resulting out-of-band peaking and to make the modulator robust to clock jitter and to out-of-band blockers.

The sixth chapter, by Hajime Shibata, Yunzhi Dong, Wenhua Yang, and Richard Schreier, contrasts the use of conventional single-loop with MASH ΣΔ modulator architectures in the context of wideband wireless applications subject to out-of-band blockers. The performance of two wideband MASH implementations in a 28 nm CMOS process are compared and discussed.

Chapter 1
WiFi Receiver Evolution in a Dense Blocker Environment

Patrick Torta, Antonio Di Giandomenico, Lukas Dörrer, and Jose Luis Ceballos

1.1 Introduction

The most demanding scenario that a receiver chain must sustain is a dense blocker environment where many contiguous receiving channels are allocated and simultaneously in use by many users. In such a scenario the signal to be converted in the band of interest can be transmitted from a far station and can therefore be very weak compared to the transmitted signal of a near user. The WiFi transceiver can be embedded into a platform which serves also other standards like cellular, GNSS, BT or FM radio in a co-running mode. As the isolation of the antenna is limited to approximately 10–12 dB, it is important to ensure that other signals do not degrade the wanted signal due to alias, folding or distortion. The analysis of all possible disturber combinations is a difficult task, in particular when the A/D clock frequency is low and its multiples can generate intermodulation products and folding effects together with the receiver chain mixer clock. Figure 1.1 shows a baseband spectrum scenario and a receiver chain composed by many consecutive filter-and-gain stages followed by a medium resolution ADC. The ADC is clocked at twice of the Nyquist rate of the wanted signal bandwidth. A very high chain gain is needed to provide sufficient signal to noise and distortion ratio (SNDR) at the ADC input: this sets stringent constraints to all the baseband blocks. In particular the chain strongly amplifies the non-idealities of the blocks immediately following the mixer. The frequency behavior of the analog active components can be strongly affected by poorly controllable parasitics and even by the configuration settings for the block bandwidth and the gain. For instance, the DC offset of the mixer as well as the I-Q skew of the direct conversion receiver need to be carefully controlled and

P. Torta (✉) • A. Di Giandomenico • L. Dörrer • J.L. Ceballos
Intel Austria GmbH, Villach, Austria
e-mail: patrick.torta@intel.com

© Springer International Publishing Switzerland 2017

A. Baschirotto et al. (eds.), *Wideband Continuous-time ΣΔ ADCs, Automotive Electronics, and Power Management*, DOI 10.1007/978-3-319-41670-0_1

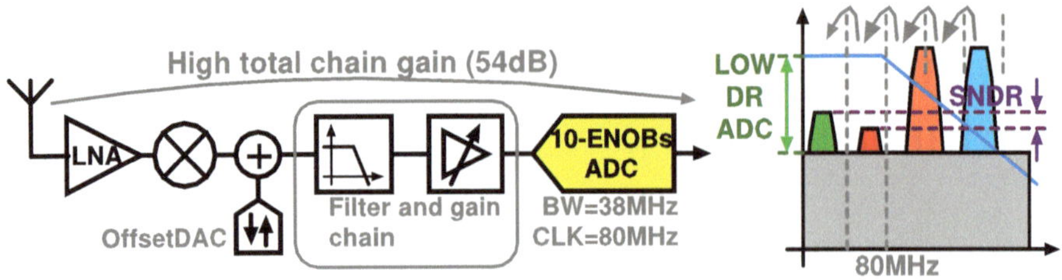

Fig. 1.1 A filter and gain receiver chain and the spectrum at the ADC input

mapped to all chain configurations to guarantee high chain resolution. A low noise DC offset cancellation may be required. This leads to complex gain and calibration schemes.

When an oversampled high resolution ADC is used, the chain can keep the required inband SNDR and, at the same time, absorb the power of the out of band signals employing a lower chain gain. This relaxes some analog blocks requirements and the amount of digital calibrations needed. The receiver lineups presented in Fig. 1.2 are minimal receiver architectures implementing high dynamic range (DR) continuous time sigma delta ADC (SD-ADC). The upper one implements only one active filter stage with trans-impedance amplifier (TIA) between the passive mixer and the SD-ADC. This structure targets to minimize area and power consumption compared to the filter-and-gain chain and is less exposed to the circuit imperfections of the blocks interconnections. It has a RF-to-baseband interface that can be easily simulated with common known techniques. A further step to reduce the area and the power consumption avoids the instantiation of the TIA stage, as shown in Fig. 1.2 (middle), and replaces the active stage by a small passive attenuation stage and a lower pole in the first stage. A high order SD-ADC is a chaotic system and it is in general not possible to use directly harmonic balance to simulate this chain performance and a linearized model can be used instead. Both low gain proposed chains have in common that the first active stage after the mixer suffers of noise gain due to the equivalent small impedance in the RF chain. The third presented lineup in Figure (bottom) uses a passive mixer in voltage mode followed by a Gm stage having a high output impedance interface to the SD-ADC.

1.2 The Receiver Chain as a Full Featured and Fair ADC Verification Environment

All presented receiver architectures are able to convert signals present in the two WiFi RF bands. The signal at the antenna is amplified by a cascaded differential low noise amplifier (LNA) and down converted to zero IF. A passive switch in

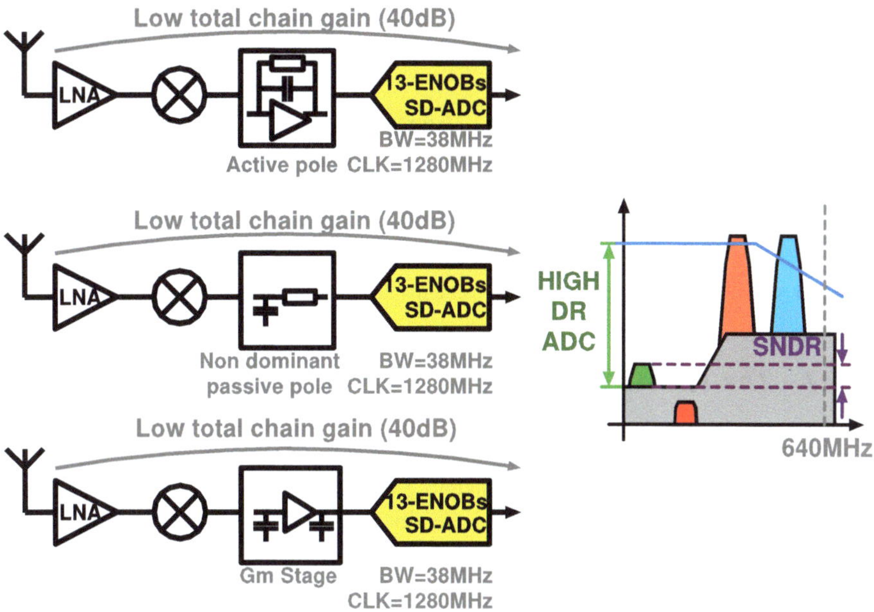

Fig. 1.2 The three minimal gain, high dynamic range lineups presented in this work and the typical PSD plot scenario in baseband. The out of band signals does not saturate the chain. Less folding in band is achieved due to the high clk frequency

series with the switching quad mixer selects the band of interest and the signal is routed to the I and Q baseband paths. The WiFi standard defines the possibility to allocate for each user a channel containing a different number of carriers. As a result all proposed lineups are able to process a baseband signals with bandwidth of 8.9, 18.3 and 38.3 MHz. At the output of the ADC a chain of decimators and digital filters processes the converted signal. The implemented analog gain control adapts the LNA gain proportionally to the signal strength detected at the ADC output. In the digital domain it is also possible to collect the undecimated SD-ADC data stream stored into an internal memory. Asides of the active pole there is a significant difference that distinguish the systems: the ADC clock can be derived from the mixer local oscillator (LO) or from a separated PLL. A LO derived clock is particularly helpful when the isolation of an active block is missing. In this case a variable rate converter needs to be added at the end of the decimation chain to interpolate the decimated samples to the fixed frequency of the system. The variable rate converter can be skipped if a multiple of the system clock is used for the ADC. The three chips have a compatible pinout; in particular the programming of the devices, the routing of the RF signals to the LNA and of the supply lines are the same and thus the same motherboard is used to measure and compare them (Fig. 1.3).

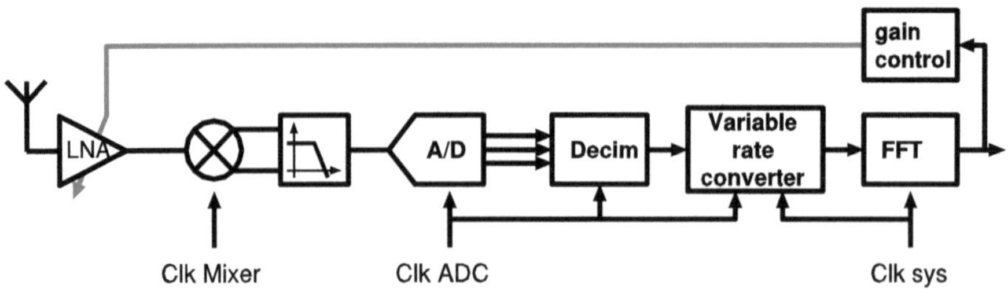

Fig. 1.3 Clock sources in the full receiver including decimators and AGC

1.2.1 The TIA-ADC Lineup

The functional block of the receiver is presented in Fig. 1.4. In this SD-ADC, as well as in the other presented SD-ADC, the chosen continuous time filter is implemented as a third-order chain of integrators in feedback. An advantage of this structure is that the resistor between TIA and ADC can be programmed to achieve an additional gain and DR boost of few dBs at the cost of added gain complexity. A block-level noise budget is used to size and design the blocks and it's calculated by replacing the antenna, the LNA and the mixer with a greatly simplified model as shown in Fig. 1.5. The model replaces the antenna by an ideal voltage source driving a noiseless voltage gain element of value G connected to a noiseless resistor R_{mix}, which represents the mixer's output impedance. This resistor value is affected by the chosen RF frontend architecture and in our design has a value of few hundred Ohms. As a further simplification the noise of all passive elements in the baseband is neglected, since the noise of the amplifiers is the one that dominates the total noise and, as a consequence, the system area and power consumption. To fairly compare the lineups we assume that both G and R_{mix} are fixed in all systems. The transfer function (TF) of the TIA and the one of the ADC are flat in the band of interest. Therefore in the noise budget model we can skip the TIA capacitor and the high frequency suppression out of band of the ADC STF. The complete ADC is thus reduced to a simple inverting amplifier: the feedback resistor R_{dac} is the equivalent resistor of the first current steering DAC, which together with the resistor R_2 sets the ADC gain. N_{op1} and N_{op2} mimic the input referred noise of the OPAMPs and N_q replaces the ADC quantization noise.

We can use the superposition principle to calculate the input referred noise of the system under the assumption that the gain of the OPAMPs is very high. The system gain from input X to output Y is given by:

$$\frac{Y}{X} = \frac{R_{dac}R_1G}{R_{mix}R_2}$$

By calculating the TF of each noise source to the output and dividing them by the system gain we obtain the input referred noise contribution equation:

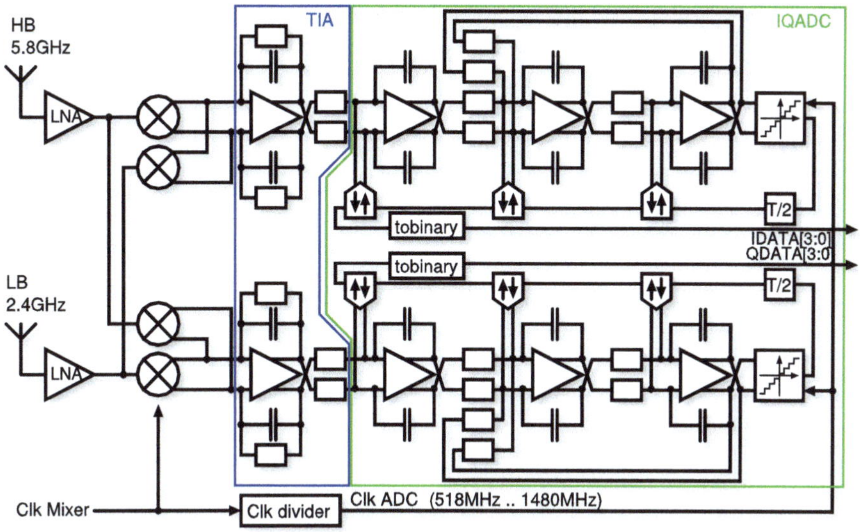

Fig. 1.4 Conceptual drawing of the first proposed low gain high dynamic range lineup. The TIA filters the high frequency signals

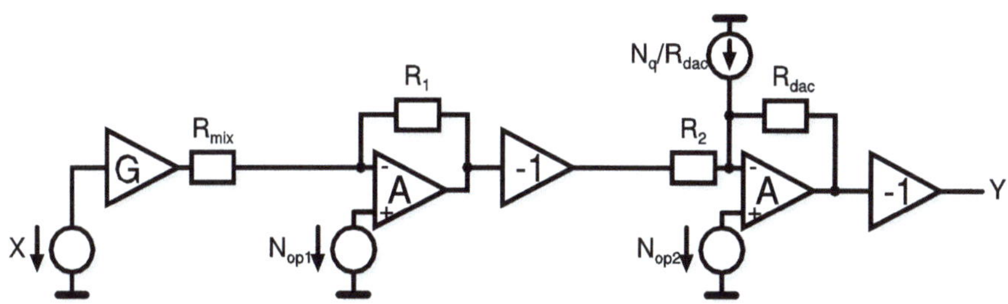

Fig. 1.5 Simplified DC noise model of the TIA and ADC cascade

$$N_{input}^2 = N_{op1}^2 \left| \frac{(R_1 + R_{mix})}{R_1 G} \right|^2 + N_{op2}^2 \left| \frac{(R_{dac} + R_2) R_{mix}}{R_{dac} R_1 G} \right|^2 + N_q^2 \left| \frac{(R_2 R_{mix})}{R_{dac} R_1 G} \right|^2$$

The value of the quantization noise N_q is driven by the clock. On one side, to reduce the power consumption the clock needs to be minimized, on the other side a lower frequency decreases the rate of the DAC feedback and reduces the loop stability. The poles of the ADC have a limited highest frequency constraint for the same reasons. As the quantization noise is added at the end of the chain it can be minimized by rising the gain through an increasing of R_1 and R_{dac}. The supply voltage limits the total gain of the first stage and sets constraints on the ADC gain R_{dac}/R_2. Thus having a flat transfer function on two stages and a small R_2 sets equivalent stringent constraints to both the TIA noise and to the ADC first stage noise. It is possible to calculate the input referred noise budget for each frequency and with

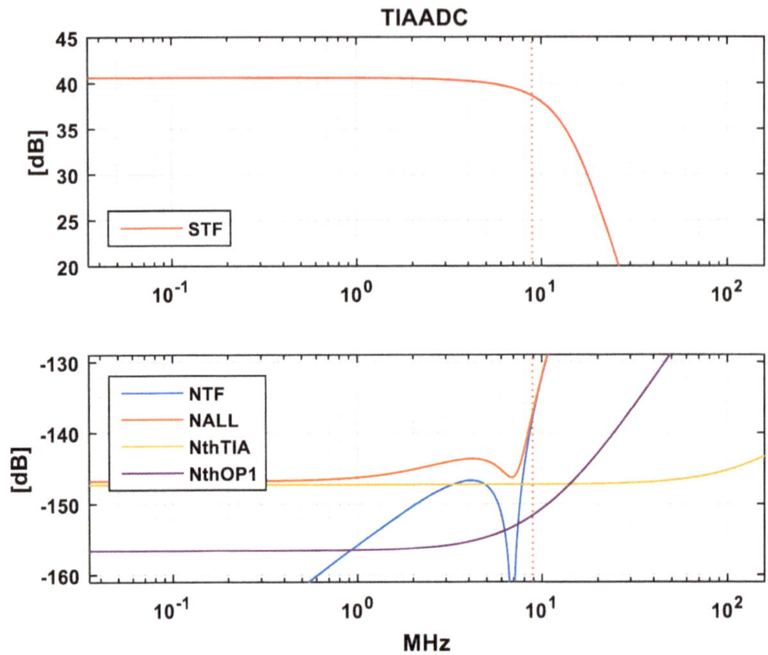

Fig. 1.6 Chain gain and STF (*top*) and input referred noise budget (*bottom*). This includes the contribution of the quantization noise, of the TIA and of the first ADC opamp noise. If R2 equals Rdac then the opamp1 noise can be greatly relaxed

less simplifications by using math tools to solve the system equations. An example for the breakdown is shown in Fig. 1.6. The area and the power consumption of these two blocks is considerable.

To reduce at a minimum the thermal noise contribution of the first stage, the multipath OPAMP structure [1] was used in the TIA stage. The differential OPAMP concept is depicted in Fig. 1.7. The g_m stages can be programmed to find a better tradeoff between the noise (g_{m1} and C_{g1}), the *GBW* and the power consumption. The two cascaded g_m stages allow to achieve very high gain at lower and medium frequencies. A single stage version of the class AB output stage as reported in [2] is used to efficiently drive the integrator feedback cap and the load.

The following OPAMPs in the design need to achieve a higher bandwidth but not much inband gain and a two stage miller compensated structure was used.

The clock of this lineup has been derived from the mixer clock. The local oscillator (LO) is divided by an integer factor which changes with the WiFi mode. The clock can thus span over a wide range, and the ADC needs to fulfill all stability and DR requirements for the lowest clock, but also needs to be able to run at the highest clock frequency. The datapath between quantizer and DACs needs to be designed to avoid the metastability region when the sample is captured in the DAC latch.

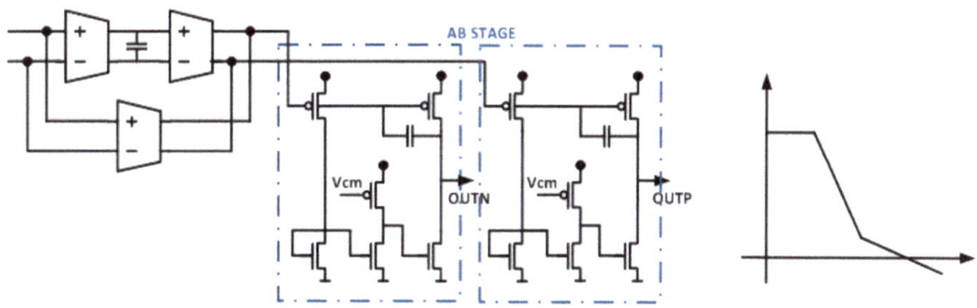

Fig. 1.7 The TIA and the first ADC opamp is a GmC multipath structure cascaded to a class AB stage (*left*). Opamp open loop frequency response (*right*)

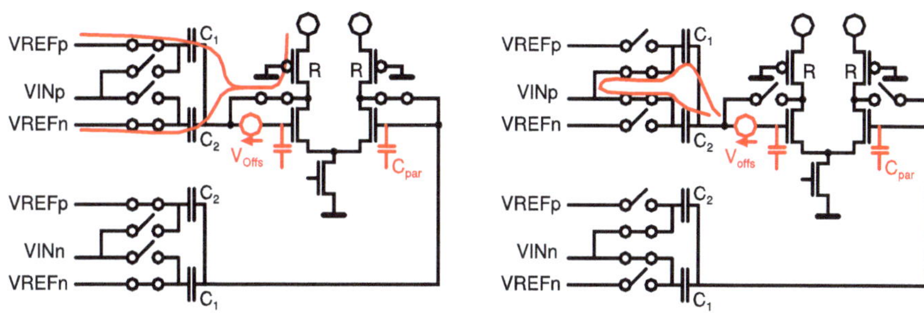

Fig. 1.8 The reference and offset sampling phase (*left*), and the input signal tracking phase (*right*) for a comparator in the quantizer

The quantizer design implements an evolution of the offset compensated comparators presented in [3]. The basic differential comparator slice circuit including the reference generation capacitive network and the preamplifier for the latch is shown in Fig. 1.8. It operates essentially in two non-overlapping phases driven by the ADC clock. In the first phase, depicted on the left, the differential pair outputs and inputs are shorted. The reference fullscale voltage, common to all comparators, is connected to the comparator capacitor array. In this way, the voltage stored in the capacitors is sampled against the operating point of the differential pair. Assuming a differential preamplifier gain G, the output voltage in this operating condition is:

$$V_{outdiff\,ph1} = \frac{-GV_{Offs}}{\mathrm{G}+1}$$

This means that also the offset voltage V_{Offs} of the comparator is stored in the capacitors. In the second phase, depicted on the right of Fig. 1.8, a switch connected to the SD-ADC filter is closed. In this way charge redistribution between C_1 and C_2 occurs. The voltage at the output of the preamplifier is then given by:

$$V_{outdiff\,ph2} = G\left(V_{indiff} + \frac{C_2 - C_1}{C_1 + C_2}V_{ref\,diff} - \frac{GV_{Offs}}{\mathrm{G}+1} + V_{Offs}\right)$$

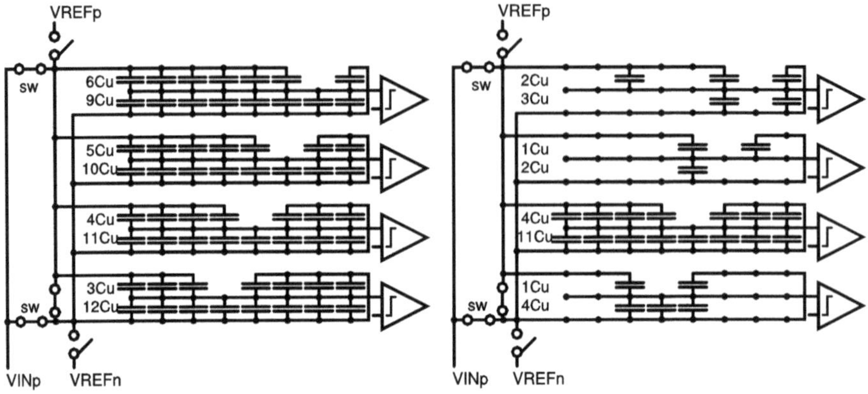

Fig. 1.9 A full quantizer array of four comparators (*left*) and an optimized version that enables a faster charge redistribution due to the reduced amount of capacitors (*right*). The VINp switches can be boosted to achieve higher frequency

If the gain is high enough then the last two terms are canceled and the offset is compensated. Although the capacitors can be altogether in the range of tenths of fF the speed of this structure is limited in phase one by the parasitic capacitors at the differential input pair. For this reason it is essential to build an extremely compact layout for this cell. The total parasitic capacitance stems from parasitics along the routed lines together with the differential pair gate capacitance and the parasitic capacitance that the resistive loads show versus the substrate.

The latter term appears due to the fact that in advanced technologies the minimum resistor width is wide compared to the transistors. In these technologies the allowed minimum distance between transistors is short compared to the one between a transistor and a poly resistor. To mitigate all these parasitic effects two long PMOS transistors driven in the ohmic region are used as load instead of the resistors. Due to the embedded offset compensation of the structure the mismatch between the two PMOS loads is not of concern. An additional speed limit comes from the charge redistribution switches in the second phase. Contrary to the reference switches, the input switches do not sit at a potential near to the supply. The issue is that V_{inp} and V_{inn} are often close to common mode voltage and this can be solved using two tricks implemented as shown in Fig. 1.9. Comparing the left and the right part: three boosted switches *sw* can be used instead of two to reduce their series resistance R_{on}. Moreover, where the comparator level allows it, the number of unit capacitor C_u can be reduced to the minimum required ratio based on ratio matching and gain requirements.

The output of the quantizer is fed to the DACs after a small delay of several hundreds of picoseconds. To increase the performance of the ADC the DAC transfer function shall be linear. As the DAC cells suffer from mismatch a background calibration has been implemented. The literature reports several ways to linearize the DAC. The scrambling of the cells used by dynamic element matching technique is very efficient in simulation, but difficult to be implemented out of the box due to the added delay in the loop and due to the common mode glitch energy, generated at the first OPAMP input. This glitch modulates with the input signal generating distortion tones. Another way to linearize the DAC is to calibrate the mismatch of the cells. It is possible to calibrate the mismatch in a digital controlled fashion [4], or it is also possible to do it within an analog loop [5]. The advantage of the digital calibration is that the observability and manipulation is well controlled in the digital domain and the refresh rate in background can be performed seldom; the implementation is complex and the residual error is limited by the implemented LSB. On the other hand the analog calibration loop needs a continuous refresh and the calibration is virtually able to reduce the residual error to a small ε.

In this ADC design the analog calibration was chosen because the complexity is reduced. The circuit idea of the calibration is shown Fig. 1.10 and the basic technique is known since very long time [6]. To decrease the amount of noise injected in the integrators the unused DAC cells are disconnected from the filter and connected to a floating node *dump*. If a cell is in use then the digital word coming from the quantizer is routed to the switches DP_x/DN_x. The DACs inject the current pulses in a non return-to-zero fashion. The current steering PMOS and NMOS are split in two parts: a wide transistor part is connected to a fixed bias voltage that grants to inject a current of about 90 % of the wanted current. Additionally a smaller transistor in parallel injects the residual calibrated current of about 10 %. Its bias voltage is stored into a local capacitor. If a cell is in calibration its quantizer data is routed to a spare cell that can take over the task of injecting the required current in the filter integrators. The cell in calibration activates the switches marked with CAL_{sw} and DC_x. In this way a reference current cell is connected and compared against the cell in calibration. As both reference cell and in-calibration current steering cell shows a high impedance output $\sim 1/gds$ the voltage at the node CAL_{sw} settles to the required voltage needed to produce the reference current. After settling is achieved the cell is reinserted back in the SD-ADC loop. As the disconnection and the re insertion of the cell can affect the stored bias voltage this operation is done in a non-overlapping fashion.

The functionality of the circuit has been tested with measurements shown in Fig. 1.11 and the circuit is able to linearize the DAC by reducing the third harmonic.

1.2.2 The RF-ADC Lineup

The lineup is shown in Fig. 1.12. As demonstrated in the previous section the presence of an active stage between mixer and ADC is tightening the noise requirements for both TIA and the first ADC stage. The idea is to assign the portion of noise generated from the TIA into the total ADC noise budget, actually relaxing its requirements.

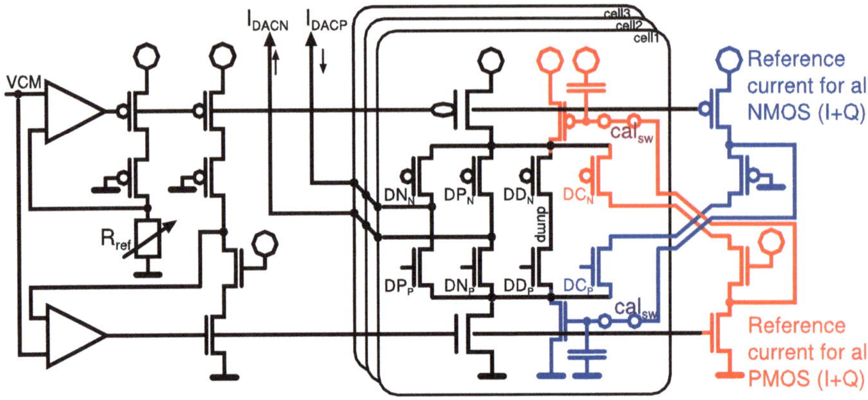

Fig. 1.10 The DAC bias and a current steering DAC cell including the switches for the operation, for the dumping and for the calibration against a reference current

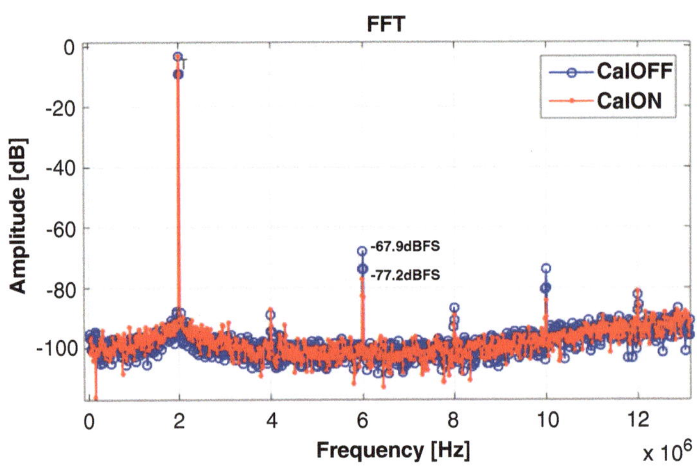

Fig. 1.11 Measured FFT spectrum performed injecting a −3dBFS tone with and without calibration. With calibration the third harmonic is suppressed by 9dBFS

Given the simplified noise model of Fig. 1.13, the chain DC gain with high gain amplifiers A is

$$\frac{Y}{X} = \frac{R_{dac}G}{R_{mix}}$$

and the input noise power

$$N_{input}^2 = N_{op1}^2 \left| \frac{(R_{mix} + R_{dac})}{GR_{dac}} \right|^2 + N_q^2 \left| \frac{R_{mix}}{GR_{dac}} \right|^2$$

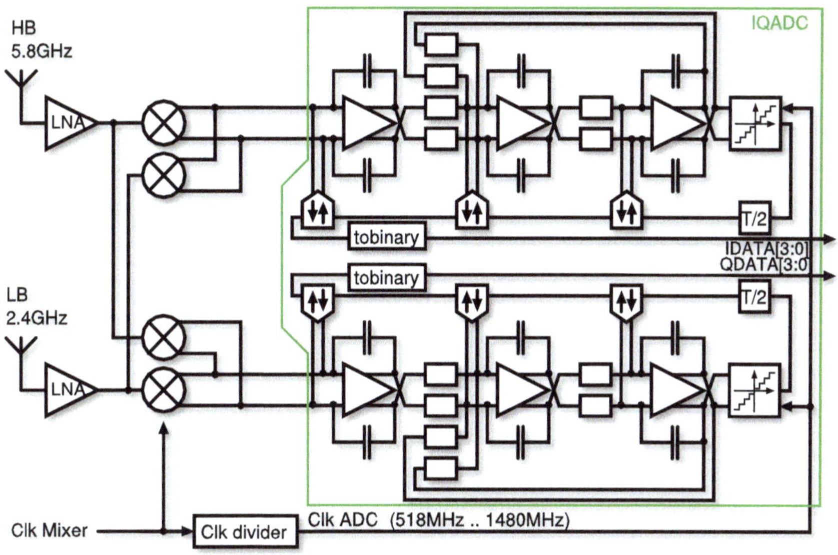

Fig. 1.12 Conceptual drawing of the second proposed low gain high dynamic range lineup. The SD-ADC and the mixer are connected directly

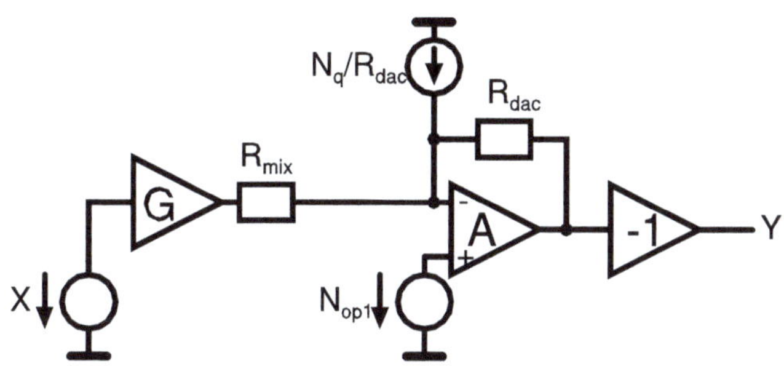

Fig. 1.13 Simplified DC noise model of the RFADC

The final goal is to reduce the overall area and power consumption. The SD-ADC coefficient set needs to fulfill the out of band signal rejection requirements: this can be partly achieved by using a passive pole in front of the ADC and a lower cut-off frequency in the first stage which limits the quantization noise. A disadvantage of this lineup is that changing the low frequency gain of the STF would require the DAC1 current and the first filter capacitor C_1, or the second stage input resistor R_2, to be reprogrammed to keep the pole position at the same frequency. The input referred noise breakdown is shown in Fig. 1.14.

The building blocks of this ADC are the same as the ones of the previous design. The interface between mixer and ADC needs careful attention. If the clock of the LO and the clock of the first DAC are not synchronized spurs can pollute the

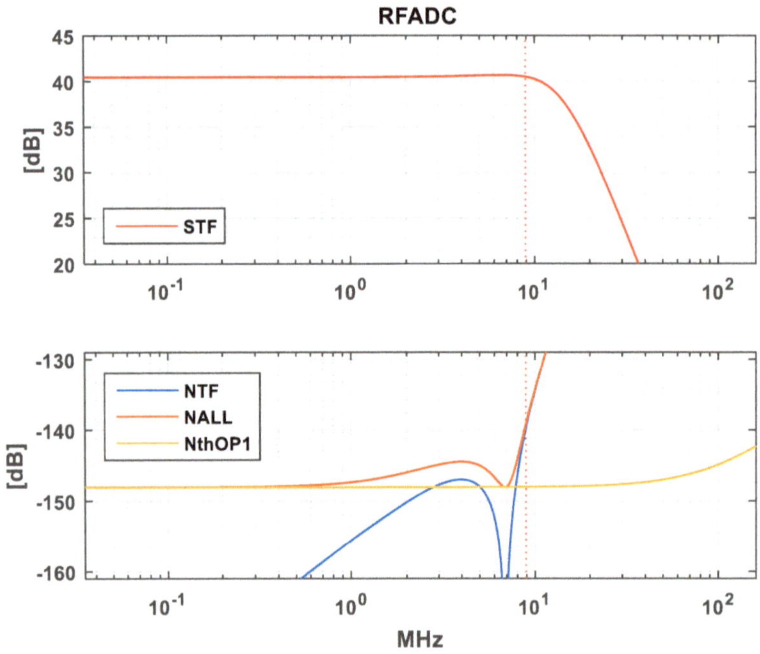

Fig. 1.14 Noise budget breakdown for the RFADC. The dominant noise is generated in the first opamp and at the quantizer

inband spectrum. The DAC acts as a differential current source, but when a cell changes its state the generation of a small common mode glitch cannot be avoided. The common mode glitch contains several harmonics at the multiples of the ADC clock. A differential signal produced by the mixer at the same frequency is down-converted to baseband: its amplitude at the digital ADC output depends on the first stage performance, which is usually limited at higher frequencies. In this lineup the clock is derived from the LO; this makes the analysis and the assessment of the potential problematic scenarios much easier. For high frequency isolation purposes it is advisable to add a small and non-dominant RC passive pole at the ADC input. A too big capacitor boosts the first stage noise.

The design of the RF components requires a good model of the baseband circuitry. Due to the coarse quantization of the signal at the output of the ADC filter, it is not granted that the datastream is periodic. This non-periodic behavior of a SDADC is a blocking point for all harmonic balance based simulators. A full transistor level linear model of the ADC is needed. For this purpose the quantizer and the DAC can be replaced by a simple voltage controlled current source as shown in Fig. 1.15. This model fits well the real STF and NTF at higher oversampling ratio and can be used to verify the overall chain performance. Good matching between simulations and measurements has been achieved. In the model it is possible to emulate the noise contribution of the quantizer. The noise generated by a resistor

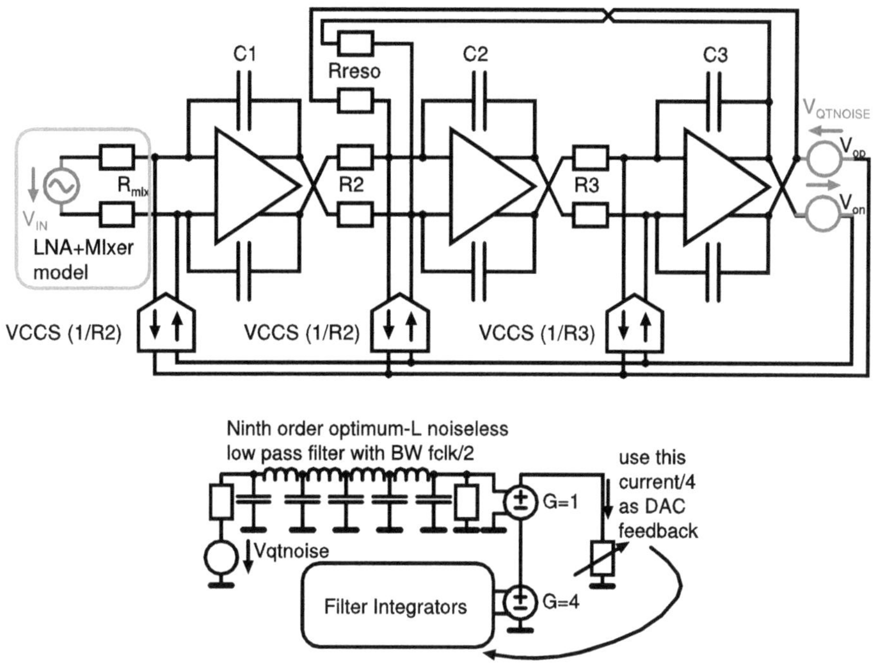

Fig. 1.15 Linearized model of the SD-ADC (*top*) and equivalent quantization noise generator for transient simulations (*bottom*)

can be sized to fit to the one determined by the number of bits and by the ADC clock frequency. The quantization noise in the real ADC is sampled at fclk and thus it is filtered at its multiples. The thermal noise of a resistor shows a flat characteristic: it is therefore advisable to strongly filter the resistor noise with an LC filter prior to the injection at the filter output.

1.2.3 *The Gm-ADC Lineup*

Having a clock frequency synchronized with the LO is simplifying the analysis of the possible blocker scenarios but requires the use of a variable rate converter in the digital domain, which is a very power hungry block. One way to avoid this additional digital block is to keep the SD-ADC clock in synch with the system and independent of the LO. In such an architecture it is very important to implement a strong anti-alias function in cascade with the ADC STF which avoids intermodulation between the mixer and the DAC. This is achieved introducing a g_m stage in the chain. In all the presented architectures the mixer is passive and composed of simple switches driven at the common mode voltage. The mixer to baseband interface in this case is not fixed at common mode as it is in the voltage domain. This is an opportunity to

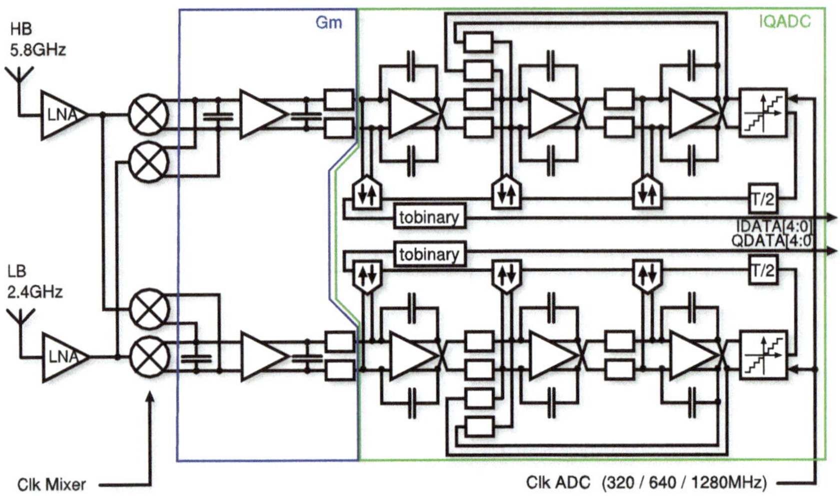

Fig. 1.16 Conceptual drawing of the third proposed low gain high dynamic range lineup. A Gm stage filters the out of band signals with a capacitor at the input and at the output

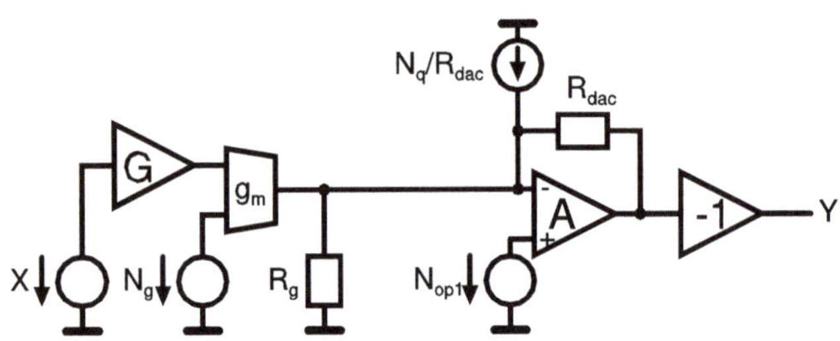

Fig. 1.17 Simplified DC noise model of the GmADC

filter the input signal with a capacitor. The g_m cell is directly feeding the ADC in current mode, and it can be seen at the input of the opamp as a high impedance input load R_g. Compared to the TIA lineup the g_m inter-stage relaxes the first OPAMP noise budget. The equivalent circuit for the basic noise budget calculation does not contain anymore the mixer impedance; just the LNA gain G (Figs. 1.16 and 1.17).

The chain gain, under the assumption of a high gain amplifiers A, is given by:

$$\frac{Y}{X} = Gg_mR_g$$

and the input referred noise is

$$N_{input}^2 = N_{op1}^2\left|\frac{\left(R_g + R_{dac}\right)}{Gg_mR_gR_{dac}}\right|^2 + N_g^2 + N_q^2\left|\frac{1}{Gg_mR_{dac}}\right|^2$$

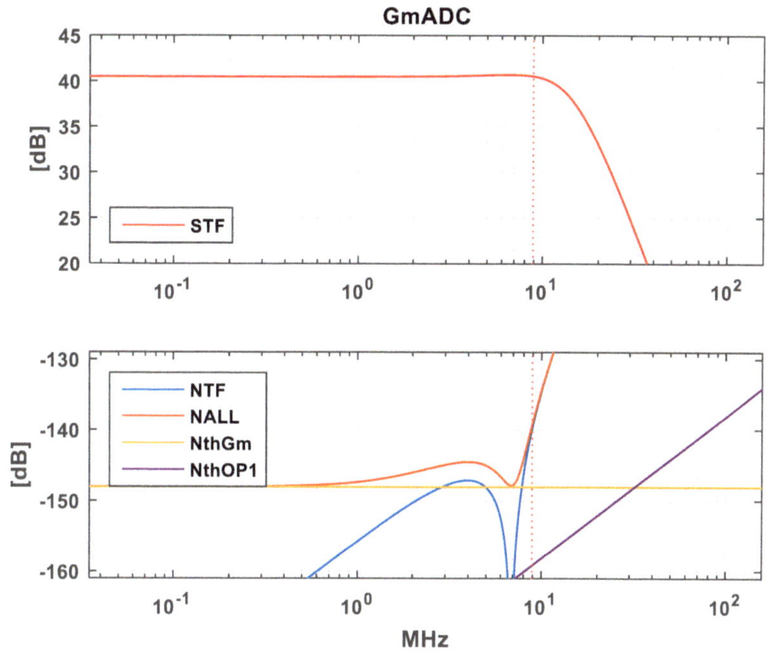

Fig. 1.18 The STF (*top*) and the input referred noise budget (*bottom*). The Gm cell noise together with the quantization noise is the dominant source

The g_m cell noise N_g is considerable and is in the same range of the TIAs noise for an equivalent power consumption.

The shaped input referred noise breakdown is shown in Fig. 1.18. Most of the savings in this structure are at a more macroscopic level: the poles of the ADC can be pushed higher in frequency if some pre-ADC filtering is implemented. This allows a lower ADC clock or a lower input referred quantization noise N_q. Lowering the ADC clock and skipping the variable rate converter at the end of the digital decimation chain introduce a considerable saving in the total power consumption: this tradeoff was favorable and a 5bit quantizer was used to support it further.

The g_m cell is able to grant a linear output when its differential peak to peak input is within a range of few hundreds of mV. The voltage interface followed by the gain exposes and amplifies any previous mismatch. The passive mixer and the path between the passive mixer and the Gm input need to be checked very carefully and must be well matched between I and Q paths. The interface is thus an opportunity to add a pole at the input, but unfortunately this pole cannot be the dominant one due to the subsequent amplification of the mismatch without any further digital IQ-mismatch calibration. A capacitor and a series resistor can be added at the $g_m$output to build a g_mC low pass filter and relax the SD-ADC first stage. Again the pole cannot be placed too low in frequency as the capacitor would affect the feedback factor of the first stage OPAMP.

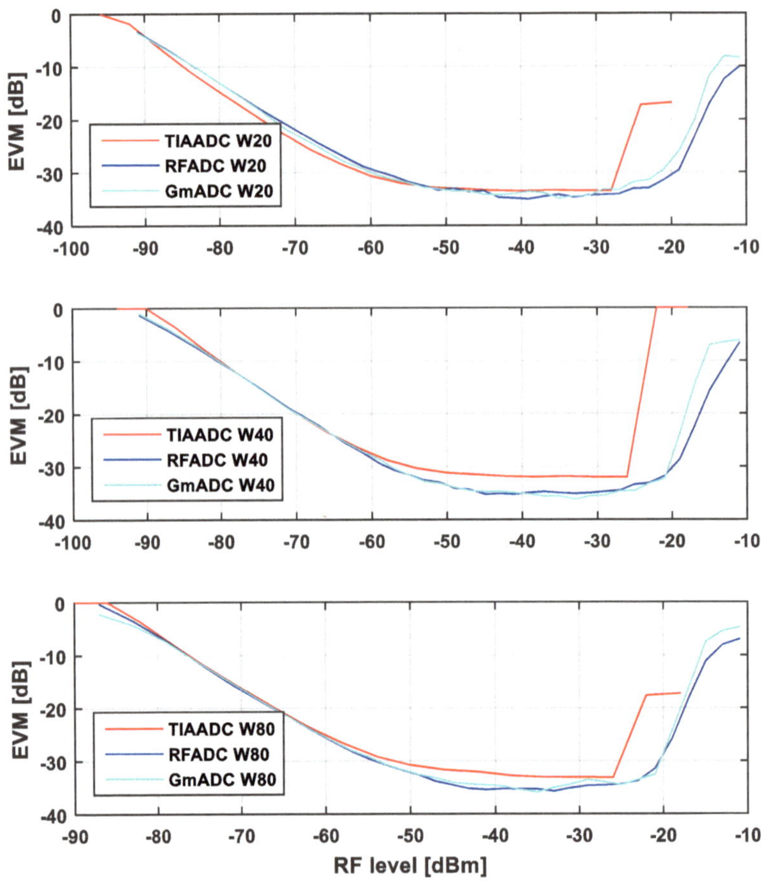

Fig. 1.19 A comparison of the EVM of the three lineups in the three modes by increasing the input signal power

1.3 The Measurement Results

Three testchips implementing the presented lineups were fabricated in a 28 nm CMOS process to directly compare the results. A WiFi OFDM pattern was applied at the antenna and the EVM performance measured in the main modes is shown in Fig. 1.19. In Fig. 1.20 and example of this EVM pattern is shown. As expected the performance was comparable allowing a fair comparison of the remaining key parameters reported in Table 1.1 together with the die micrograph of the GmADC.

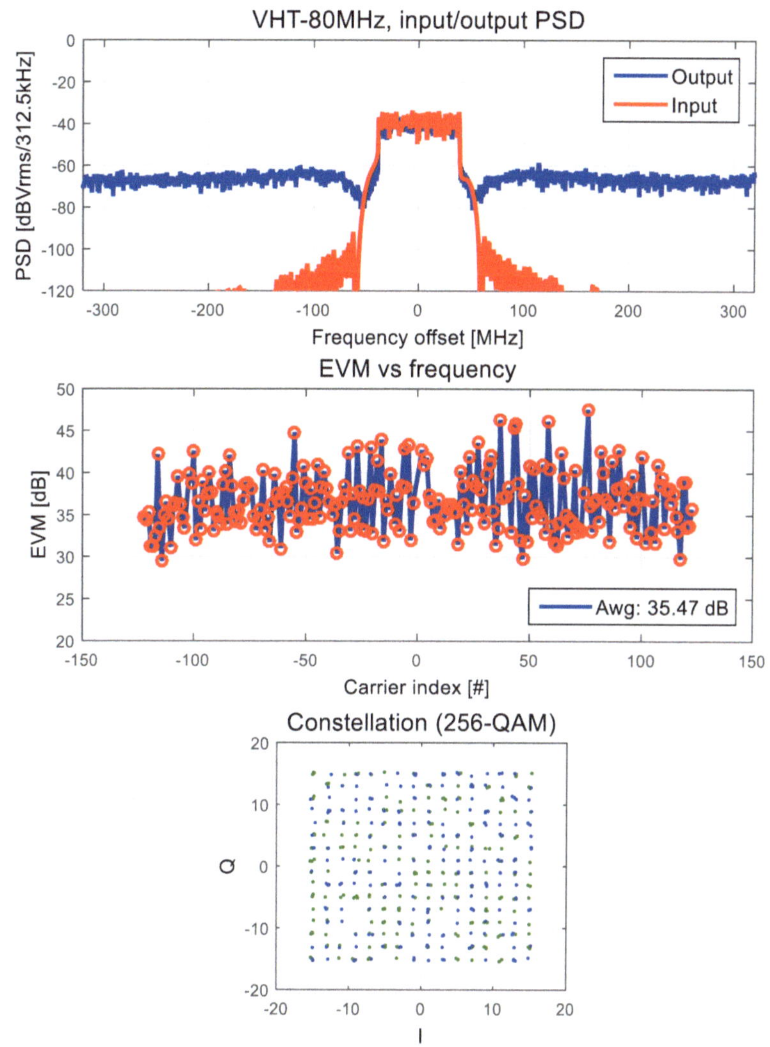

Fig. 1.20 Performance of the RFADC lineup in the 80 MHz with a 256QAM VHT80 pattern

1.4 Conclusions

The presented low gain receiver lineups make use of a high dynamic range SD-ADC to reduce the amount of blocks, the area and the power consumption. If the area needs to be optimized the RFADC lineup shall be preferred. In terms of out of band rejection and power consumption the GmADC is preferred because it's reduced and fixed clock frequency leads to major power savings in the digital domain.

Table 1.1 Summary of the measured performances

Parameter	TIA-ADC	RFADC	GmADC
Clock frequency (MHz)	518–720 (W20)1036–1280	518–720 (W20)1036–1280	320 (W20) 640 (W40) 1280 (W80)
Total current consumption for the BB blocks in WiFi20/40/80 (mA)	16/20/23	11/15/19	11/16/17 (w/o digital)
Max EVM in WiFi20/40/80 (dB)	33.4/31.8/33.0	35.0/35.2/35.7	34.9/36.1/35.8
Area compared to TIAADC (%)	100	61	72
Supply (V)	1.1	1.1	1.1
Technology (nm)	28	28	28

References

1. R.G.H. Eschauzier, L.P.T. Kerklaan, J.H. Huijsing, A 100 MHz 100 dB operational amplifier with multipath nested Miller compensation structure, in *Solid-State Circuits Conference, IEEE Journal of, 1992. Digest of Technical Papers. 39th ISSCC, 1992 IEEE International*, 19–21 Feb 1992, pp. 196–197
2. X. Peng, W. Sansen, AC boosting compensation scheme for low-power multistage amplifiers. IEEE J. Solid State Circuits **39**(11), 2074–2079 (2004)
3. L. Dorrer, F. Kuttner, P. Greco, P. Torta, T. Hartig, A 3-mW 74-dB SNR 2-MHz continuous-time delta-sigma ADC with a tracking ADC quantizer in 0.13-μm CMOS. IEEE J. Solid State Circuits **40**(12), 2416–2427 (2005)
4. J.G. Kauffman, P. Witte, M. Lehmann, J. Becker, Y. Manoli, M. Ortmanns, A 72 dB DR, CT $\Sigma\Delta$ modulator using digitally estimated, auxiliary DAC linearization achieving 88 fJ/conv-step in a 25 MHz BW. IEEE J. Solid State Circuits **49**(2), 392–404 (2014)
5. M. Clara, W. Klatzer, B. Seger, A. Di Giandomenico, L Gori, A 1.5 V 200MS/s 13b 25 mW DAC with randomized nested background calibration in 0.13 μm CMOS, in *Solid-State Circuits Conference, 2007. ISSCC 2007. Digest of Technical Papers. IEEE International*, 11–15 Feb 2007, pp. 250–600
6. D.W.J. Groeneveld, H.J. Schouwenaars, H.A.H. Termeer, C.A.A. Bastiaansen, A self-calibration technique for monolithic high-resolution D/A converters. IEEE J. Solid State Circuits **24**(6), 1517–1522 (1989)

Chapter 2
High-Resolution Wideband Continuous-Time $\Sigma\Delta$ Modulators

Lucien Breems and Muhammed Bolatkale

2.1 Introduction

This paper is organized as follows. In Sect. 2.2 some key architectural choices and trade-offs are discussed when defining the optimal noise-transfer function (NTF) of a wideband and high-resolution delta-sigma modulator. Among the architectural choices are type and order of the loopfilter, ADC resolution, sampling frequency and number of stages. Important limiting factors related to the practical implementation of wideband delta-sigma converters are discussed in Sect. 2.2. Examples of such non-idealities are metastability, excess loop delay (ELD), non-linearity, and phase noise. These aspects can be taken into account on the architectural level to mitigate their impact. In Sect. 2.4 recent designs are presented that have pushed the envelope with respect to bandwidth and linearity. The conclusions are drawn in Sect. 2.5.

2.2 Architectural Choices

There are several important choices to be made when defining the architecture and the optimal noise transfer function of a delta-sigma modulator for given bandwidth and resolution specifications. Figure 2.1 shows the basic block diagram of a single-loop delta-sigma modulator that consists of a loopfilter, A/D converter (ADC) and feedback D/A converter (DAC). The main degrees of freedom are the position of the sampler, the loopfilter (LF) type, order and oversampling ratio (OSR) and the accuracy of the ADC and DAC.

L. Breems (✉) • M. Bolatkale
NXP Semiconductors, Eindhoven, The Netherlands
e-mail: lucien.breems@nxp.com

© Springer International Publishing Switzerland 2017

A. Baschirotto et al. (eds.), *Wideband Continuous-time ΣΔ ADCs, Automotive Electronics, and Power Management*, DOI 10.1007/978-3-319-41670-0_2

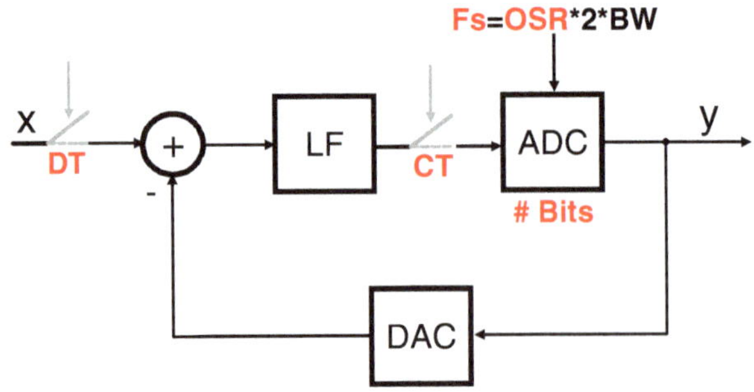

Fig. 2.1 Basic block diagram of a delta-sigma modulator

2.2.1 *Sampler*

The sampler is typically placed either at the input of the delta-sigma modulator or at the input of the ADC. The first configuration represents a discrete-time delta-sigma modulator which can be implemented with switched-capacitor circuits while the latter configuration is a continuous-time modulator with a continuous-time loopfilter implementation. Since a discrete-time modulator has the sampler directly at the input it suffers from aliasing of signals that may be present close to (multiples of) the sampling frequency. Therefore, an anti-alias filter is required in front of the delta-sigma modulator to adequately suppress aliased signals. In case of a continuous-time modulator, the loopfilter acts as an inherent anti-alias filter which relaxes or even eliminates the anti-alias filter in front of the modulator. In case of a switched-capacitor filter implementation, the filter coefficients are defined as capacitor ratios which are very accurate and do not require calibration. Also, the bandwidth of a discrete-time modulator can be easily scaled by changing the sampling frequency without the need for adapting the filter coefficients. This is different from a continuous-time filter implementation where filter coefficients are determined by e.g. RC time constants that are subject to significant spread ($\sim$40 %) which also need to be tuned in case of sampling rate scaling. A drawback of switched-capacitor circuits is that the resolution depends on the settling accuracy of charge transfers which requires high-speed amplifiers, especially if the sampling frequency is high. Continuous-time filters do not have the settling requirement and usually require lower bandwidth amplifiers. Therefore, in the bandwidth range of several tens of MHz and above, many continuous-time modulators have been reported in literature [1–7].

2.2.2 *Filter*

The noise transfer function of the delta-sigma modulator is determined by the loopfilter order, filter type and the choice of filter coefficients. As an example, Fig. 2.2 shows the magnitude plots of two different NTF curves (both second-order).

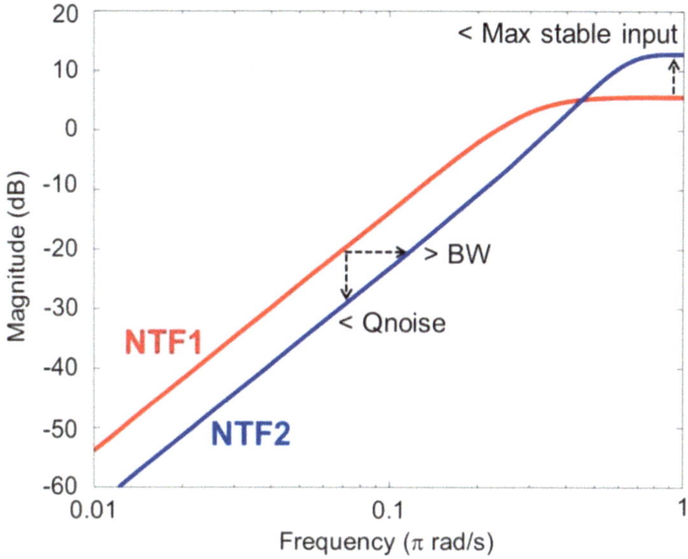

Fig. 2.2 Magnitude plot of second-order NTFs with different OOB gains

The blue transfer (NTF2) has lower low-frequency NTF gain compared to the red transfer (NTF1) but as a result it has higher OOB gain compared to NTF1. From Fig. 2.2 it becomes clear that a higher OOB gain leads to lower quantization noise (Qnoise) in a certain target (low-frequency) bandwidth or to an increased bandwidth (BW) for a certain target noise suppression. Besides the improved low-frequency quantization noise suppression and/or increased bandwidth, a negative implication of a higher OOB gain is that the maximum stable input level of the modulator is lower. The impact of a smaller maximum input signal level is that the modulator will become more sensitive to other noise sources like circuit noise. Further increasing the OOB NTF gain will ultimately lead to an inherently instable modulator even without an input signal. Therefore, the design of the NTF will be a compromise between the maximum stable input level, bandwidth and quantization noise suppression.

Besides optimizing the NTF OOB gain as described above, the bandwidth and/or noise suppression can be improved by increasing the order of the loopfilter. This is shown in Fig. 2.3 where a third-order NTF is compared with a fifth-order NTF. At low frequencies the fifth-order NTF provides much better quantization noise suppression. Notice that at a certain frequency (F_x), the NTF curves cross and the fifth-order NTF becomes worse compared to the third-order NTF. As this crossing frequency will move to lower frequencies for higher-order filters it does not pay off to go to very high filter orders when the oversampling ratio is low (if $F_x <$ bandwidth), which is usually the case for wideband delta-sigma modulators.

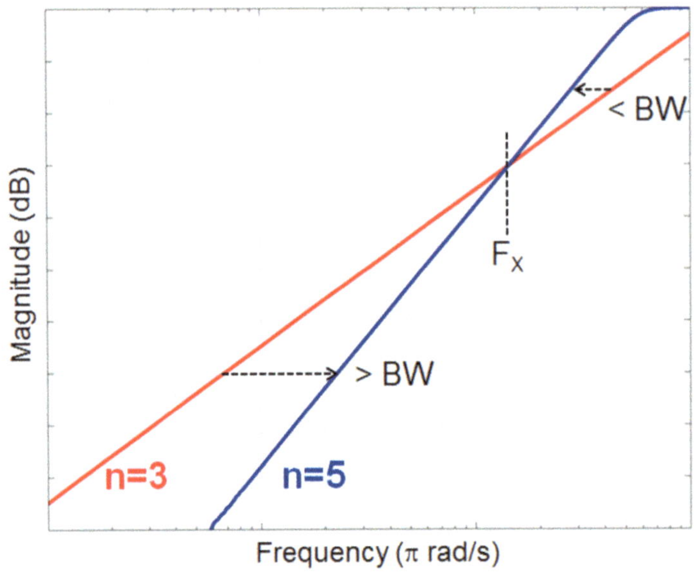

Fig. 2.3 Magnitude plot of third-order and fifth-order NTFs

Further optimization can be done by optimally distributing the NTF zeros in the signal band of interest as shown in Fig. 2.4. Spreading the NTF zeros in the signal band gives much better overall suppression of the quantization noise compared to the situation where all NTF zeros are at DC.

2.2.3 *ADC*

The quantization noise of the ADC is filtered by the NTF as was described in the previous section. Due to the noise-shaping and oversampling of the modulator, the intrinsic resolution of the ADC can be as low as 1-bit, while the accuracy in a specific bandwidth can still be very high. Moreover, employing a 1-bit ADC has the advantage that only a 1-bit DAC is required in the feedback path of the delta-sigma modulator, which is inherently linear. In wideband delta-sigma modulators, usually a 1-bit ADC is not enough to achieve the bandwidth and resolution requirements. Typically a 4-bit ADC is utilized [3–6] which seems to be a good compromise between ADC complexity, speed and performance. Increasing the number of bits in the ADC can reduce the quantization noise and increase the bandwidth in two ways. Firstly, the quantization noise itself will be smaller as a result of more bits in the ADC. Secondly, as the quantization noise in the delta-sigma loop becomes smaller, the maximum OOB gain of the NTF can be increased while still having a sufficiently large maximum stable input level. Increasing the OOB gain helps to further reduce

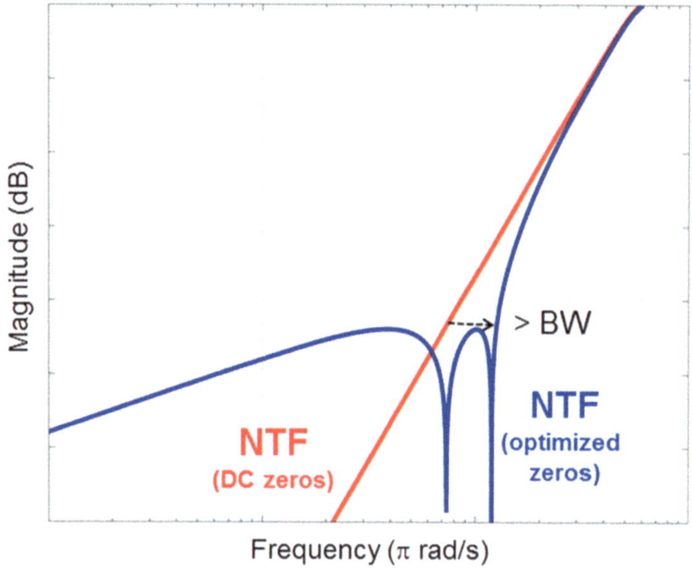

Fig. 2.4 Magnitude plot of fifth-order NTFs with DC zeros and optimized zeros

the in-band quantization noise as was shown in Fig. 2.2. As a result of the NTF optimization, the quantization noise floor in the signal band can be improved more than the improvement in intrinsic ADC resolution when increasing the number of bits.

2.2.4 *Multi-Stage Noise-Shaping (MASH)*

In the previous sections an overview of the main optimization opportunities of a delta-sigma modulator was given. The key parameters are the filter order, number of ADC bits and oversampling ratio. The degrees of freedom in this design space are more than adequate to design a high-resolution delta-sigma ADC with low to medium bandwidth. Recently published delta-sigma designs however demonstrate a trend towards much higher bandwidths, far exceeding 100 MHz which is being pushed by new wireless standards. On the architectural level the design parameters of the delta-sigma modulator are constrained to practical limitations [3]. As a delta-sigma modulator is a feedback system, the loop stability requirements put constraints on the maximum number of ADC bits, the maximum sampling frequency and hence the oversampling ratio and the maximum loopfilter order. This is explained in more detail in Sects. 2.3 and 2.4.

An architectural solution to break the bandwidth barrier and enable wider bandwidths at low oversampling ratios within the constraint of the maximum sampling frequency is the cascaded or multi-stage noise-shaping (MASH) delta-sigma modulator. The MASH delta-sigma modulator originated as a solution to implement a stable higher-order modulator with multiple low-order stages [8]. The concept of a (2-1-1) MASH modulator is shown in Fig. 2.5. For simplicity, the DACs are not shown in Fig. 2.5. It consists of N (3 in this example) cascaded delta-sigma modulators where each consecutive stage digitizes the quantization error of the previous stage. This way redundancy is created in the system of the quantization errors of the first N-1 stages. With this redundancy, the quantization errors of the first N-1 stages can ideally be cancelled in the digital domain and what remains is the quantization error of the Nth stage which is shaped by the NTFs of all stages combined. As an example, consider the output of the first (2nd-order) delta-sigma modulator stage of Fig. 2.5. The output y_1 contains the input signal x with two delays and the quantization noise q_1 shaped by a second-order NTF. The quantization error q_1 is extracted from the first stage by subtracting the ADC1 input and ADC1 output (DAC omitted) and fed through interstage gain g_1 to the input of the second (first-order) delta-sigma stage. The output bitstream y_2 of the second stage contains the amplified quantization noise $g_1 \cdot q_1$ with one delay and the quantization error q_2 of ADC2 shaped with a first-order NTF. Both outputs y_1 and y_2 contain q_1, but with different transfer functions. The transfers of q_1 can be equalized by delaying y_1 with one clock period while filtering y_2 with the NTF of stage 1 and dividing by interstage gain g_1. After subtraction, the combined output y_{12} does not contain quantization error q_1. A similar analysis can be done for q_2 and the third stage of the MASH modulator. The overall output y does not contain q_1 and q_2 while the quantization error q_3 of ADC3 is filtered with a 4th-order NTF and suppressed by interstage gains g_1 and g_2 combined. As the overall 4th-order NTF is a combination of low-order (1st and 2nd) NTFs, it provides better quantization noise suppression compared to the NTF of a single-stage 4th-order delta-sigma modulator. In addition to that, the application of interstage gains (g_i) between the stages can further suppress the quantization error. Figure 2.6 compares the NTF of a 2nd order delta-sigma modulator with a 2-1 and 2-1-1 MASH.

The effectiveness of the digital noise cancellation is limited by the matching inaccuracy between the analog and digital filters. In practice the analog loop filters will suffer from mismatch. Therefore, the quantization noise contributions of the individual stages will not be perfectly cancelled which puts a limit on the maximum noise suppression that can be achieved. Traditionally, MASH delta-sigma modulators have been solely implemented with switched-capacitor (SC) circuits due to the fact that the coefficients of SC filters are set by capacitor ratios which have very high accuracy and match very well with the coefficients of the digital noise-cancellation filters. In [9] the feasibility of a continuous-time implementation of a MASH delta-sigma modulator has been demonstrated. Recently, the application of continuous-time MASH delta-sigma modulators has pushed the bandwidth from tens of MHz to far beyond 100 MHz [6].

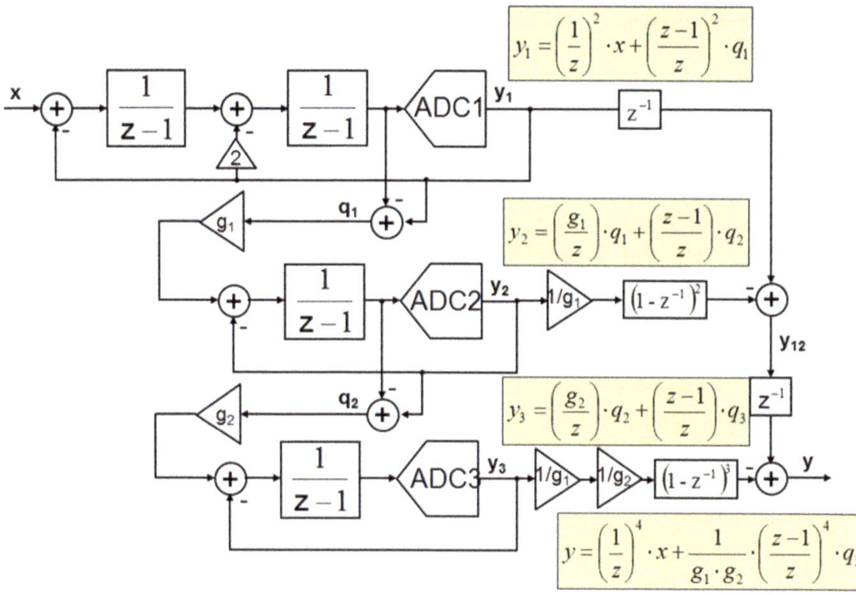

Fig. 2.5 Example of a 2-1-1 MASH delta-sigma modulator

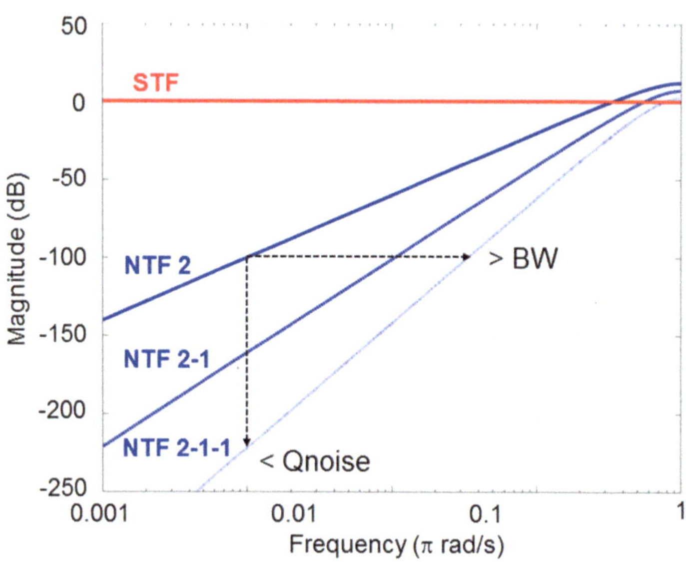

Fig. 2.6 NTFs of 2nd-order, 2-1 MASH and 2-1-1 MASH

2.3 Non-Idealities

Previously it was mentioned that the design space of a high-speed wideband delta-sigma modulator is constrained due to loop stability requirements which is related to the speed and accuracy limitations of the used technology. As an example, the ADC in the delta-sigma loop of Fig. 2.1 requires a certain amount of time to convert the analog input signal into a digital representation. This latency of the ADC is inside the feedback loop and directly impacts the stability. In particular if the ADC is clocked at a very high (GHz) frequency, the loop becomes very sensitive to latency. Another aspect of loop stability is for example related to the parasitic loading of the loopfilter output with the ADC input capacitance. In high-speed wideband delta-sigma modulators, typically low-latency multi-bit flash ADCs are employed, which introduce a significant load capacitance for the loopfilter [3]. This results in an additional parasitic pole in the system that hampers stability. These and other non-idealities related to high-speed delta-sigma modulators will be discussed in the following sections.

2.3.1 Metastability

A metastability error occurs if the ADC does not have enough time to generate a fully settled digital output for a certain input signal. Usually the digital decision is made by means of a regenerative latch with a positive feedback loop. Due to the exponential behavior of the latch, the delay of the latch becomes large for example in case of a very small latch input signal or an input signal with certain dynamics [10]. In case of insufficient gain, the metastable output state of the ADC will propagate through the feedback loop to the DAC and will result in an incomplete DAC feedback charge. The impact of metastability can be easily verified when modelling the ADC in the delta-sigma loop (Fig. 2.1) with finite quantizer gain, while the metastable output y of the delta-sigma modulator is converted into an ideal digital signal by an ideal digitizer with infinitely large gain. Figure 2.7 shows two simulated output spectra in case of infinite quantizer gain (blue) and a finite quantizer gain of 80 dB. The red spectrum shows a flat highly elevated noise floor in the low-frequency band that degrades the maximum resolution that can be achieved.

There are two approaches to reduce the probability of metastable errors. Either the amount of gain within the ADC time budget must be increased by e.g. decreasing the time constant of the regenerative latch, or the time budget for the ADC should be increased. Improving the time constant is a delicate task and can be realized by e.g. increasing the bias current or reducing the capacitive load of the latch. Increasing the time budget leads to larger latency that impacts the stability of the loop. A dedicated excess loop delay compensation loop can allow for more delay while maintaining stable operation of the modulator loop which is described next.

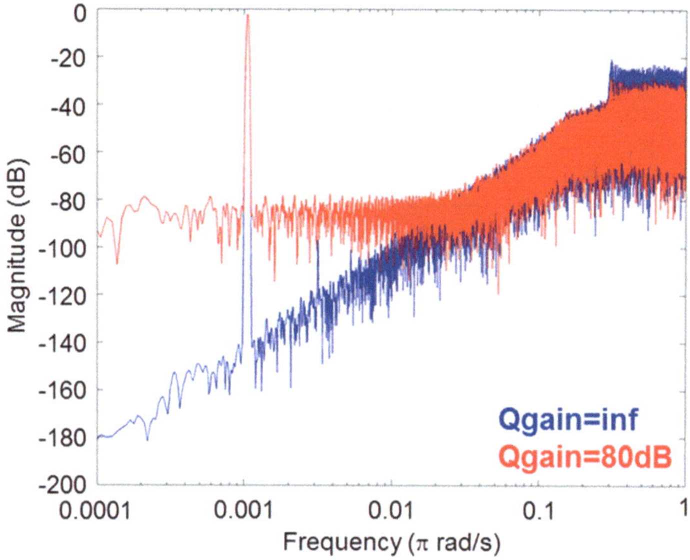

Fig. 2.7 Simulated output spectra of a 1-bit delta-sigma modulator with infinite quantizer gain (*blue*) and with finite quantizer gain of 80 dB (*red*)

2.3.2 Excess Loop Delay (ELD)

Metastability and excess loop delay are tightly coupled as the metastability probability requirement dictates the ADC time budget for a given latch time constant. Other delay contributions origin from e.g. propagation delays of digital circuitry and DACs inside the delta-sigma feedback loop. An ELD compensation loop [11] can allow for a clock period delay without compromising stability or the shape of the NTF. The concept of the ELD loop is shown in Fig. 2.8. In this example the quantizer latency is modelled as a clock period delay. An additional feedback path d around the quantizer delay stabilizes the loop in the presence of the extra delay and the NTF can be mapped to the ideal second-order NTF (in the example of Fig. 2.8 the NTF mapping is achieved for coefficients $a_1 = 1$, $a_2 = 3$ and $d = 2$). ELD compensation is widely used in high-speed wideband delta-sigma modulators. Excess phase, e.g. due to a parasitic pole at the output of the loopfilter as a result of the ADC input capacitance, can be compensated in a similar way.

2.3.3 Non-Linearity

The focus has been so far on the (quantization) noise aspects of the delta-sigma modulator. Besides noise, distortion is another critical performance parameter of

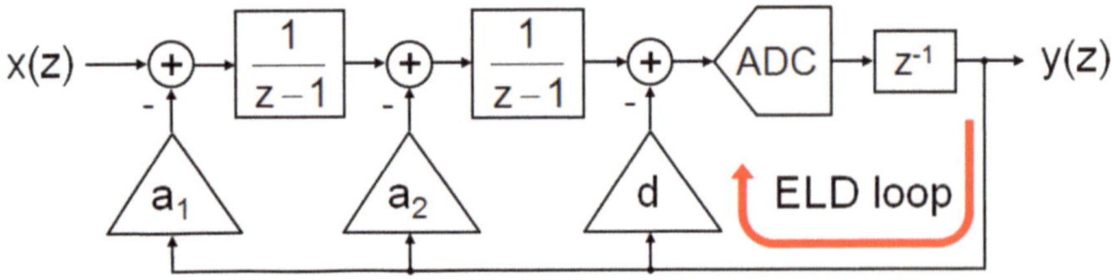

$$NTF_{ideal} = \frac{(z-1)^2}{z^2} \quad NTF_{ELDC} = \frac{(z-1)^2}{z^2 + \underbrace{z\cdot(d-2) + (1-2d+a_2) + (d-a_2+a_1)/z}_{=0}}$$

Fig. 2.8 Second-order delta-sigma modulator with one period ELD compensation

the ADC. Due to non-linearity, distortion components at harmonic frequencies can occur as well as an increased noise level in case of large input signals. A dominant contributor to distortion is typically the feedback DAC. As mentioned before, many wideband delta-sigma modulators incorporate multi-bit ADCs and multi-bit feedback DACs. Due to the low number of bits, the DACs are usually implemented as a thermometer DAC with e.g. unit current sources. Due to mismatch between the unit current sources, distortion components at even and odd harmonic frequencies will occur and high frequency quantization noise will leak in the low-frequency signal band via intermodulation distortion (Fig. 2.9). It is not uncommon that the peak signal-to-noise-and-distortion ratio (SNDR) of a wideband multi-bit delta-sigma modulators drops 5–10 dB relative to the dynamic range (DR) due to non-linearity effects. For some applications this may be ok but for e.g. multi-channel receivers (a typical application for high-resolution wideband ADCs) such a loss is unacceptable, as the presence of a strong in-band blocker signal leads to desensitization of the receiver for weak wanted signals. Besides static mismatch errors, also dynamic errors like intersymbol interference (ISI) will degrade the performance of the delta-sigma modulator. In literature, many techniques have been reported to improve the static errors of a multi-bit DAC via analog calibration [12], digital post correction [13–15], dynamic element matching algorithms such as data weighted averaging (DWA) [16–21] as well as static errors like ISI [22] via return-to-zero (RTZ) coding, ISI shaping [23], etc.

2.3.4 Phase Noise

For continuous-time delta-sigma modulators the purity of the sampling clock is very important. This is due to the fact that the amount of feedback current from the DAC during a clock period is directly depending on the clock timing. Noise on

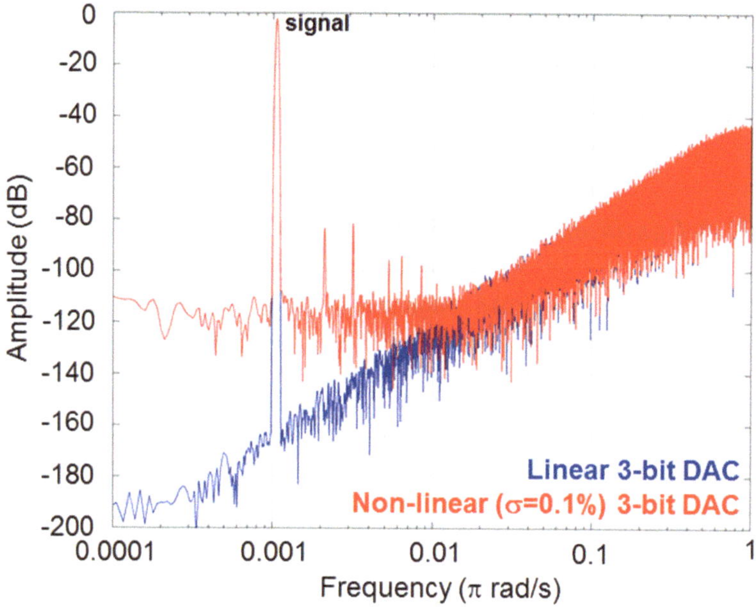

Fig. 2.9 Simulated output spectra of a 3-bit delta-sigma modulator with a linear DAC (*blue*) and 0.1 % mismatch DAC (*red*)

the clock edges directly reflects into noise injection at the input of the modulator. The wideband phase noise of the clock leaks into the signal band via reciprocal mixing with the OOB quantization noise. This is in particular a limitation in 1-bit delta-sigma modulators that have high OOB quantization noise. For multi-bit DACs the OOB noise is lower which relaxes the far-off phase noise requirements. An effective solution to mitigate the effect of reciprocal mixing is to filter the DAC signal with a finite impulse response (FIR) filter [24] to reduce the OOB quantization noise. Mismatch between the FIRDAC coefficients results in a slightly altered filter transfer but does not introduce non-linearity. However, a FIRDAC can introduce significant delay in the feedback path which needs to be compensated to maintain loop stability [1].

2.4 High-Resolution Wideband Delta-Sigma Modulator Designs

In this paragraph some implementation examples are described that have pushed the envelope for wideband delta-sigma modulators with respect to architecture, bandwidth and linearity. The first section presents an implementation of a continuous-time MASH delta-sigma modulator. The second example is a GHz rate modulator with >100 MHz bandwidth. The last design demonstrates a high-resolution wideband delta-sigma modulator that realizes <−100 dB in 25 MHz bandwidth.

2.4.1 A Continuous-Time MASH Delta-Sigma ADC

In [9] a first implementation of a continuous-time 2-2 MASH delta-sigma modulator has demonstrated 67 dB DR in 10 MHz at an oversampling ratio of only 8×. A quadrature configuration of the 2-2 MASH was presented in [25]. The design of the quadrature 2-2 MASH is shown in Fig. 2.10. The ADC consists of two different channels to handle complex in-phase (I) and quadrature phase (Q) signals. The loopfilter of each delta-sigma stage consists of two RC integrator stages (e.g. R1C1, R2C2 in the top left corner) with a feedforward capacitor (e.g. Cff1) for loop stabilization. The 2nd-order loopfilters of the quadrature sigma-delta stages have complex cross-coupling resistors (e.g. Rfb1, Rfb2). Depending on the sign of the cross-coupling connections the NTF zeros can be freely distributed either in the positive or negative frequency band. All resistors in the loopfilters can be calibrated with 1 % accuracy to match the analog loopfilter to the digital noise cancellation filters (not shown). The delta-sigma modulators incorporate 4-bit flash quantizers with offset calibration and 4-bit SC feedback DACs. The quantization errors of the first quadrature sigma-delta stages are duplicated by means of resistors R3 and DAC2 that are connected to the input virtual ground nodes of the second quadrature stages. The topology of the second quadrature stages is identical to the first. Figure 2.11 shows the measured complex quantization noise spectrum of the quadrature 2-2 MASH. All NTF zeros are positioned in the positive frequency band from 0.5 to 20.5 MHz. In this measurement, a −1 MHz image tone has been applied to the input. The image leakage appears at +1 MHz and is 58 dB down. The quantization noise at the edge of the image band at -20 MHz is about 40 dB higher than the in-band noise between 0.5 and 20.5 MHz. The quadrature 2-2 MASH realized 77 dB DR and 69 dB peak SNDR in 20 MHz zero-IF (0.5–20.5 MHz) bandwidth at a sampling frequency of 340 MHz and 56 mW power consumption from a 1.2 V supply (90 nm CMOS). Recent publications of wideband continuous-time MASH delta-sigma architectures have demonstrated 85 dB DR/74.6 dB SNDR in 50 MHz (3-1 MASH) [5], 90 dB DR/72.6 dB SNDR in 53 MHz (0-3 MASH) [2] and 67 dB DR/64.7 dB SNDR in 465 MHz (1-2 MASH) [6].

2.4.2 A 4 GHz CT Delta-Sigma ADC with 125 MHz Bandwidth

In this section, a 4 GHz rate continuous-time delta sigma (CT$\Sigma\Delta$) ADC is presented [3]. The ADC achieves 70 dB dynamic range and −74 dBFS total harmonic

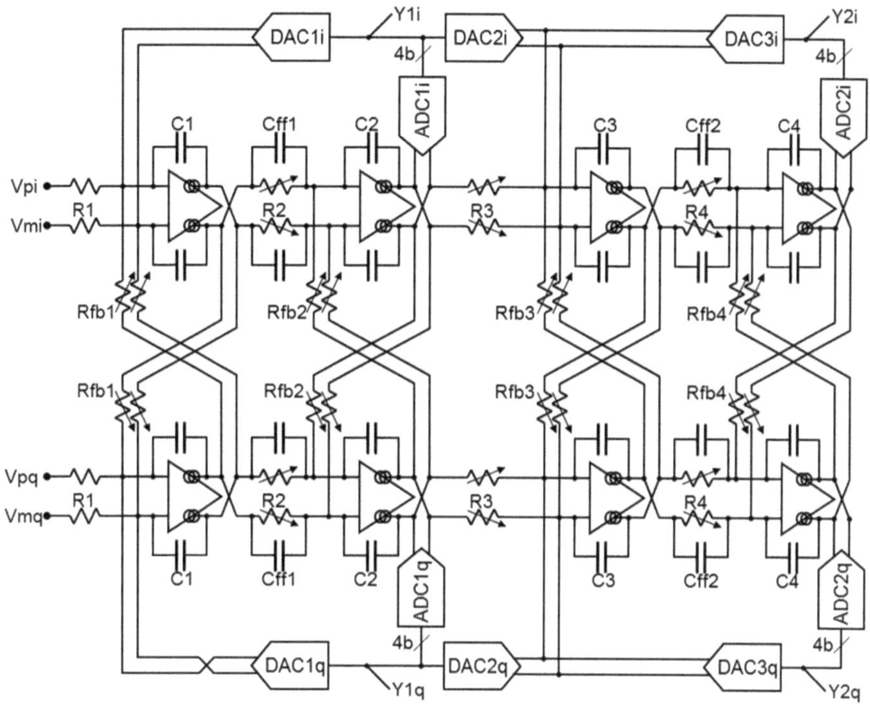

Fig. 2.10 Continuous-time quadrature 2-2 MASH

distortion (THD) in 125 MHz. Such large bandwidth and high linearity is achieved by employing a high-speed loop filter topology, in combination with a low latency 4-bit Quantizer and DAC architecture.

Figure 2.12 shows the block level diagram of the high-speed CTΣΔ modulator, which uses a capacitive feedforward loopfilter architecture. A feedforward loopfilter architecture requires a summation node for its feedforward coefficients. The summation node should not introduce additional ELD. A summation node can be implemented by using an active summing amplifier. When the modulator is clocked at GHz sampling rates, the limited gain-bandwidth product of the summing amplifier limits the SQNR performance of the modulator. The limited gain of the summing amplifier acts as a fixed attenuation in the delta-sigma loop and reduces the effective gain of the loop filter. Moreover, the limited BW of the summing amplifier acts as an additional pole in the loop and degrades the phase margin of the loop filter. For a design targeting 4 GHz sampling rate, an active feedforward summation node requires an amplifier with a gain-bandwidth product in excess of 20 GHz [26]. In addition to its limited GBW, the summation amplifier needs to drive a multi-bit quantizer, which further increases the GBW requirement of the amplifier. Therefore, the summing amplifier is one of the major bottlenecks that limits the maximum sampling speed of the modulator.

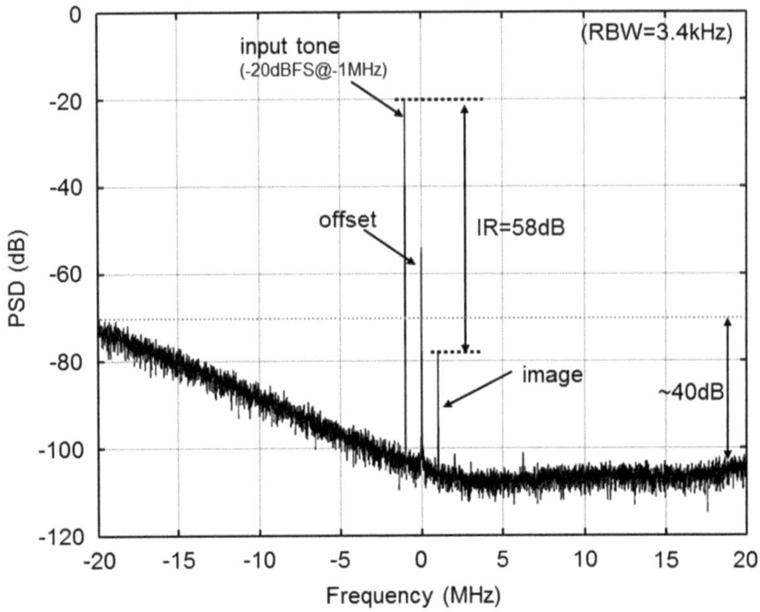

Fig. 2.11 Measured in-band spectrum of the CT quadrature 2-2 MASH ADC

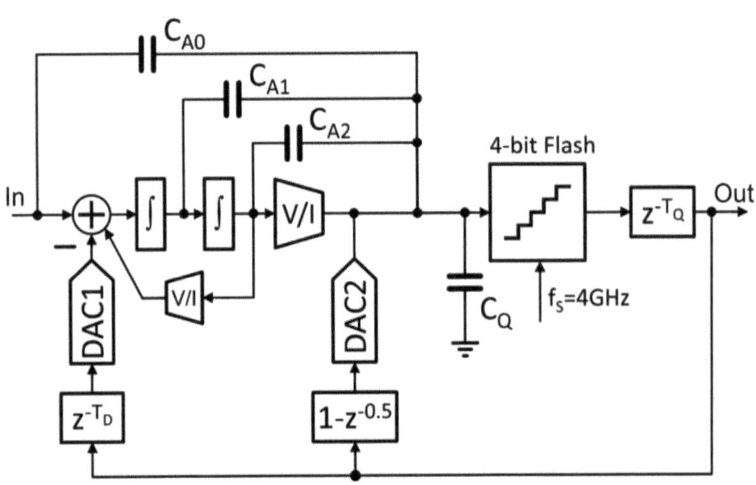

Fig. 2.12 Block diagram of the 3rd order CTΣΔmodulator

Instead of an active summation node, we have implemented a wideband passive summation node (CQ) for differentiated signals in current domain. The last stage of the loop filter is implemented as an gmC integrator, the input capacitance of the multi-bit quantizer can be used to realize of the loop filter poles. The ELD DAC (DAC2), which is required to create a high speed path around the quantizer,

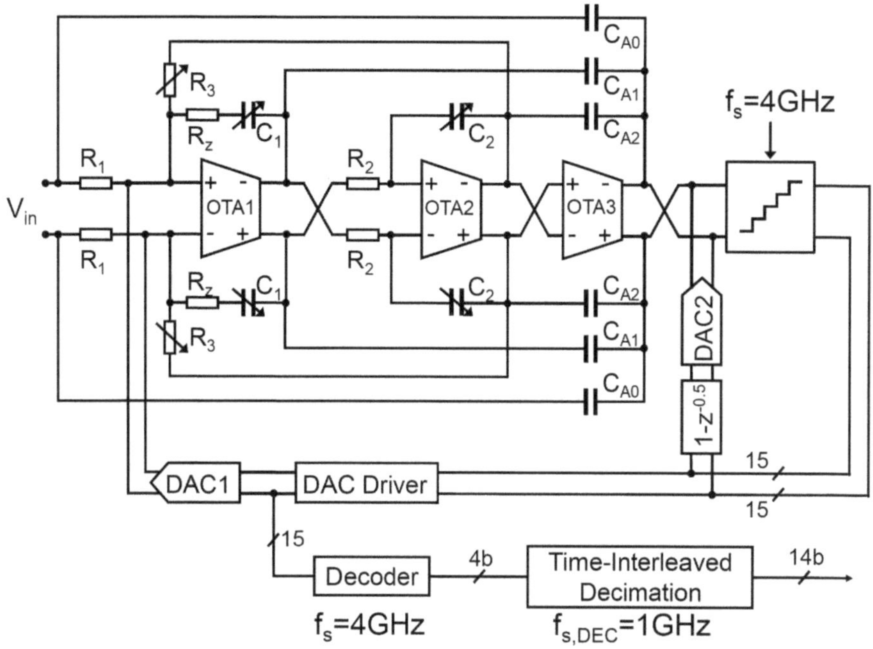

Fig. 2.13 Measured spectra with a 0.85 V@2.5 MHz and 0.85 V@7.67 MHz signal

integrates a digital differentiator [27]. Furthermore, an overall feed-forward path is implemented by CA0 to relax the requirements on the loop filter's linearity [28] and to reduce the peaking in the signal transfer function of the modulator at the cost of lower anti-alias filtering.

In Fig. 2.13 shows the top level architecture of the 3rd order CTΣΔ modulator. The first two integrators are implemented as RC integrators since they can meet the linearity requirement of −70 dBFS THD. Moreover, the zero introduced due to the limited transconductance (gm) of the first integrator, a resistor (Rz = 1/gm) in series with the integration capacitor (C1) is employed. The first two operational-transconductance-amplifiers (OTAs) are implemented as two-stage amplifiers with feedforward frequency compensation [29]. The last integrator is a g_mC integrator since it has relaxed linearity requirements. It is implemented as a resistively degenerated folded-cascode amplifier. Since, the modulator uses a capacitive feedforward architecture, the last OTA is not in the speed critical path thus relaxing its gain-bandwidth requirements. The power dissipation of the last integrator is negligible compared to first two integrators.

To compensate RC spread, the capacitors C1&C2, and resistor R3 can be separately adjusted via five-bit networks. The tuning range covers 0.5-2× of the nominal RC time constant. In order to reduce calibration overhead, the nominal bias current of the gmC integrator can also be varied 0.5-2× to calibrate the pole frequency of the third integrator of the loopfilter (ω3∝gm). The feedforward

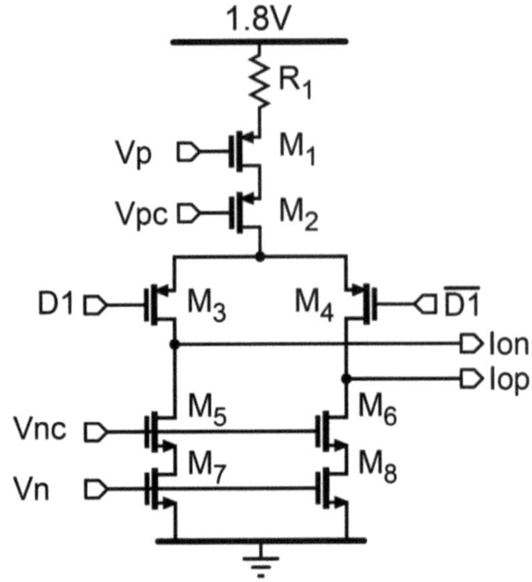

Fig. 2.14 Schematic of a unit element of 4-bit DAC1

coefficients of the loopfilter are fixed by the ratios of the capacitors (CA1 and CA2). The quantizer is implemented as a 4-bit flash ADC. To meet the stability requirement of the modulator, the quantizer should generate a valid signal in less than one clock period. In this design, we have allocated half a sampling-clock period (125 ps) to meet the stability requirements. The remaining half of the sampling period is reserved for propagation of signals and re-clocking of DAC input signals.

DAC1 is a 4-bit current-steering DAC designed for 11-bit intrinsic matching. Achieving this with MOS current sources consumes too much area and results in poor high-frequency linearity. Figure 2.14 shows one-unit element of DAC1. Therefore, DAC1 uses resistively degenerated PMOS current sources. By using a higher supply voltage for DAC1 (1.8 V), R_1 can be made larger, effectively reducing the noise contribution of DAC1 and reducing the ADC's overall power consumption. Since the voltage drop on R_1 is about 0.7 V, $M1-8$ can still be implemented using thin-oxide transistors. At high sampling rates, the unequal rise and fall time of the output of DAC1 can cause inter-symbol interference (ISI) [22, 30]. To minimize this, DAC1 employs a fully differential architecture and DAC1 driver's D-FF and switch drivers are dimensioned to achieve a signal-to-noise ratio (SQNR) of better than 80 dB.

The ADC has been fabricated in 45 nm baseline LP-CMOS and has an active area of 0.9 mm^2. The ADC including the decimation filter dissipates 256 mW from a 1.1 V supply and 3.2 mW from a 1.8 V supply, which is the supply voltage of the DAC1 as described before. Figure 2.15 shows an FFT of the measured-decimated output of the $\Sigma\Delta$ ADC with no input signal. The ADC's noise floor[1] is flat in the

[1]The noise floor is the average of four measurements.

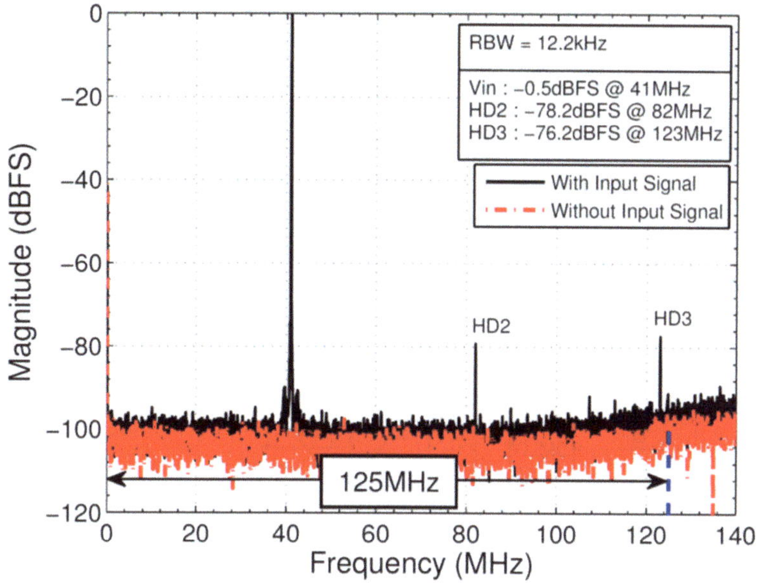

Fig. 2.15 FFT of measured output for an input signal of −0.5 dBFS at 41 MHz

signal BW of 125 MHz and rises slightly at higher frequencies due to the presence of out-of-band quantization noise. To measure the ADC's distortion, sinusoidal input signals with a maximum input voltage of 2.0-Vp − p differential were supplied to the ADC. The decimated output for a 41 MHz input signal at −0.5 dBFS has been captured in real-time; its FFT is shown in Fig. 2.15. The THD is −74 dBFS.

2.4.3 A 25 MHz BW CT Delta-Sigma ADC with <−100 dB THD

In [7] a continuous-time delta-sigma ADC is presented that achieves <−100 dB THD and 77 dB SNDR in 25 MHz bandwidth. The modulator comprises of a 1-bit feedback DAC, which is highly insensitive to process spread and mismatch, and a wideband high precision voltage regulator to mitigate dynamic errors of the 1-bit DAC.

Figure 2.16 shows the model block diagram of the delta-sigma ADC. It has a 4th-order loopfilter with two resonance filters (ω_1/s-ω_2/s-d_1 and ω_3/s-ω_4/s-d_2), two internal feedforward paths (c_2 and c_3), a signal feedforward path (c_1) and three 1-bit feedback DACs. The main feedback DAC (DAC1) and the reference circuit are critical for the linearity of the ADC. Although theoretically a 1-bit DAC is inherently linear, any signal- or data-dependent residue on the DAC reference voltage will lead to distortion, spurious tones and increased in-band noise. To achieve the noise and

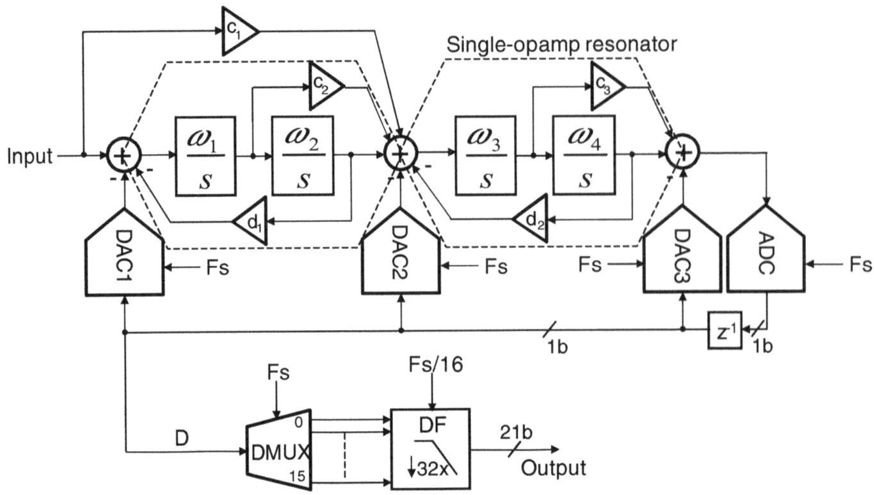

Fig. 2.16 Block diagram of the 4th-order delta-sigma ADC

distortion performance a resistive return-to-open DAC is used. During open state the data dependent charge in the parasitic capacitance at the switch side of the DAC resistor must be sufficiently discharged via the DAC resistor to the loopfilter virtual ground nodes to avoid distortion due to memory effects. The loopfilter model is mapped to a cascaded implementation of two single-opamp resonators [28, 31]. The resonator filter implementation has an inherent zero. Therefore, no additional summing amplifiers are needed which saves power as only two amplifiers are used to implement the complete 4th-order loopfilter. The pseudo differential amplifier consists of three inverter stages with Miller compensation. The common-mode of the amplifier input nodes is controlled via resistors by an inverter-based common-mode amplifier sensing the output common-mode. The bias current of the first resonator filter is dictated by linearity requirements and the current of the second filter by speed requirements.

The choice for a partial feedforward and feedback topology of the loopfilter is a trade-off between power, noise and minimal out-of-band (OOB) peaking of the signal transfer function (STF). STF peaking is highly undesired as it results in loss of dynamic range for OOB interferers. Signal feedforward path c_1 has been added to reduce STF peaking close to the signal band to ~4 dB. Because of the continuous-time loopfilter, no extra anti-alias filter is required. The 1-bit quantizer is dithered and has a local excess loop delay (ELD) compensation DAC (DAC3) to allow for one clock period delay in the feedback loops. The quantizer and DACs are clocked at 2.2 GHz and the time budget for the quantizer latch is maximized to half a clock cycle to minimize meta-stability related errors. The remaining time budget accounts for latency of buffer and retiming circuits. The 2.2 GHz 1-bit output data is de-multiplexed and fed to a decimation filter (DF) that decimates by 32× and outputs 21-bit data at 68.75 MS/s.

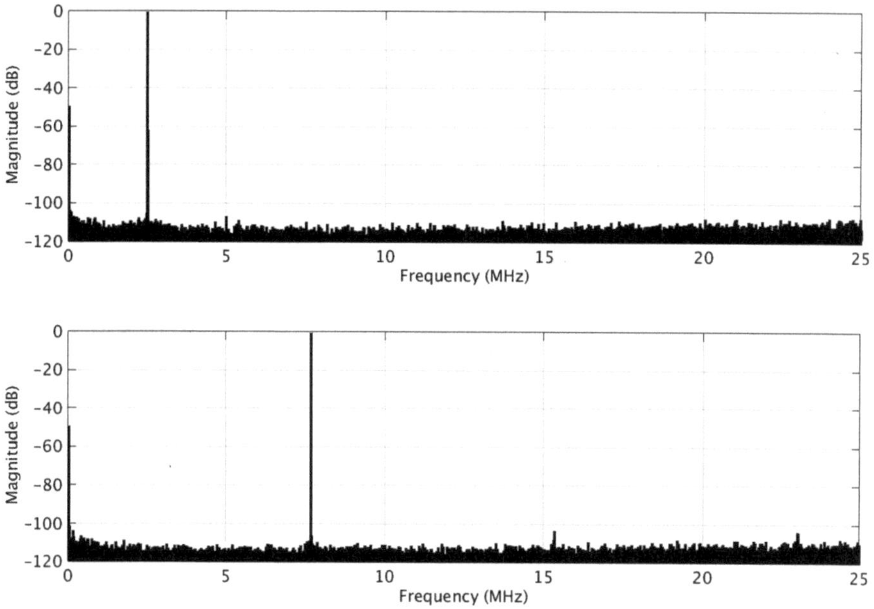

Fig. 2.17 Measured spectra with a 0.85 V@2.5 MHz and 0.85 V@7.67 MHz signal

The ADC has been fabricated in a TSMC 65 nm CMOS process. The modulator and the 1.2 V supply regulators measure 0.25 mm^2 and 0.35 mm^2 respectively. Figure 2.17 shows the output spectra (RBW 880 Hz) with 0.85 V peak differential input signals at 2.5 MHz and 7.667 MHz respectively. In both measurements HD2 and HD3 are below -100 dBc. The dynamic range and peak SNDR are 77 dB in 25 MHz bandwidth. The modulator power consumption including clock distribution is 41.4 mW.

2.5 Conclusions

This paper describes design aspects of high resolution and wideband continuous-time ΣΔ modulators for wireless applications. The optimal noise transfer function and the choice of architecture are the two most important design steps required to achieve the resolution in a given bandwidth. First of all, in a single loop delta-sigma modulator, the main design parameters are the loopfilter order, oversampling ratio and the accuracy of the ADC and DAC. Since a delta-sigma modulator is a feedback system, the loop stability requirements limit the maximum sampling rate in practical implementations, thus the bandwidth of the modulator. In that case, multi-stage noise-shaping (MASH) delta-sigma modulator architecture enables wider bandwidth by eliminating the maximum sampling rate limitation.

The design space of single-loop and MASH delta-sigma modulator architectures is limited by speed and accuracy limitations of the available process technologies. Recently published delta-sigma modulators are clocked at GHz sampling rates, the non-idealities associated with the internal ADC and DAC of the modulator limit their performance. Metastability and ELD limit the maximum sampling rate, whereas the non-linearity and phase noise of the sampling clock limit the achievable dynamic range.

Three implementations of CT delta-sigma modulators are presented which advanced the state-of-the-art envelope in terms of architecture, bandwidth and linearity. The first implementation introduced a continuous-time 2-2 MASH delta-sigma modulator architecture which achieves 77 dB DR in 20 MHz and 56 mW power consumption. The second implementation pushed the sampling speed of single loop delta-sigma modulators to 4 GHz by employing a high-speed capacitive feedforward loop filter architecture and achieved the 70 dB DR in 125 MHz. The third implementation achieves <-100 dB THD and 77 dB SNDR in 25 MHz bandwidth. Such high linearity is enabled using a 1-bit feedback DAC and integrated high precision voltage regulator.

References

1. P. Shettigar, S. Pavan, A 15mW 3.6GS/s CT-ΣΔ ADC with 36MHz bandwidth and 83dB DR in 90nm CMOS. *ISSCC Digest of Technical Papers*, pp. 156–157, Feb 2012
2. Y. Dong, R. Schreier, W. Yang, S. Korrapati, A. Sheikholeslami, A 235mW CT 0-3 MASH ADC achieving -167dBFS/Hz NSD with 53MHz BW. *ISSCC Digest of Technical Papers*, pp. 480–481, Feb 2014
3. M. Bolatkale, L.J. Breems, R. Rutten, K. Makinwa, A 4GHz CT ΣΔ ADC with 70dB DR and -74dBFS THD in 125MHz BW. *ISSCC Digest of Technical Papers*, pp. 470–471, Feb 2011
4. H. Shibata, R. Schreier, W. Yang, A. Shaikh, D. Paterson, T. Caldwell, D. Alldred, P.W. Lai, A DC-to-1GHz tunable RF ΣΔ ADC achieving DR=74dB and BW=150MHz at f_0=450MHz using 550mW. *ISSCC Digest of Technical Papers*, pp. 150–151, Feb 2012
5. D. Yoon, S. Ho, H. Lee, An 85dB-DR 74.6dB-SNDR 50MHz-BW CT MASH ΣΔ Modulator in 28nm CMOS. *ISSCC Digest of Technical Papers*, pp. 272–273, Feb. 2015
6. Y. Dong, J. Zhao, W. Yang, T. Caldwell, H. Shibata, R. Schreier, Q. Meng, J. Silva, D. Paterson, J. Gealow, A 930mW 69dB-DR 465MHz-BW CT 1-2 MASH ADC in 28nm CMOS. *ISSCC Digest of Technical Papers*, pp. 278–279, Feb 2016
7. L. Breems, M. Bolatkale, H. Brekelmans, S. Bajoria, J. Niehof, R. Rutten, B. Oude-Essink, F. Fritschij, J. Singh, G. Lassche, A 2.2GHz continuous-time ΣΔ ADC with -102dBc THD and 25MHz BW. *ISSCC Digest of Technical Papers*, pp. 272–273, Feb 2016
8. Y. Matsuya, K. Uchimura, A. Iwata, T. Kobayashi, M. Ishikawa, T. Yoshitome, A 16-bit oversampling A-to-D conversion technology using triple-integration noise-shaping. IEEE J. Solid-State Circuits, **22**(6), 921–929 (1987)
9. L. Breems, A cascaded continuous-time ΣΔ modulator with 67dB dynamic range in 10MHz bandwidth. *ISSCC Digest of Technical Papers*, pp. 72–73, Feb 2005
10. J. Cherry, W. Snelgrove, Clock jitter and quantizer metastability in continuous-time delta-sigma modulators. IEEE Trans. Circuits Syst. II **46**, 661–676 (1999)
11. P. Benabes, M. Keramat, R. Kielbasa, A methodology for designing continuous-time sigma-delta modulators, *in Proceedings of European Design and Test Conference, ED TC'97*, pp. 46–50, Mar 1997

12. P. Witte, J. Kauffman, J. Becker, Y. Manoli, M. Ortmanns, A 72 dB-DR CT ΣΔ modulator using digitally estimated auxiliary DAC linearization achieving 88 fJ/conv in a 25 MHz BW. *ISSCC Digest of Technical Papers*, 2012, pp. 154–156, 2012
13. J. Kauffman, P. Witte, J. Becker, M. Ortmanns, An 8.5 mW continuous-time ΣΔ modulator with 25 MHz bandwidth using digital background DAC linearization to achieve 63.5 dB SNDR and 81 dB SFDR. IEEE J. Solid State Circuits **46**(12), 2869–2881 (2011)
14. Y.-S. Shu, J.-Y. Tsai, P. Chen, T.-Y. Lo, P.-C.Chiu, A28 fJ/convstep CT delta sigma modulator with 78 dB DR and 18 MHz BW in 28 nm CMOS using a highly digital multibit quantizer. *ISSCC Digest of Technical Papers*, pp. 268–269, 2013
15. T. Cataltepe et al., Digitally corrected multi-bit ΣΔ data converters. *ISCAS*, pp. 647–650, May 1989
16. M.J. Story, Digital to analogue converter adapted to select input sources based on a preselected algorithm once per cycle of a sampling signal. U.S. Patent 5 138 317, 11 Aug 1992
17. B.H. Leung, S. Sutarja, Multi-bit sigma-delta A/D converter incorporating a novel class of dynamic element matching techniques. IEEE Trans. Circuits Syst. II **39**, 35–51 (1992)
18. R. Schreier, B. Zhang, Noise-shaped multi-bit D/A converter employing unit elements. Electron. Lett. **31**(20), 1712–1713 (1995)
19. R.T. Baird, T. Fiez, Linearity enhancement of multi-bit ΣΔ A/D and D/A converters using data weighted averaging. IEEE Trans. Circuits Syst. II **42**, 753–762 (1995)
20. T. Kwan, R. Adams, R. Libert, A stereo multi-bit ΣΔ D/A with asynchronous master-clock interface. *ISSCC Digest of Technical Papers*, vol. 39, Feb 1996, pp. 226–227
21. I. Galton, Spectral shaping of circuit errors in digital-to-analog converters. IEEE Trans. Circuits Syst. II **44**(10), 808–817 (1997)
22. R.W. Adams, Design and implementation of an audio 18-bit analog-to-digital converter using oversampling techniques. J. Audio Eng. Soc. **34**(3), 153–166 (1986)
23. L. Risbo, R. Hezar, B. Kelleci, H. Kiper, M. Fares, A 108 dB DR, 120 dB THD and 0.5 Vrms output audio DAC with intersymbol interference shaping algorithm in 45 nm. *ISSCC Digest of Technical Papers*, pp. 484–485, 2011
24. B. Putter, ΣΔ ADC with finite impulse response feedback DAC. *ISSCC Digest of Technical Papers*, pp. 76–77, Feb 2004
25. L. Breems, R. Rutten, R.H.M. van Veldhoven, G. van der Weide, A 56mW continuous-time quadrature cascaded ΣΔ modulator with 77dB DR in a near zero-IF 20MHz band. *IEEE J. Solid-State Circuits*, pp. 2696–2705, Dec 2012
26. M. Bolatkale, L. Breems, K. Makinwa, *High Speed and Wide Bandwidth Delta-Sigma ADCs* (Springer, Dordrecht, 2014)
27. G. Mitteregger et al., A 20-mW 640-MHz CMOS continuous-time ADC With 20 MHz signal bandwidth, 80-dB dynamic range and 12-bit ENOB. IEEE J. Solid State Circuits **41**(12), 2641–2649 (2006)
28. H. Chae, J. Jeong, G. Manganaro, M. Flynn, A 12mW low-power continuous-time bandpass ΣΔ modulator with 58dB SNDR and 24MHz bandwidth at 200MHz IF. *ISSCC Digest of Technical Papers*, pp. 148–149, 2012
29. J. Silva, U. Moon, J. Steensgaard, G. Temes, Wideband low distortion delta-sigma ADC topology. Electron. Lett. **37**(12), 737–738 (2001)
30. E. van der Zwan, E. Dijkmans, A 0.2-mW CMOS ΣΔ modulator for speech coding with 80 dB dynamic range. IEEE J. Solid State Circuits **31**(12), 1873–1880 (1996)
31. S. Zeller et al., A 9th-order continuous time ΣΔ-ADC with X-coupled differential single-opamp resonators *MWSCAS*, pp. 1–4, Aug 2011

Chapter 3
Sigma-Delta ADCs with Improved Interferer Robustness

Rudolf Ritter, Jiazuo Chi, and Maurits Ortmanns

3.1 Introduction

The Sigma-Delta ($\Sigma\Delta$) ADC is a widely used topology for analog-to-digital conversion from slow to fast conversion rates as well as from low to high resolutions [1, 2]. In recent years, it has become one of the favorable ADCs for wireless applications, where it is affected by strong interferer signals. This is obvious for receivers in which beside the signal of interest, also called the inband signal, at least the interfering adjacent channels are received. From this scenario, a common challenge arises: it is often not the inband signal itself which defines the overall required dynamic range, but the interfering signals, which can be orders of magnitude larger than the inband signal and thus they define the dynamic range, the linearity as well as the jitter requirements [3–5].

In order to illustrate such requirements of a $\Sigma\Delta$ ADC, a simplified homodyne receiver block diagram is given in Fig 3.1 from the antenna to the digital signal processor. The signals entering the antenna at the input are illustrated as a weak received signal of interest (RX) and the surrounding—possibly strong—interferers. The antenna is often reused for the transmitted signal (TX), which is subsequently separated by a duplexer towards the receive chain. The TX cannot be removed completely and thus the TX is often one of the dominant input signals at this point of the receiver signal chain and becomes consequently a strong interferer. A band select filter (BSF) passes the band with all channels, which might be of interest, and blocks all others. A low noise amplifier (LNA) including an automatic gain control (AGC), which is mostly the first integrated stage of the receiver, amplifies the remaining signal to a maximum level without overload and reduces consequently the dynamic range requirements of the subsequent stages. An IQ-demodulator converts

R. Ritter • J. Chi • M. Ortmanns (✉)
Institute of Microelectronics, University of Ulm, Ulm, Germany
e-mail: maurits.ortmanns@uni-ulm.de

© Springer International Publishing Switzerland 2017
A. Baschirotto et al. (eds.), *Wideband Continuous-time $\Sigma\Delta$ ADCs, Automotive Electronics, and Power Management*, DOI 10.1007/978-3-319-41670-0_3

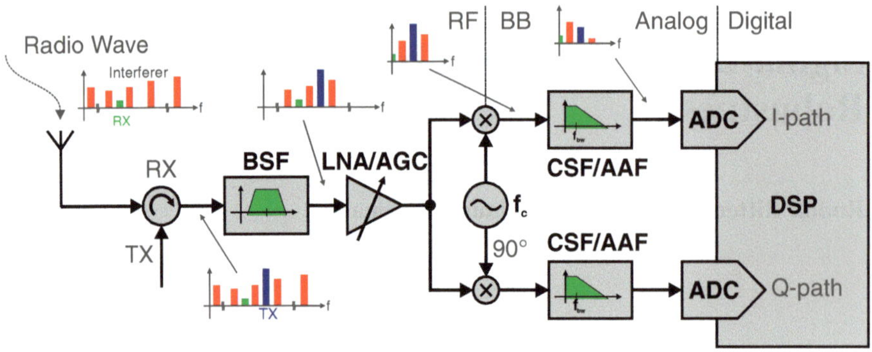

Fig. 3.1 Homodyne receiver block diagram including signal levels. Source: [6]

the inband signal to base-band (BB). A last filter stage acts as channel select filter (CSF), to separate the wanted channel from the others, and as anti aliasing filter (AAF), such that the preceding ADC can work without overload or distortion.

Implementing the ADC in such a receiver by a ΣΔ ADC is advantageous due to its increased interferer robustness compared to Nyquist rate ADCs; this basic advantage is outlined in Sect. 3.2 reviewing the advantage of implicit signal filtering, anti-alias filtering and oversampling. More recently, ΣΔ architectures were investigated to allow even improved features regarding the robustness to interferers which are summarized in Sect. 3.3. A short conclusion is given in Sect. 3.4.

3.2 ΣΔ Modulators in an Interferer Environment

The ΣΔ ADC is very well suited for applications including interferers, for mainly two reasons. Due to the oversampling nature of the ΣΔ ADC, the interferers close to the inband don't have to be attenuated below the inband signal level in order to avoid aliasing as in Nyquist rate ADCs, but only to a level such that no overload in the ΣΔ modulator occurs. The filtering is done in the digital domain which is especially advantageous together with technology scaling [7]. The signal band, which generates distortion by aliasing, is typically far enough from the inband, such that simple AAF filtering can be used. This is even improved with continuous time (CT)-ΣΔ ADCs which feature an inherent AAF making an explicit AAF needless [8]. Secondly, the ΣΔ modulator features a signal transfer function (STF), which enables an intrinsic filtering of the interferers; this allows a reduction of the requirements of the preceding stages or even enables their elimination. Therefore, the ADC in Fig. 3.1 can be implemented with a filtering STF, which then partly or even completely replaces an explicit CSF [4, 5, 9–14].

3.2.1 RF ΣΔ Modulators

Modern wireless devices often embed many different receivers with different requirements, for example a smart-phone receives Bluetooth, GPS, UMTS, LTE, WLAN, etc. This led many years ago to the idea of a software defined radio, which—due to the high flexibility of digital circuitry and its advantages in modern CMOS technologies—realizes as many as possible components of the receiver signal chain by digital circuitry; consequently, the required ADC is shifted as close as possible towards the antenna and the ADC is ideally directly connected to the antenna [15, 16]. Such an ADC would consequently work on radio frequency input signals, but with speed and resolution requirements which are not—and presumably will never become—feasible to be implemented efficiently for consumer devices.

A more common solution, which is one step closer to this idea than the classical receiver in Fig. 3.1, is the application of bandpass (BP) ΣΔ modulators as sketched in Fig. 3.2a [17–19]. This ΣΔ modulator features a BP STF and its required loop filter works completely on RF signals—in contrast to the lowpass (LP) ΣΔ modulator, which works on baseband (BB) signals. Therefore, this technique allows to shift the mixer and the IQ-path separation into the digital domain, in which all their implementation issues are more efficiently manageable. The major drawback of this technique is the loop filter which is built in the RF domain as bandpass filter,

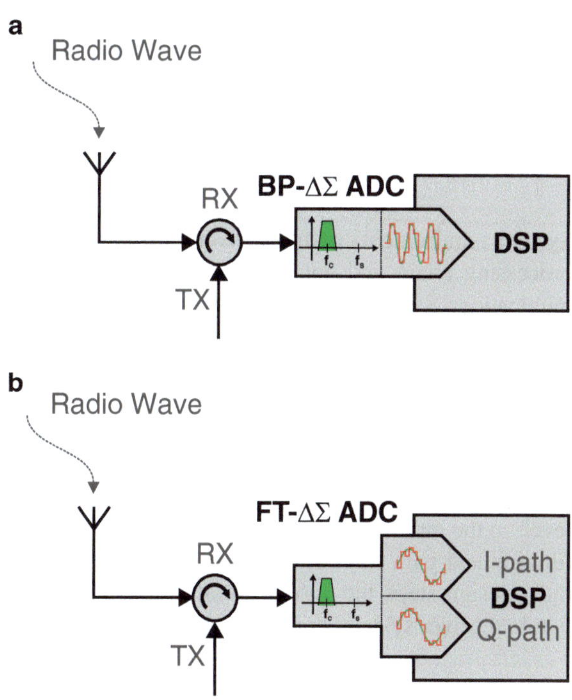

Fig. 3.2 Homodyne receiver block diagram applying RF ΣΔ modulators. (**a**) BP-ΣΔ ADC. (**b**) FT-ΣΔ modulator. Source: [6]

and whose implementations are—up to today—by far not as competitive as LP ΣΔ modulators as stated throughout the state-of-the-art (SoA) [20].

Another solution is the frequency translating (FT) ΣΔ modulator [21, 22], which works on RF input signals as the BP architecture, but whose digital output signal is similar to the LP ΣΔ modulator as sketched in Fig. 3.2b. This is achieved with internal analog mixing within the loop filter of the FT ΣΔ modulator. This is advantageous as major parts of the ΣΔ modulator can be built in the LP domain, which is more efficiently done compared to the bandpass implementation. In contrast, the mixer and IQ signal processing cannot be omitted as in the BP ΣΔ modulator, but they are located at nodes with non-ideality suppression within the loop filter. This topology at first sight seems quite advantageous compared to the BP ΣΔ modulator, but due to an additionally required mixer for up-conversion within the loop filter feedback path and thus at a most sensitive signal node, the advantages are compromised.

Both techniques are promising for future radio receivers, but cannot compete with the power efficiency of LP ΣΔ modulators in classical receiver implementations [20]. This is due to the implementation challenges in the BP domain which are—beyond others—the low Q-factors of integrated resonators and the increased sensitivity to jitter. Moreover, as the bandwidth of the RF ΣΔ modulators is limited, they need to be highly reconfigurable to enable the demand on the large flexibility of multi standard receivers.

This article will thus focus—without loss of generality—on LP CT-ΣΔ modulators and their behavior and improvements in the presence of OOB interferers, while most of the statements and principles are transferable to BP and FT ΣΔ modulators.

3.2.2 ΣΔ Modulators DR Requirements

System requirements of transceivers are typically specified for the whole signal processing chain and not directly for the ADC. The ETSI (European Telecommunications Standards Institute) provides for example the specifications for LTE receiver systems, where the most important values are plotted in Fig. 3.3 exemplary for the 20 MHz mode [5, 23]. Thereby, the minimum and maximum inband as well as worst case interferer signal levels are specified at the antenna. Combinations of them are defined, in which the receiver is required to work fully functional with a specified maximum bit error rate (BER). As the ΣΔ modulator is usually not directly connected to the antenna, but subsequently to the mixer or a CSF, the signal levels at the input of the ΣΔ modulator must be adopted according to the preceding filtering and gain stages in order to identify its requirements. Furthermore, during most design steps the ΣΔ modulator is simulated standalone and not within the whole receiver system; hence, its own SNR instead of the overall BER is evaluated

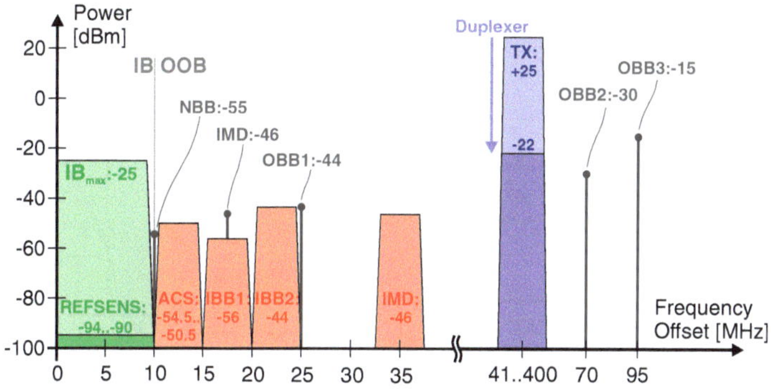

Fig. 3.3 Signal levels for a 20 MHz LTE receiver system. Source: [6]

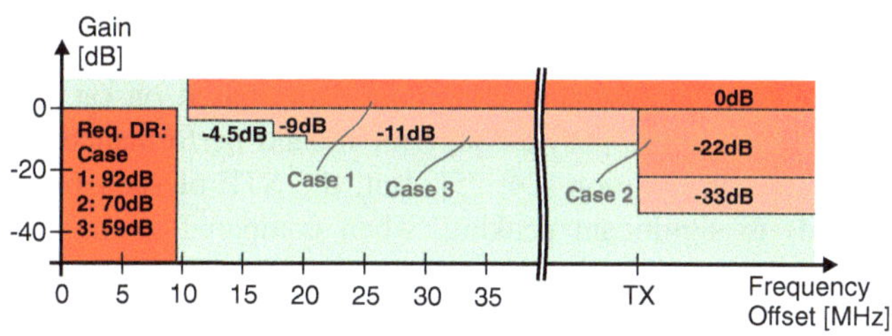

Case 1: No explicit nor implicit filtering
Case 2: Explicit or implicit filtering at TX
Case 3: Strong explicit or implicit filter at TX location and weak filter at close to IB frequencies

Fig. 3.4 Exemplary filter masks for the CSF and $\Sigma\Delta$ modulator in series

which does not fit to the receiver specifications. It is in the following generally assumed, that an inband SNR of typically 20 dB is sufficient for the digital baseband.

This translates in filter masks, for which 3 are exemplary given in Fig. 3.4 [5] for the 20 MHz LTE case; such filter masks then need to be fulfilled by the $\Sigma\Delta$ modulator in series with the preceding CSF. As seen from Fig. 3.4, a $\Sigma\Delta$ modulator without any explicit CSF would require a DR of more than 92 dB for proper operation without overload in all close to IB test cases. On the other hand, a strong preceding filter, whether explicitly realized in the CSF or implicitly in the $\Sigma\Delta$ modulator, reduces the required IB DR to only 59 dB, which is easily achievable; but this comes at the cost of a rather complex filter of at least third order. A compromised solution requires medium filtering at the TX location with -22 dB attenuation, coming with relaxed filtering requirements and a more reasonably achieved DR $\approx$ 70 dB.

The loop filter of a $\Sigma\Delta$ modulator, as illustrated in Fig. 3.5a, can be implemented in various ways whereas the major differences can be characterized into feed-forward (FF) and feedback (FB) compensated architectures. It is shown in

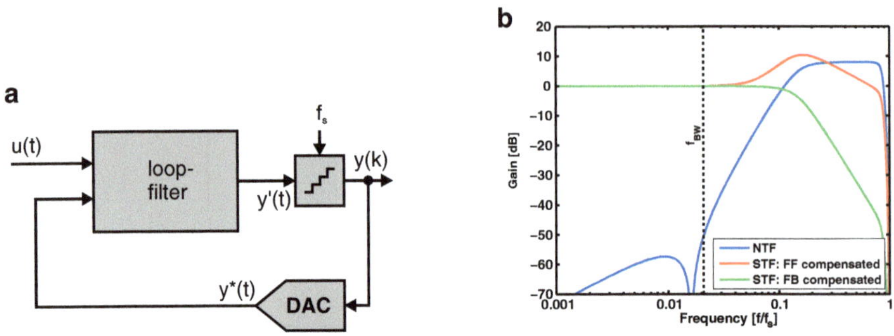

Fig. 3.5 FF vs. FB compensated ΣΔ modulators. (**a**) ΣΔ modulator block diagram. (**b**) NTF and STF of a third order modulator with 4 bit internal quantization, OOB NTF gain of 8 dB and OSR = 24

many SoA designs, that wideband ΣΔ modulators based on (at least partially) FF compensated loop filters provide superior inband performance with the most competitive power consumption [24, 25]. But, the STF of such FF-compensated modulators tends to significant peaking, when compared to the STF of a FB-compensated modulator with an otherwise equivalent NTF. Thus, as illustratively calculated in Fig. 3.5b, interferers are thereby not attenuated by the STF, but they are even amplified.

Thus—to continue our example from Fig. 3.4—gain at OOB frequencies even increases the required IB DR beyond the stated 92 dB (without attenuation in case 1) which obviously needs to be avoided.

Consequently, it can be stated that ΣΔ modulators with an OOB peaking STF require significantly better inband dynamic range and linearity than modulators with flat or attenuating STF; this needs to be considered in comparing various implementations, which is usually not reflected in Figures-of-Merit or comparison tables. This can be concluded as a quite general shortcoming of published state of the art.

3.2.3 ΣΔ Modulators Internal Dynamics

Designing the STF of a ΣΔ modulator according to a specified filter mask, as outlined through Sect. 3.2.2 and exemplary illustrated in Fig. 3.4 reduces the required IB DR without clipping and distorting the ΣΔ modulator's output; this is variously stated throughout the state of the art, and techniques to improve the overall signal transfer function have been reported.

But obviously, the loop filter also features internal physical nodes which are not allowed to clip as well [5, 26]. The exemplary fourth order FB compensated CT-ΣΔ modulator as sketched in Fig. 3.6, which fulfills the second filter mask in Fig. 3.4, is calculated in Fig. 3.7a, where not only the overall STF, but also the internal

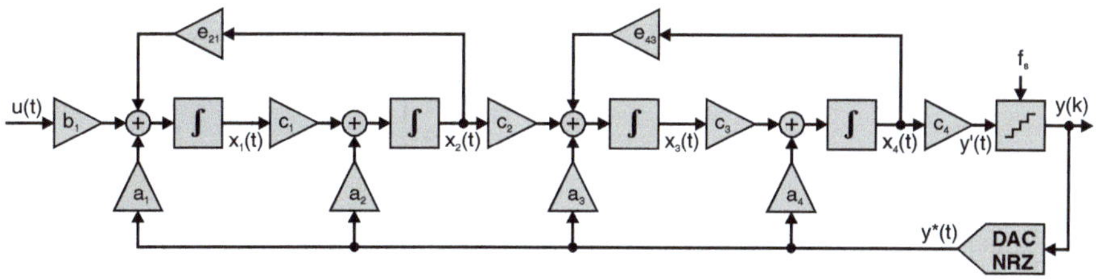

Fig. 3.6 Fourth order FB compensated ΣΔ modulator. Source: [6]

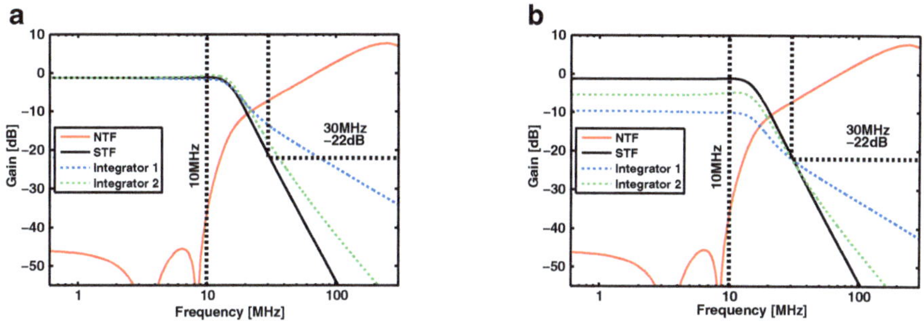

Fig. 3.7 Fourth order ΣΔ modulator fulfilling case 2 in Fig. 3.4 [5]. (**a**) Scaled for IB. (**b**) Scaled according filter mask

transfer functions are shown. The internal transfer functions are thereby calculated according to [5], a technique which will also be outlined at the end of this section. The integrators in this example were scaled for the overall NTF and STF fulfilling the design case 2 in Fig. 3.4; but now the first and the second integrator output violate the filter mask without any further countermeasures. Thus, proper operation in the presence of interferers is not guaranteed. This is because the signal transfer functions to the first and a second integrator outputs feature only a first and a second order LP filter, respectively, even though the input-to-output STF features a third order roll-off. Hence, the weights within the loop filter must be adopted such that the overall STF and NTF are unaltered, but the gain factors are changed according to the defined filter mask such as to change the dynamics at the internal states. This leads to the transfer functions in Fig. 3.7b—the process is called rescaling: by using this procedure, the loop gain in the early stages is reduced and shifted to the later stages, thereby the non-idealities and noise attenuation of the early stages is disadvantageously reduced.

Even though the process of rescaling for ΣΔ modulators is often reported in publications and is at least for DT-ΣΔ modulators also available by using the delta-sigma design toolbox [27], those approaches generally miss one important point: due to the generation of aliases and due to the superposition of those signals, the internal swings can become worse as it would be expected from individual signal transfer functions [5].

Thus, in order to be able to rescale the ΣΔ modulator according to a filter mask, the transfer functions to the internal nodes need to be accurately calculated

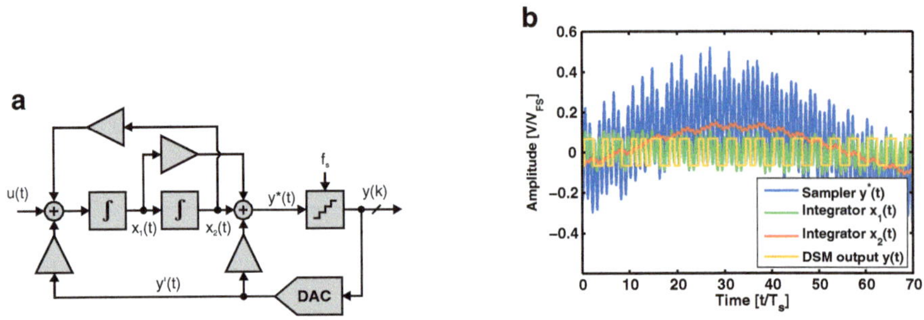

Fig. 3.8 Internal swings of a ΣΔ modulator [5]. (**a**) ΣΔ modulator block diagram. (**b**) Transient signals

or estimated. This is not easily achieved as illustrated with Fig. 3.8 and more extensively discussed in [5]. In this simulation a large input signal is applied with a frequency close to the sampling rate. The modulator output toggles only around one LSB due to the implicit AAF of the CT-ΣΔ modulator and the strong input-output attenuation of the signal close the f_S.

But from this simulation, it is also obviously noticed that the input signal of the quantizer $y^*(t)$ is not quiet at all, which is in contrast to what the output signal would imply. This is because the continuous-time signal strongly moves between the sampling events, and only at the sampling instants the quantizer input signal is truly close to zero. Thereby, a low frequency component can be observed at the output of the second integrator in Fig. 3.8b, which is not at the same frequency as the input signal. This frequency is a replica, which is generated by the sampler and which is fed back over the feedback DAC.

Consequently, neither pure CT nor pure DT calculations can be applied to approximate the internal transfer functions accurately enough. This is since the DT calculation does not take the signal movement between the sampling instants into account, while the CT calculation misses the aliases generated by sampling. Hence, a method was presented in [5] which allows the calculation of worst case transfer functions to the internal nodes of a CT-ΣΔ modulator, more precisely of the worst case envelopes of the fundamental signal and all superposed aliases. This method has been used to rescale the modulator coefficients from Fig. 3.7b which yielded the modulator transfer functions calculated in Fig. 3.7b, i.e. all internal transfer functions now fulfill the mask. Moreover, note that the shown transfer functions to the internal nodes in Fig. 3.7 are calculated upon exactly this worst case approach from [5].

Concludingly, it can be stated that the internal dynamics of ΣΔ modulators have to be paid the same attention as the overall input-output transfer functions in the case of OOB interferers. Thereby, not only the fundamental of the input signal must be considered, but also worst case superposition of the fundamental with its aliases and the internal states need to be rescaled for this.

3.2.4 ΣΔ Modulators Jitter Issues

Another major challenge in the presence of strong interferers, which can be accounted for on system level, is the robustness with respect to clock jitter. The sampler of CT-ΣΔ modulators is, in contrast to its DT counterpart, at the least sensitive position within the loop filter, and sampling jitter can thus be usually ignored. As has been stated from its very early publications on, the high jitter sensitivity of CT-ΣΔ modulators originates from the feedback DAC, and jitter errors caused within this DAC are injected into the input and consequently they are not attenuated towards the output.

This jitter induced error can be separated into two parts: the first error source is timing jitter in the feedback DAC together with the (strong OOB) quantization noise, sketched in Fig. 3.9a as a random error signal $e_J(t)$. This is the most commonly regarded error source, and many approaches have been proposed to counteract it, e.g. by a reduction of the quantizer step size which is realized by using an internal multi-bit quantizer with a larger number of quantization levels [28], or also shaped feedback DAC waveforms as exponential decaying [29], raised cosine [30] or FIR feedback [31, 32].

The second error source of jitter in the feedback DAC, which behaves more similar to aperture jitter of a sampler, is visualized in Fig. 3.9b. In this case, the distortion $e_J(t)$ is not caused by the quantization steps, but by the feedback signal converted by the DAC. Thereby, the jitter induced error is increased with larger signal amplitudes and faster signal changes. This effect is already known from RZ feedback DACs and it becomes the most dominant error in case of non-white clock jitter, e.g. in the case of accumulated clock jitter [33]. This jitter, originating from colored phase noise, creates skirts around CW signals, and thus—in the presence of strong OOB interferers—causes strong noise leakage into the inband.

This second error source of jitter is often not regarded in the literature, as there is also no obvious countermeasure by e.g. feedback pulse shaping. Nonetheless, it is obvious to understand, that architectures with good interferer robustness should

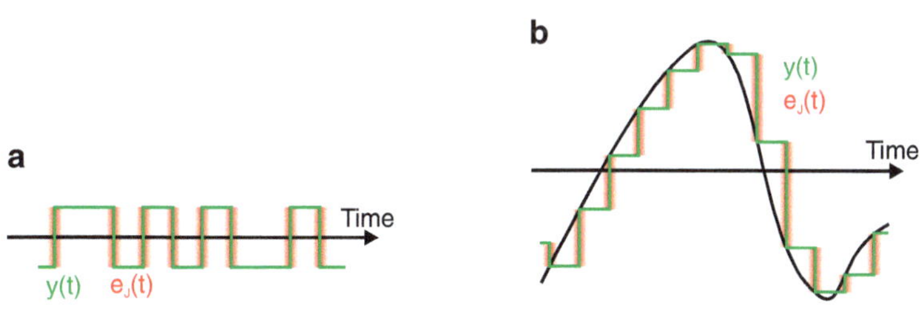

Fig. 3.9 Jitter in ΣΔ modulators (**a**) Together with quantization noise. (**b**) Together with feedback signals. Source: [6]

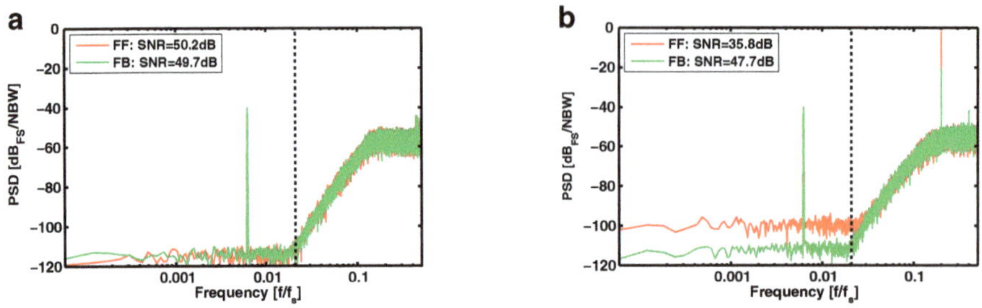

Fig. 3.10 Jitter effect in FF vs. FB compensated ΣΔ modulators. (**a**) Without interferer. (**b**) With interferer

also be more jitter tolerant: e.g. a FB compensated modulator should be better than the FF compensated modulator, as the strong OOB interferers are attenuated in the FB compensated case [34] and thus less jitter noise should leak into the inband.

This effect is demonstrated in Fig. 3.10, where third order, OSR = 24, 4 bit NRZ FB and FF compensated CT-ΣΔ modulators with the NTF and STFs as in Fig. 3.5b are simulated in the presence of white jitter noise ($\sigma_J = 0.001/f_s$). The ΣΔ modulators are firstly simulated without interferers in Fig. 3.9a in which no differences are observable; they are simulated again with a strong interferer signal at $f_{int} = 0.2\,f_s$ and an amplitude of $S_{int} = -12\,dB_{FS}$. The inband performance of the FB compensated ΣΔ modulator is almost not affected as it was expected due to its implicit attenuation of the interferer at this high frequency; this is in contrast to the amplification within the STF of the FF compensated ΣΔ modulator, which consequently shows an interferer amplified to almost $0\,dB_{FS}$, which yields a strong jitter noise leakage into the inband.

Concluding, jitter tolerance in CT-ΣΔ modulators is another important factor which needs to be considered in the presence of interferers, and as a general statement good interferer tolerance also leads to improved jitter tolerance. Thus, an interferer tolerant CT-ΣΔ modulator can not only save power by relaxing or omitting the CSF, but also by relaxing the jitter requirements and thus PLL implementation.

3.3 State-of-the-Art Filtering ΣΔ Modulator Techniques

Over the last years, the STFs of ΣΔ modulators received an increased attention throughout the literature, whereas traditional architectures have been modified for improved STF behavior, or new architectures have been proposed with quasi user selectable interferer robustness. The drawback of peaking STF in FF compensated ΣΔ modulators can e.g. be eliminated or reduced without loosing most of their advantages as discussed in [10, 35, 36]. This is achieved by systematically selecting loop filter scaling coefficients of the FB, FF and feed-in paths, for which no additional circuitry is required; this makes those approaches simple in implementation. Such modifications of the loop filter even enable steep OOB roll-off in the filter function as shown in [11, 37–39]. Improved performance was also achieved by employing a symbiosis of the ΣΔ modulator with the CSF in which both circuits benefit from each other as demonstrated in [4, 9, 13, 40]. Another approach is to digitally modify the feedback signal of the ΣΔ modulator to achieve enhanced performance together with strong interferers [14, 41]. These techniques are shortly illustrated in the following.

3.3.1 Improved STF Behavior Based on Classical ΣΔ Modulator Loop Filters

The ΣΔ modulator architecture shown in Fig. 3.11 is a combination of a FB compensated ΣΔ modulator together a FF compensated one [35]. The second FB path of a fully FB compensated modulator is omitted and replaced by a FF path, bypassing the second stage. This combination enables a larger gain in the loop filter's first stage and a strong reduction of the non-idealities injected by subsequent stages, and it features low inband signal swing at the output of the first integrator, which leads to an even reduced distortion caused from non-linearity. This architecture therefore enables a low power consumption filter implementation. The STF of this structure still shows peaking, but when compared to the all FF compensated ΣΔ modulator, the peaking is reduced and the modulator in Fig. 3.11 features good attenuation at far OOB frequencies. The benefits of this architecture are also reflected by its implementation in several recent SoA prototypes with among the highest power-performance efficiency. Nonetheless, this architecture still requires a strong CSF in front of the ΣΔ modulator featuring a steep OOB roll-off, which needs to compensate the still present peaking in the STF especially close to the IB, which otherwise would require a very large IB DR requirement of the ΣΔ modulator, as discussed in Sect. 3.2.2.

In [10] an alternative technique was proposed which enables a flat, non-peaking STF within a purely FF compensated CT-ΣΔ modulator; this is achieved by systematically scaling feed-in paths as shown in Fig. 3.12. This architecture, was

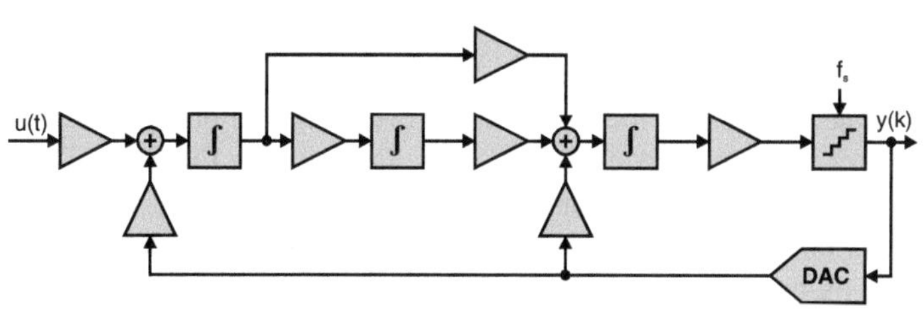

Fig. 3.11 ΣΔ modulator with improved OOB performance [35]

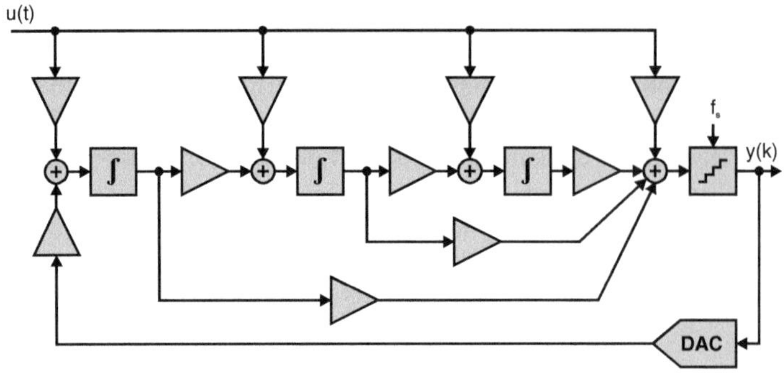

Fig. 3.12 FF compensated ΣΔ modulator featuring a flat STF [10]

successfully adopted in a recent SoA implementation [36]. Advantageously, the first stage of Fig. 3.12 features large gain and thus all subsequent disturbances are heavily attenuated, which enables a very power efficient implementation common for FF compensated modulators. Minor drawbacks of this architecture are slightly increased swing at the internal nodes compared to the FF compensated modulator, and the lack of current mode input signals due to the feed-in paths. Additionally, a preceding CSF and AAF are required, since the STF is completely flat.

The STF of the architecture in Fig. 3.12 could even be improved to feature a steep OOB roll-off as shown in [11] and [37], which would relax or even omit the need for the preceding AAF and CSF. But it was shown that the STF featuring a steep OOB roll-off in this topology is at the same time very sensitive to coefficient mismatch, which makes it unattractive as long as no calibration is applied to center the resulting STF.

Another ΣΔ architecture was presented in [38] and is shown in Fig. 3.13. This modulator is a combination of FB and FF compensation with still steep but more robust Butterworth like OOB roll-off. With this topology only two feedback paths are required, i.e. the outermost feedback and the one to the input of the second integrator. Disadvantageously, the gain within the first integrator is similarly low as in a purely FB compensated ΣΔ modulator and the attenuation of the non-idealities of the subsequent stage is consequently inferior, which is a drawback for the IB performance compared to a purely FF compensated ΣΔ modulator. Due to the advantages of the FF compensation in the remaining stages, this topology is still advantageous compared to a purely FB compensated ΣΔ modulator as it can achieve exactly the same STF.

A technique to improve the STF roll-off even above a purely FB compensated ΣΔ modulator was introduced in [39] and is illustrated in Fig. 3.14. There, several integrators are bypassed in order to place notches within the STF, whereas the NTF remains unchanged. While a purely FB compensated ΣΔ modulator allows to design STFs comparable to Chebyshev filters of type I—passband ripple—the modification with the bypass path enables STFs analog to Cauer filter functions. It should be noticed that firstly these additional STF notches are not seen at all

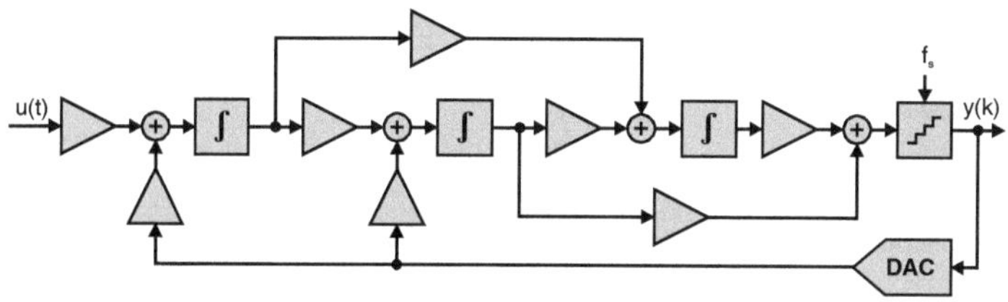

Fig. 3.13 Mixed FB and FF compensated ΣΔ modulator [38]

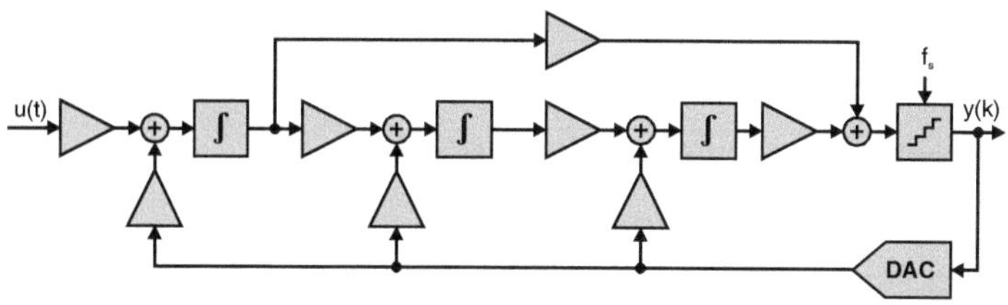

Fig. 3.14 FB compensated ΣΔ modulator with superior OOB roll-off [39]

internal nodes but only for the overall input-output STF, which makes—according to the discussion in Sect. 3.2.3—an in-depth analysis of the worst case internal swings essential. Secondly, each realized STF notch reduces the filter attenuation at far OOB frequencies, which is usually not a severe drawback but needs to be considered. As this technique comes with low implementation effort it is still recommended for FB compensated ΣΔ modulators.

3.3.2 *Improved STF Behaviour Based on Enhanced ΣΔ Modulator Loop Filters*

A more recent methodology to improve the overall performance of receiver ADCs is given by merging an explicit CSF with a CT-ΣΔ modulator.

This can firstly be achieved by embedding the ΣΔ modulator into the loop filter of a CSF as sketched in Fig. 3.15a [4, 40]; the idea is that the output of a filter is digitized and the feedback signal again converted to the analog domain, which would not change the filter function, but implicitly digitize the filter output. An example, similar to the structure presented in [4], is illustrated in Fig. 3.15b. The ΣΔ modulator is embedded into the feedback loop of a CSF featuring an inverse-follow-the-leader structure. The ΣΔ modulator is thereby located at the output of the CSF, which is the location with the lowest sensitivity regarding distortion. The noise and linearity requirements and consequently the power consumption of the embedded ΣΔ modulator are thus strongly reduced, while the design and specifications of the CSF are quasi unaltered by the embedded CT-ΣΔ modulator. Altogether, the resulting structure resembles a high order CT-ΣΔ modulator, but with a well defined

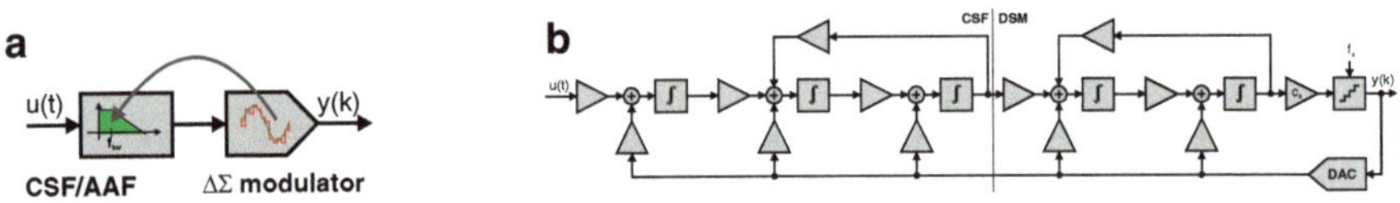

Fig. 3.15 Embedded ΣΔ modulator [6, 40]. (**a**) Basic Idea. (**b**) Example

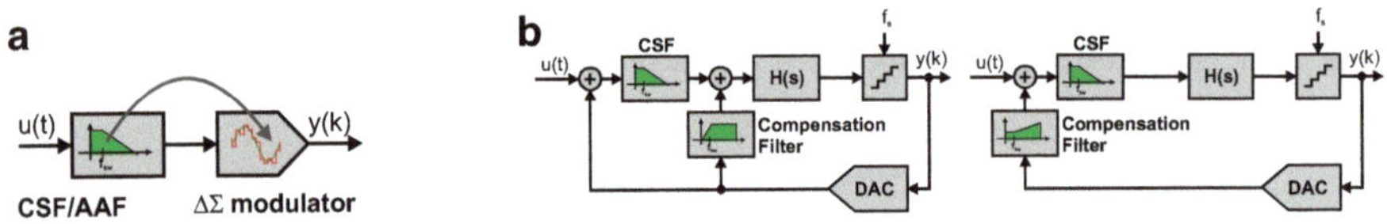

Fig. 3.16 Embedded CSF [6, 9]. (**a**) Basic Idea. (**b**) Example

implicit filter structure. This makes the method either an embedding of a CT-ΣΔ modulator into a CSF, or more simply speaking a design methodology for high order, filtering ΣΔ modulators.

Vice versa, a strategy to embed the function of the CSF into the ΣΔ modulator's loop filter can be considered as sketched in Fig. 3.16a and proposed in [9, 13]. The requirements on the CSF are reduced by the preceding gain of the ΣΔ modulator loop filter, and simultaneously the STF of the ΣΔ modulator can be modified at will. As the inclusion of the CSF destabilizes the original control loop of the ΣΔ modulator, compensation is required: in [9] two possibilities are described and illustrated in Fig. 3.16b. The CSF attenuates the high frequency components of the control loop and a high pass filter in the feedback either has to compensate this attenuation in series or in parallel to the embedded CSF. The SoA of these implementations [9, 13] successfully achieved this compensation and thereby achieved to flatten the STF of an originally peaking ΣΔ modulator as in Fig. 3.11.

It should though be noted, that both SoA implementations in [9, 13] embedded the CSF behind the first integrator stage of the ΣΔ modulator. This relaxed the requirements of the CSF, but it also made the first filter stage more demanding: instead of the original frontend CSF, the ΣΔ modulator input stage has to handle the full interferer signal now, and thus it requires higher dynamic range and linearity in presence of the large interferer signal. Nonetheless, [13] showed in a direct comparison that this could be advantageously handled in their implementation.

In contrast to all previously discussed architectures for filtering ΣΔ modulators, the technique presented in [14] applies a digital modification to improve the interferer robustness. Thereby, a digital control loop with a transfer function Cr(z) is added around the ΣΔ modulator as sketched in Fig. 3.17a. The intention of this control loop is to extract the critical interferer from the digital output of the ΣΔ modulator and to cancel it at its input. This should then avoid the interferer from entering the analog loop filter of the ΣΔ modulator at full power level, and thus it would consequently reduce the DR requirements.

Disadvantageously, any distortion caused by this new outermost feedback signal is directly seen in the ΣΔ modulators output, which makes the implementation of the control filter Cr(z) and the additional DAC challenging. This was solved by realizing the control filter in a Δ-structure, as shown in Fig. 3.17b, which allows a power efficient filter implementation; at the same time truncation noise shaping enabled a low resolution DAC, whereas OOB noise of the digital noise shaper is implicitly filtered by the analog loop filter within Fig. 3.18a. This technique was used in order to increase the OOB dynamic range of the overall system [14], where a digitally reconfigurable resonator Cr(z) was used to realize a digitally reconfigurable notch in the overall STF.

A corresponding example is given in Fig. 3.18b. The STF of a FB compensated ΣΔ modulator as in Fig. 3.18a is modified with a digital feedback filter Cr(z) implemented as resonator at the interferer frequency. This yields to a enhanced STF, which features a strong attenuation exactly at the resonance frequency. Due to its digital implementation, the notch frequency can also be digitally tuned.

As discussed in Sect. 3.2.4, jitter performance of ΣΔ modulators can be dominated by the strongest signal, which is often not the inband signal but an OOB interferer. When the digital filter is used to extract the interferer and cancel it at the modulator input in order to realize an improved STF, the feedback signal contains the strong interferer in full amplitude, which makes the discussed technique superior concerning OOB dynamic range, but not advantageous concerning jitter sensitivity in the DAC. Therefore, a similar technique was introduced in [41],

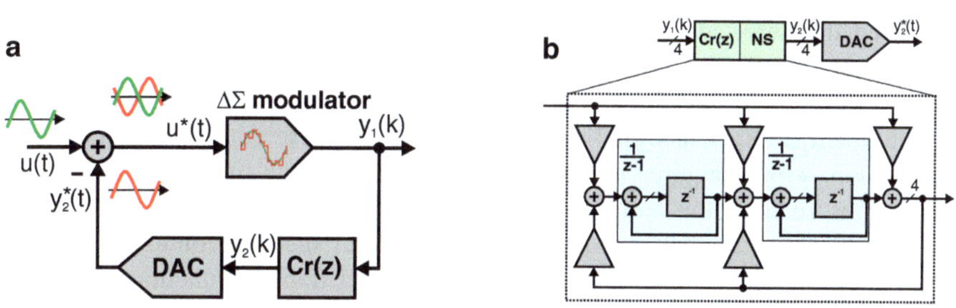

Fig. 3.17 Digitally modified feedback. (**a**) Interferer cancellation. (**b**) Noise shaping IIR filter implementation. Source: [6]

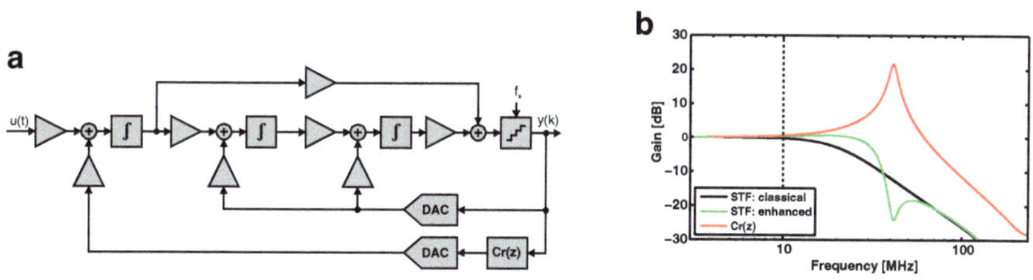

Fig. 3.18 FB compensated ΣΔ modulator with digitally modified feedback. (**a**) Block diagram [14, 41]. (**b**) Transfer functions with Cr(z) = Resonator. Source: [6]

where—instead of canceling the interferer at the input of the ΣΔ modulator—the digital feedback filter was applied to cancel the interferer before entering the outermost DAC. This can be achieved by using a digital low pass filter in the feedback path Cr(z) in Fig. 3.18a, which attenuates the strongest blocker signal in front of the outermost DAC. Since the distortion, which is caused by the outermost DAC, is seen unattenuated at the output of the ΣΔ modulator, the intention of this technique is thus the reduction of the strongest OOB interferer and thus the distortion caused by the DAC and e.g. nonlinearity and jitter noise.

The transfer functions in Fig. 3.19a show an exemplary digital LP filter within the digital feedback filter Cr(z) in the modulator in Fig. 3.18a. It can be seen, that the modification only marginally changes the STF, but the transfer function to the DAC (STF · Cr(z)) features a strong OOB attenuation. A simulation including jitter is given in Fig. 3.19b, in which a strong OOB dual-tone causes strong distortion together with the clock jitter when the classical ΣΔ modulator is simulated without the digital feedback filter. Including the digital filter, the IB distortion by jitter is nicely relaxed.

It must be noted that the techniques from [14] and [41] cannot be jointly applied and the selection depends on the character of the problems, e.g. overload or jitter.

3.4 Conclusion

This contribution gives an overview of the state of the art of ΣΔ ADCs with improved interferer robustness and discusses their requirements, advantages and application selection. It is shown that the ADC's dynamic range is—in many applications—not defined by the inband signal, but by the interferer signals, which can be much stronger in signal power. This is even worse with the trend of early digitization, which requires the ADC to move towards the receiver frontend and which thus reduces the amount of preceding interference filtering prior to the ADC. The connection between an exemplary receiver specification to the required ΣΔ ADC's dynamic range dependent on its signal transfer function was given to visualize the need of the large interference robustness. This emphasized that a

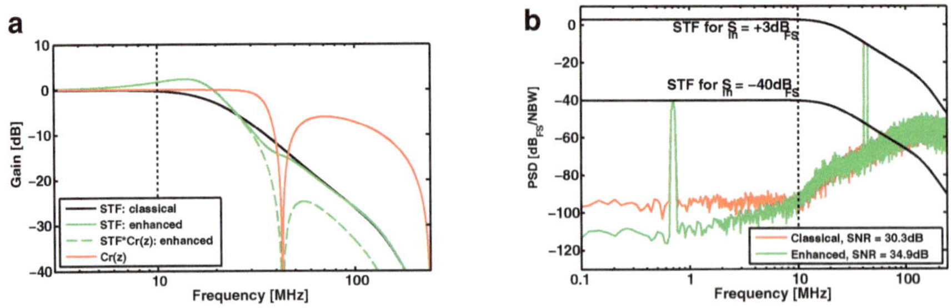

Fig. 3.19 FB compensated ΣΔ modulator with digitally low pass filtered feedback. (**a**) Transfer functions with Cr(z) = lowpass. (**b**) Simulation results (Jitter: $\sigma_J = 0.001/f_s$). Source: [6]

filter function realized by the CSF or intrinsically by the $\Sigma\Delta$ modulator reduces the required ADC dynamic range. It was furthermore shown, that this robustness needs to be guaranteed not only for the input-output behaviour of the $\Sigma\Delta$ ADC, but the interferer is simultaneously not allowed to internally overload the $\Sigma\Delta$ ADC, even when worst case superposed aliases are taken into account. Furthermore, the problem of jitter has been discussed as third major issue to be considered in the presence of interferers.

Consequently, the SoA of $\Sigma\Delta$ modulator techniques with improved filter characteristic is reviewed, each of them showing its distinct advantages and disadvantages, which allows the reader to select the most suitable implementation for a given specification and application.

References

1. J. de la Rosa, Sigma-delta modulators: tutorial overview, design guide, and state-of-the-art survey. IEEE Trans. Circuits Syst I Regul. Pap. **58**(1), 1–21 (2011)
2. J. de la Rosa, R. Schreier, K.-P. Pun, S. Pavan, Next-generation delta-sigma converters: trends and perspectives. IEEE J. Emerging Sel. Top. Circuits Syst. **5**(4), 484–499 (2015)
3. R. Saad, D. Aristizabal-Ramirez, S. Hoyos, Sensitivity analysis of continuous-time $\Sigma\Delta$ ADCs to out-of-band blockers in future SAW-less multi-standard wireless receivers. IEEE Trans. Circuits Syst. I Regul. Pap. **59**(9), 1894–1905 (2012)
4. M. Andersson, M. Anderson, L. Sundstrom, S. Mattisson, P. Andreani, A filtering $\Sigma\Delta$ ADC for LTE and beyond. IEEE J. Solid State Circuits **49**(7), 1535–1547 (2014)
5. R. Ritter, M. Ortmanns, Continuous-time delta-sigma ADCs with improved interferer rejection. IEEE J. Emerging Sel. Top. Circuits Syst. **5**(4), 500–513 (2015)
6. R. Ritter, Delta-Sigma modulators with enhanced interferer cancellation in receiver applications. Ph.D. dissertation, University of Ulm (2016)
7. K. Kim, Silicon technologies and solutions for the data-driven world, in *2015 IEEE International Solid- State Circuits Conference - (ISSCC)* (2015), pp. 1–7
8. M. Keller, A. Buhmann, J. Sauerbrey, M. Ortmanns, Y. Manoli, A comparative study on excess-loop-delay compensation techniques for continuous-time sigma-delta modulators. IEEE Trans. Circuits Syst. I Regul. Pap. **55**(11), 3480–3487 (2008)
9. K. Philips, P. Nuijten, R. Roovers, A. van Roermund, F. Chavero, M. Pallares, A. Torralba, A continuous-time sigma delta ADC with increased immunity to interferers. IEEE J. Solid State Circuits **39**(12), 2170–2178 (2004)
10. J. De Maeyer, J. Raman, P. Rombouts, L. Weyten, Controlled behaviour of STF in CT $\Sigma\Delta$ modulators. Electron. Lett. **41**(16), 19–20 (2005)
11. M. Ranjbar, O. Oliaei, Continuous-time feedforward SD modulators with robust lowpass STF. Electron. Lett. **43**(24), 1340–1341 (2007)
12. M. Sosio, A. Liscidini, R. Castello, F. De Bernardinis, A complete DVB-T/ATSC tuner analog base-band implemented with a single filtering ADC, in *2011 Proceedings of the ESSCIRC (ESSCIRC)* (2011), pp. 391–394
13. R. Rajan, S. Pavan, Design techniques for continuous-time $\Sigma\Delta$ modulators with embedded active filtering. IEEE J. Solid State Circuits **49**(10), 2187–2198 (2014)
14. R. Ritter, J. Kauffman, J. Becker, M. Ortmanns, A 10MHz bandwidth, 70dB SNDR continuous time delta-sigma modulator with digitally improved reconfigurable blocker rejection. IEEE J. Solid State Circuits, **51**(3), 660–670 (2016)
15. J. Mitola, Software radio architecture: a mathematical perspective. IEEE J. Select. Areas Commun. **17**(4), 514–538 (1999)

16. A. Abidi, The path to the software-defined radio receiver. IEEE J. Solid State Circuits **42**(5), 954–966 (2007)
17. H. Shibata, R. Schreier, W. Yang, A. Shaikh, D. Paterson, T. Caldwell, D. Alldred, P.W. Lai, A DC-to-1 GHz tunable RF $\Sigma\Delta$ ADC achieving DR = 74 dB and BW = 150 MHz at f_0 = 450 MHz using 550 mW. IEEE J. Solid State Circuits **47**(12), 2888–2897 (2012)
18. H. Chae, J. Jeong, G. Manganaro, M. Flynn, A 12mW low-power continuous-time bandpass $\Sigma\Delta$ modulator with 58dB SNDR and 24MHz bandwidth at 200MHz IF, in *2012 IEEE International Solid-State Circuits Conference Digest of Technical Papers (ISSCC)* (2012), pp. 148–150
19. J. Harrison, M. Nesselroth, R. Mamuad, A. Behzad, A. Adams, S. Avery, An LC bandpass $\Sigma\Delta$ adc with 70dB SNDR over 20MHz bandwidth using CMOS DACs, in *2012 IEEE International Solid-State Circuits Conference Digest of Technical Papers (ISSCC)* (2012), pp. 146–148
20. B. Murmann, ADC performance survey 1997–2015. Available: http://web.stanford.edu/~murmann/adcsurvey.html
21. M. Englund, K. Ostman, O. Viitala, M. Kaltiokallio, K. Stadius, K. Koli, J. Ryynanen, A programmable 0.7-2.7 GHz direct $\Sigma\Delta$ receiver in 40 nm CMOS. IEEE J. Solid State Circuits **50**(3), 644–655 (2015)
22. P. Chopp, A. Hamoui, A 1-V 13-mW single-path frequency-trans-lating $\Sigma\Delta$ modulator with 55-dB SNDR and 4-MHz bandwidth at 225 MHz. IEEE J. Solid State Circuits **48**(2), 473–486 (2013)
23. ETSI, *LTE; Evolved Universal Terrestrial Radio Access (E-UTRA); User Equipment (UE) radio transmission and reception (Release 11)*, 3GPP TS 36.101, 04 2013, v11.4.0
24. C.-Y. Ho, C. Liu, C.-L. Lo, H.-C. Tsai, T.-C. Wang, Y.-H. Lin, A 4.5mW CT self-coupled $\Sigma\Delta$ modulator with 2.2MHz BW and 90.4dB SNDR using residual ELD compensation, in *2015 IEEE International Solid- State Circuits Conference - (ISSCC)* (2015), pp. 1–3
25. Y.-S. Shu, J.-Y. Tsai, P. Chen, T.-Y. Lo, P.-C. Chiu, A 28fJ/conv-step CT $\Sigma\Delta$ modulator with 78dB DR and 18MHz BW in 28nm CMOS using a highly digital multibit quantizer, in *2013 IEEE International Solid-State Circuits Conference Digest of Technical Papers (ISSCC)* (2013), pp. 268–269
26. H. Geddada, C.-J. Park, H.-J. Jeon, J. Silva-Martinez, A. Karsilayan, D. Garrity, Design techniques to improve blocker tolerance of continuous-time $\Sigma\Delta$ ADCs. IEEE Trans. Very Large Scale Integr. Syst. **23**(1), 54–67 (2015)
27. R. Schreier, G. Temes, *Understanding Delta-Sigma Data Converters* (Wiley, New York, 2004)
28. R. Adams, K. Nguyen, A 113dB SNR oversampling DAC with segmented noise-shaped scrambling. IEEE J. Solid State Circuits **33**(12), 1871–1878 (1998)
29. M. Ortmanns, F. Gerfers, Y. Manoli, A continuous-time $\Sigma\Delta$ modulator with reduced sensitivity to clock jitter through SCR feedback. IEEE Trans. Circuits Syst. I Regul. Pap. **52**(5), 875–884 (2005)
30. N. Beilleau, H. Aboushady, F. Montaudon, A. Cathelin, A 1.3V 26mW 3.2GS/s undersampled LC bandpass $\Sigma\Delta$ ADC for a SDR ISM-band receiver in 130nm CMOS, in *2009 IEEE Radio Frequency Integrated Circuits Symposium. RFIC 2009* (2009), pp. 383–386
31. O. Oliaei, Sigma-delta modulator with spectrally shaped feedback. IEEE Trans. Circuits Syst. II Analog Digit. Signal Process. **50**(9), 518–530 (2003)
32. P. Shettigar, S. Pavan, Design techniques for wideband single-bit continuous-time $\Sigma\Delta$ modulators with FIR feedback DACs. IEEE J. Solid State Circuits **47**(12), 2865–2879 (2012)
33. M. Ortmanns, F. Gerfers, Y. Manoli, Fundamental limits of jitter insensitivity in discrete and continuous-time sigma delta modulators, in *Proceedings of the 2003 International Symposium on Circuits and Systems, 2003. ISCAS '03*, vol. 1 (2003), pp. I–I
34. R. Saad, S. Hoyos, Sensitivity analysis of pulse-width jitter induced noise in continuous-time delta-sigma modulators to out-of-band blockers in wireless receivers, in *2011 IEEE International Symposium on Circuits and Systems (ISCAS)* (2011), pp. 1636–1639
35. F. Munoz, K. Philips, A. Torralba, A 4.7mW 89.5dB DR CT complex $\Sigma\Delta$ ADC with built-in LPF, in *2005 IEEE International Solid-State Circuits Conference, 2005. Digest of Technical Papers. ISSCC*, vol. 1 (2005), pp. 500–613

36. S. Loeda, J. Harrison, F. Pourchet, A. Adams, A 10/20/30/40 MHz feed-forward FIR DAC continuous-time $\Sigma\Delta$ ADC with robust blocker performance for radio receivers, in *2015 Symposium on VLSI Circuits (VLSI Circuits)* (2015), pp. C262–C263
37. T. Brueckner, C. Zorny, W. Mathisy, M. Ortmanns, A single DAC CT sigma-delta modulator with butterworth STF characteristic, in *2011 IEEE 54th International Midwest Symposium on Circuits and Systems (MWSCAS)* (2011), pp. 1–4
38. M. Ranjbar, O. Oliaei, R. Jackson, A robust STF 6mW CT $\Sigma\Delta$ modulator with 76dB dynamic range and 5Mhz bandwidth, in *2010 IEEE Custom Integrated Circuits Conference (CICC)* (2010), pp. 1–4
39. R. Ritter, M. Lorenz, M. Ortmanns, Anti-aliasing filter improvement in continuous-time feedback sigma-delta modulators. *2015 IEEE International Symposium on Circuits and Systems (ISCAS)* (2015), pp. 325–328
40. M. Sosio, A. Liscidini, R. Castello, A 2G/3G cellular analog baseband based on a filtering ADC. IEEE Trans. Circuits Syst. II Express Briefs **59**(4), 214–218 (2012)
41. R. Ritter, P. Torta, L. Dorrer, A. Di Giandomenico, S. Herzinger, M. Ortmanns, A multimode CT $\Sigma\Delta$-modulator with a reconfigurable digital feedback filter for semi-digital blocker/interferer rejection, in *2015 - 41st European Solid-State Circuits Conference (ESSCIRC)* (2015), pp. 225–228

Chapter 4
Design Considerations for Filtering Delta Sigma Converters

Shanthi Pavan and Radha Rajan

4.1 Introduction

The direct conversion architecture is perhaps the most prevalent in contemporary radio receivers. Figure 4.1 shows a simplified block diagram of such a receiver. It consists of a low noise amplifier (LNA), mixers, variable gain amplifiers (VGA), low pass filters (LPF) and analog to digital converters (ADC). The role of the LPF is two-fold. Since the received signal can be potentially dominated by large interferers, the LPF attenuates them and thereby reduces the dynamic range (DR) required of the ADC. It also acts as an anti-alias filter and prevents noise from aliasing into the signal band. There is a trade off between the performance requirements of the filter and ADC. For instance, when a Nyquist rate ADC is used, its sampling rate is typically chosen to be about 4 times the bandwidth of the baseband signal. A lower sampling rate would necessitate a higher order filter, making its design particularly challenging.

Continuous-time $\Sigma\Delta$ modulators (CT$\Sigma\Delta$Ms) are a compelling choice for implementing the ADCs in wireless receivers. The block diagram of a CT$\Sigma\Delta$M is shown in Fig. 4.2a. $L_o(s)$ and $L_1(s)$ denote the transfer functions from the input $u(t)$ and the DAC waveform $v(t)$ to the loop filter's output respectively. Assuming that quantization noise is additive, the functional block diagram of the CT$\Sigma\Delta$M is shown in Fig. 4.2b. The input $u(t)$ can be thought of as being filtered up front by a linear time invariant filter whose transfer function, denoted by $STF(f)$, is called the signal transfer function (STF). The signal component of the modulator's output $v[n]$ is the sampled version of the output of the STF "filter". Needless to say, the STF depends on $L_o(s)$, $L_1(s)$ and the DAC pulse shape. It turns out that if the modulator's loop

S. Pavan (✉) • R. Rajan
Department of Electrical Engineering, Indian Institute of Technology, Madras, Chennai 600 036, India
e-mail: shanthi@ee.iitm.ac.in

© Springer International Publishing Switzerland 2017

A. Baschirotto et al. (eds.), *Wideband Continuous-time ΣΔ ADCs, Automotive Electronics, and Power Management*, DOI 10.1007/978-3-319-41670-0_4

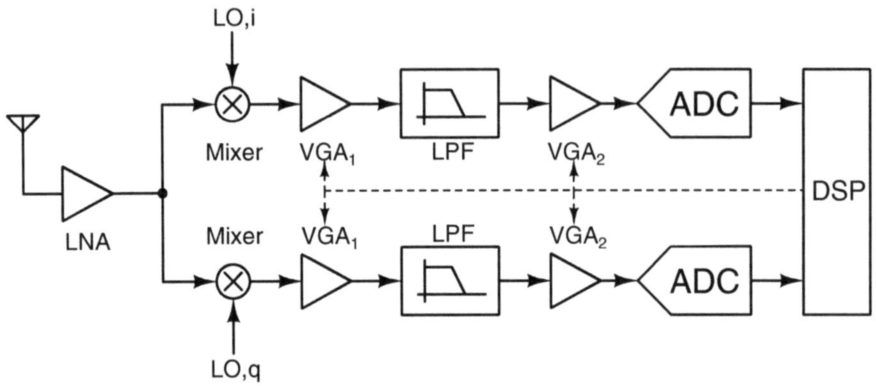

Fig. 4.1 Simplified block diagram of the front end of a typical wireless receiver

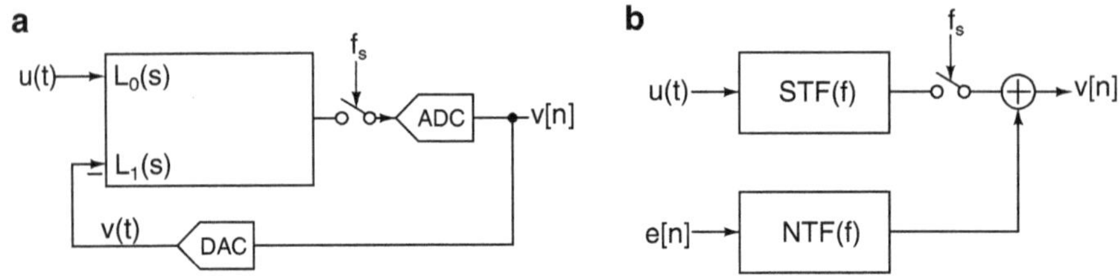

Fig. 4.2 (**a**) Block diagram of a continuous-time ΣΔ modulator and (**b**) its functional equivalent

filter is time invariant, the STF has nulls at multiples of the sampling frequency f_s. Thus, CTΣΔMs possess the inherent anti-aliasing property, which renders noise folding a non-issue. Quantization noise, denoted by the sequence $e[n]$ appears in $v[n]$ after being shaped, or high pass filtered by the noise transfer function (NTF). The aim of the design process is to choose the NTF so that the in-band quantization noise is sufficiently small. The decimation filter that follows the CTΣΔM eliminates out of band quantization noise.

When a CTΣΔM is used as the ADC in the radio receiver of Fig. 4.1, the role of the LPF is to eliminate interferers rather than address the aliasing of noise into the signal band. A CTΣΔM has several other advantages—it has a resistive input impedance, it is easy to drive and results in a simplified signal chain with reduced power dissipation. When compared to ΣΔ ADCs employing switched capacitor techniques, the opamps used in CTΣΔMs have reduced slew rate requirements, thereby resulting in better linearity for a given speed and power dissipation. CTΣΔMs are also immune to substrate noise, since sampling occurs after the signal has been processed by the loop filter. Several high performance ΣΔ modulators with signal bandwidths of several 10's of MHz (keeping communication applications in mind) and about 11-12 bit resolution have been reported recently [1–3]. However, a filter is still needed to attenuate interferers, without which the dynamic range of the ADC would need to increase.

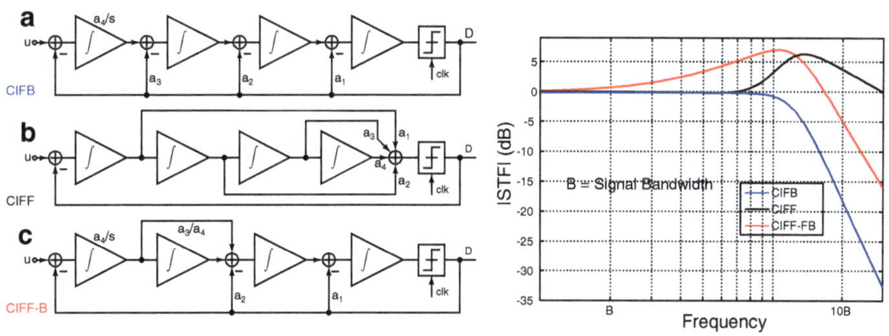

Fig. 4.3 (**a**) CIFB, (**b**) CIFF, (**c**) CIFF-B loop filter realizations for a fourth order modulator and their STFs

Since a CTΣΔM has a built-in filter (its STF), it is natural to wonder if the explicit filter of Fig. 4.1 can now be dispensed with. This, however, is problematic as explained below. Conventional CTΣΔM architectures do not allow independent control over the shape and bandwidth of the STF, as the STF is simply a by-product of NTF synthesis. The shape of the signal transfer function depends on the architecture of the loop filter used as illustrated below. Figure 4.3 shows commonly used (fourth order) modulator topologies—the cascade of integrators in feedback (CIFB), cascade of integrators in feedforward (CIFF) and cascade of integrators in feedforward-feedback (CIFF-B) and their corresponding STFs. *B* denotes the signal bandwidth. Keeping in mind a single-bit quantizer, the NTF is assumed to have an out-of-band gain of 1.5. In a CIFB structure (Fig. 4.3a), the STF has fourth order roll-off at high frequencies. However, such a structure is not power efficient, as the gain of the first integrator (a_4/s) is small in the signal band. Thus noise and distortion from succeeding integrators can be significant when referred to the input. Using a CIFF loop filter (Fig. 4.3b) avoids this problem, since the gain of the first integrator is large in the signal band. However, the STF now peaks outside the desired signal band and has only first order roll-off at high frequencies. The advantages of the CIFF and CIFB structures are combined in the CIFF-B topology (Fig. 4.3c) [3, 4]. This architecture results in a large in-band gain for the first integrator, and an STF that has third order roll-off at high frequencies. The STF still peaks at out-of-band frequencies. From the STFs shown in Fig. 4.3, it is clear that irrespective of the loop filter topology used, a CTΣΔM cannot attenuate nearby interferers—there are simply not enough degrees of freedom to specify the shape and bandwidth of the STF, *and* achieve the desired NTF at the same time. Without an explicit filter to reject interferers, the dynamic range of the CTΣΔM should be increased, like with any other ADC.

Peaking in the STF further increases the in-band dynamic range needed in the modulator due to the following. To understand this, we consider an example of a single bit CTDSM. With an NTF whose out of band gain is 1.5, the maximum stable amplitude (MSA) for low frequency inputs is typically about −3 dBFS. The MSA at a frequency f_{in} is $\approx 0.7V_{FS}/STF(f_{in})$, where V_{FS} is the full scale voltage. A 6 dB STF peak restricts the CTΣΔM input to $(1/2)0.7V_{FS}$, necessitating a 6 dB smaller in-band noise to get the same in-band SNR, which increases power dissipation. A straightforward way of attenuating nearby interferers and addressing STF peaking at the same time is to use a filter up front. The filter, however, occupies additional area, consumes power, adds noise and potentially degrades linearity of the signal chain.

Alternatively, the filter can be incorporated into the ΣΔ loop by moving it beyond the first integrator—this way, the noise from the filter is reduced by the in-band gain of the first integrator. The system noise budget, usually split between the filter and CTΣΔM, can now be allocated to the first integrator of the modulator. Alternatively, the filter and CTΣΔM can be combined so that the overall performance improves, for a given power dissipation. The authors of [5] embed a second order CTΣΔM inside a Rauch biquad. The design of [6] embeds a third order CTΣΔM in an active-RC biquad. Another version [7] by the same authors embeds a second order modulator into a third order active-RC filter. Several recent CTΣΔMs [8] also embed modulators into a filter.

This work reviews techniques to combine a filter and CTΣΔM, and gives experimental results of a design where a second order active filter is embedded in a CIFF-B CTΣΔM with the aim of more effectively attenuating close by interferers [9, 10]. We show that this has the same functionality as having the filter up front, but achieves better linearity and noise performance (for the same power dissipation) when compared to a filter-CTΣΔM cascade. The rest of this paper is organized as follows. Section 4.2 reviews techniques to modify the STF of a CTΣΔM with the aim of reducing or eliminating peaking. Section 4.3 explains the technique of embedding a filter inside a CTΣΔM. Noise, linearity, power consumption and active area of a CTΣΔM with a filter up front (H_1-ΣΔ) are compared with those in a filtering CTΣΔM (Δ-H_1-Σ). Architectural design considerations and the use of chopping to eliminate offset and flicker noise are given in Sect. 4.4. Section 4.5 discusses circuit details and measurement results of these designs, both of which incorporate a second order Butterworth filter in their STFs. The filter has a cutoff frequency which is twice the desired signal bandwidth. Measurements show that embedding the filter improves linearity and reduces area for the same power dissipation. Section 4.6 concludes the paper.

4.2 Signal Transfer Function Tailoring

Before embarking on a effort to find the best way of making a filtering CTΣΔM, it is useful to review methods that address the problem of STF peaking by introducing feed-in paths from the input to the various internal states of the modulator. Referring to the generic block diagram of Fig. 4.2a, the STF is given by

$$STF(j\omega) = L_o(j\omega)NTF(e^{j\omega}). \tag{4.1}$$

The aim of STF tailoring is to find ways of choosing $L_o(s)$ so that

a. Peaking of the STF is eliminated.
b. The order of the loop filter is not increased.
c. The NTF is not affected.

We illustrate with the "standard" second order CIFF modulator with $NTF(z) = (1 - z^{-1})^2$ shown in Fig. 4.4a. When k_1 and k_2 are zero,

$$L_o(s) = \frac{1.5s + 1}{s^2}. \tag{4.2}$$

The resulting STF is

$$STF(j\omega) = \frac{1 + j1.5\omega}{-\omega^2}(1 - e^{-j\omega})^2 \tag{4.3}$$

As shown in Fig. 4.4b, the STF peaks to about 2.2 (7 dB), which is not desirable. By adding feed-in paths with gains k_o and k_1, $L_o(s)$ is modified to

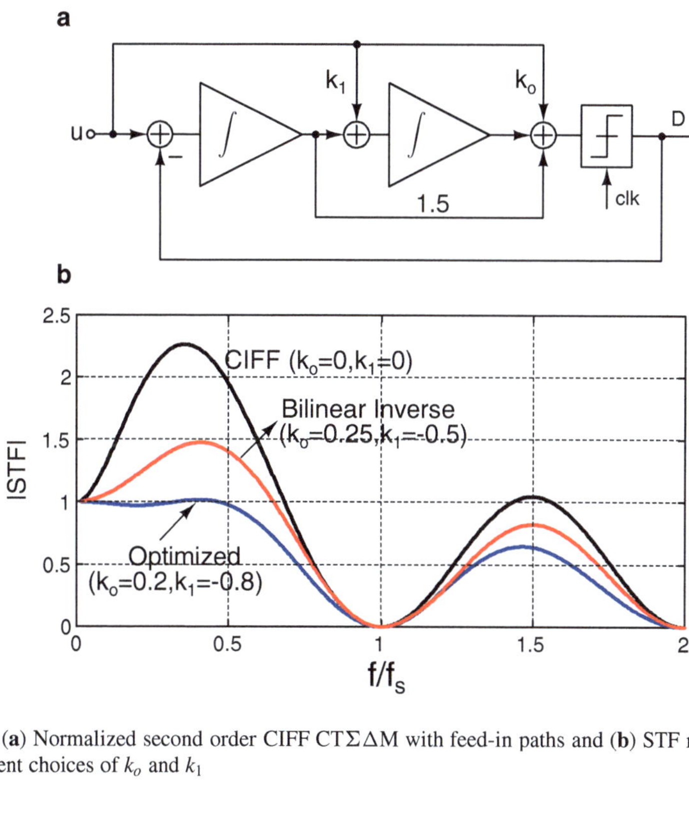

Fig. 4.4 (**a**) Normalized second order CIFF CTΣΔM with feed-in paths and (**b**) STF magnitude for different choices of k_o and k_1

$$\hat{L}_o(s) = \frac{k_o s^2 + (1.5 + k_1)s + 1}{s^2}. \tag{4.4}$$

How should k_o and k_1 be chosen? One method is to use $k_0 = 0$ and $k_1 = -1.5$. This will cause the STF to resemble that of a CIFB loop, which is desirable. However, the output swing of the first integrator increases, just like in the CIFB case, causing noise and distortion problems. To avoid this, one could compromise—rather than demand that the STF have a filtering characteristic, we would be satisfied if peaking is eliminated. The aim, therefore, would be to choose k_o and k_1 so that

$$STF(j\omega) = \frac{-k_o\omega^2 + (1.5 + k_1)j\omega + 1}{-\omega^2} NTF(e^{j\omega}) \tag{4.5}$$

has unity magnitude for $0 < \omega < \pi$. The difficulty is that $|NTF(e^{j\omega})|$ cannot be expressed as a ratio of polynomials in ω^2. An approach to circumvent this [11] is to use the so called bilinear inverse of the NTF, where $e^{j\omega}$ is approximated as

$$e^{j\omega} \approx \frac{1 + \frac{j\omega}{2}}{1 - \frac{j\omega}{2}}. \tag{4.6}$$

Equivalently, the NTF can be approximated (for low frequencies) as a polynomial in s, by replacing z with s according to

$$z \rightarrow \frac{1 - (s/2)}{1 + (s/2)} \tag{4.7}$$

Applying this to our second order modulator yields $k_o = 0.25$ and $k_1 = -0.5$. The STF peaking is reduced, as Fig. 4.4b shows, but is still about 3 dB. A more careful optimization yields $k_o = 0.2$ and $k_1 = -0.8$, which results in a largely flat STF. However, the designer has no control over its bandwidth. From the discussion above, it is seen that adjusting feed-in path gains, while useful, cannot help with attenuating close in interferers.

4.3 Embedding a Filter into a CT$\Sigma\Delta$M

The conceptually straightforward approach to control the STF is to use a filter up front. As mentioned earlier, this approach adds noise and distortion to the signal chain. Alternatively, the filter can be moved into the modulator. The first attempt at this was made in [12]. In that work, a first order passive RC filter was embedded inside a single bit CIFF CT$\Sigma\Delta$M to "tame" the STF peak characteristic of a CIFF loop.

Figure 4.5a illustrates the basic idea. A first order low pass filter formed by R_1 and C_1 is placed after the first integrator of a CIFF CT$\Sigma\Delta$M. Introducing the filter into the $\Sigma\Delta$ loop, in all likelihood, will destabilize it. To avoid this, a compensation

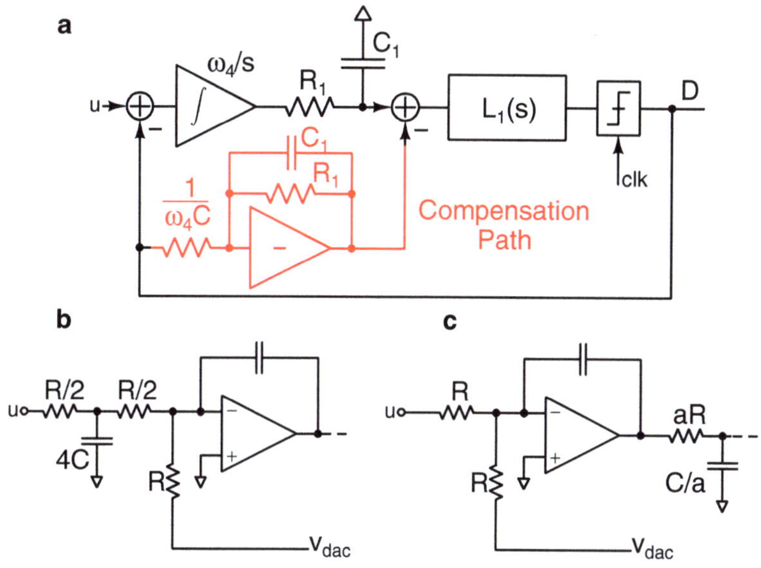

Fig. 4.5 (**a**) Basic idea behind embedding a first order filter inside a CTΣΔM (with a CIFF loop). (**b**) Input stage of a filter+CTΣΔM with filtering upfront and (**c**) Filtering CTΣΔM with the filter moved after the first integrator

path, shown in red, is introduced. Thanks to this path, the transfer function from $v(t)$ to the input of $L_1(s)$ remains unchanged, and the NTF is not affected. Thus, the bandwidth of the low pass filter, which depends on the the R_1C_1 product, can be chosen independent of the NTF. The STF, on the other hand, is the product of transfer function of the low pass filter and the STF of the original CIFF CTΣΔM. Figure 4.5b, c show the front-ends of CTΣΔMs with the filter outside and inside the loop respectively. The filter bandwidth is $1/(2\pi RC)$. When the filter is embedded, its impedance can be scaled ($a > 1$ in Fig. 4.5c), thereby reducing area.

What are the difficulties with such a structure? The first order filter offers limited selectivity and the active compensation path consumes extra power. Though not fundamental to the technique, the design of [12] used a single-bit quantizer. Due to the single bit DAC, the design is sensitive to clock jitter and the first opamp has to dissipate a lot of power to achieve the desired linearity. Another issue is the change in the NTF due to mismatch between the embedded low pass filter and the compensation path. Figure 4.6 shows Monte-Carlo simulations of the NTF with 2 % random mismatch. For these simulations, the oversampling ratio (OSR) was set to 64, with the 3 dB bandwidth of the low pass filter being twice the signal bandwidth. As with all ΣΔ loops employing 1-bit quantizers, the out-of-band gain (OBG) of the NTF was chosen to be 1.5.

Simulations show that the variations of the NTF are more (a) when the bandwidth of the embedded filter is smaller and (b) when NTF's OBG is higher. These results make intuitive sense due to the following. The low pass filter adds delay into

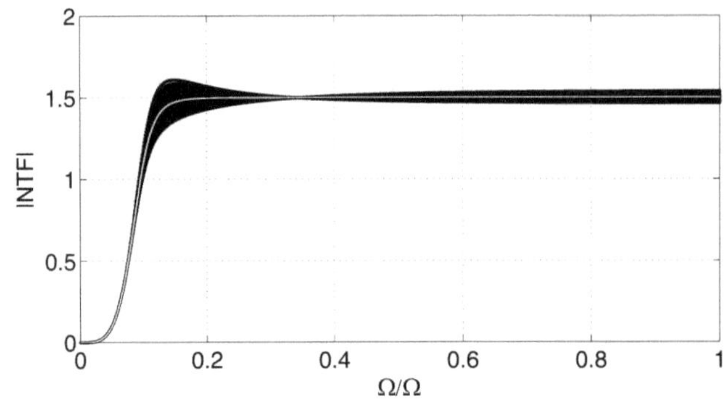

Fig. 4.6 Effect of 2 % mismatch between the embedded filter and the compensation path

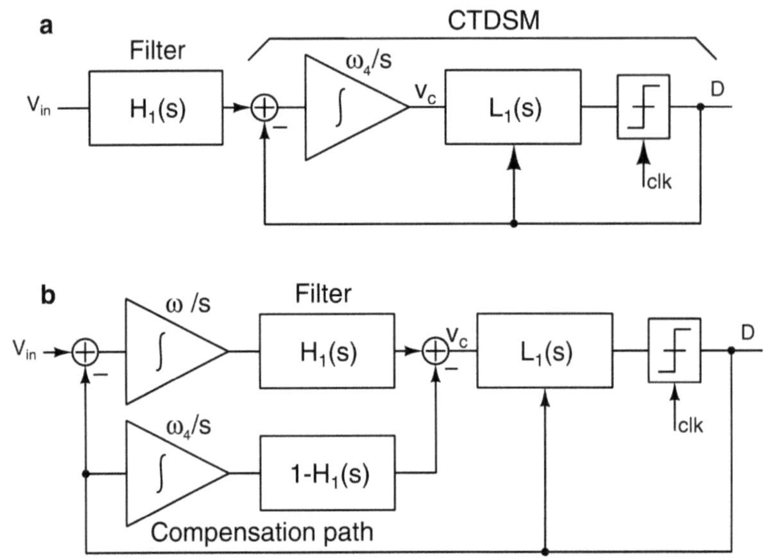

Fig. 4.7 (**a**) Block diagram of a modulator with a filter $H_1(s)$ up front, (**b**) $H_1(s)$ is moved beyond the first integrator in the $\Sigma\Delta$ loop. The compensation path is necessary to restore loop stability

the $\Sigma\Delta$ loop, and the compensation network attempts to mitigate it. Mismatch between the two affects the efficacy of excess delay compensation. With a lower filter bandwidth, the excess delay mismatch is inherently larger. With a higher OBG, the NTF is more sensitive to a given excess delay.

To improve the rejection of close-in interferers, a higher order filter needs to be used. For a sharp transition band, this filter needs to be active. As in the first order case, this filter can be moved into the $\Sigma\Delta$ loop, leading to the filtering CT$\Sigma\Delta$M architecture. Its evolution is shown in Fig. 4.7. Part (a) of the figure shows a modulator preceded by a low pass filter with transfer function $H_1(s)$. ω_4/s denotes the transfer function of the first integrator in the loop filter of the modulator. $H_1(s)$ can be moved beyond ω_4/s in the $\Sigma\Delta$ loop so that the noise from the filter gets divided by the gain of the first integrator when referred to the input (Fig. 4.7b). Doing this modifies the NTF, which can be restored by a compensation path with transfer function $(1 - H_1(s)) \cdot (\omega_4/s)$, as shown in the figure. It is straightforward to see that the modulators of Fig. 4.7a, b have the same STF and NTF, and therefore, the same functionality. Specifically, note that the voltage v_c at the input of $L_1(s)$ will be the same in both realizations.

Assume that $H_1(s)$ is of second order, with transfer function

$$H_1(s) = \frac{1}{1 + \frac{s}{\omega_o Q} + \frac{s^2}{\omega_o^2}}. \tag{4.8}$$

Since the dc gain of $H_1(s)$ is 1, $(1 - H_1(s))$ has a zero at dc, which cancels the integrator pole in the compensation path. The response of the compensation path, therefore, is given by

$$H_c(s) = \frac{\frac{\omega_4}{\omega_o Q} + \frac{s\omega_4}{\omega_o^2}}{1 + \frac{s}{\omega_o Q} + \frac{s^2}{\omega_o^2}}. \tag{4.9}$$

The filter and the compensation path have identical denominator polynomials— and can be realized using the same network, as shown in Fig. 4.8. In this figure, calculations show that the transfer function from B to C is of the form required for the compensation path. Incorporating this into Fig. 4.7b yields Fig. 4.9a, which shows a CTDSM with an embedded second order filter, with no extra hardware needed for compensation. The output of the filter v_c in Fig. 4.9a, consists of two components: one through the main feedback path and one through the filter compensation path. Assuming the dc gain of the filter is 1 ($R_{f3} = R_{f1}$), the transfer function from the modulator output D to v_c can be shown to be the following.

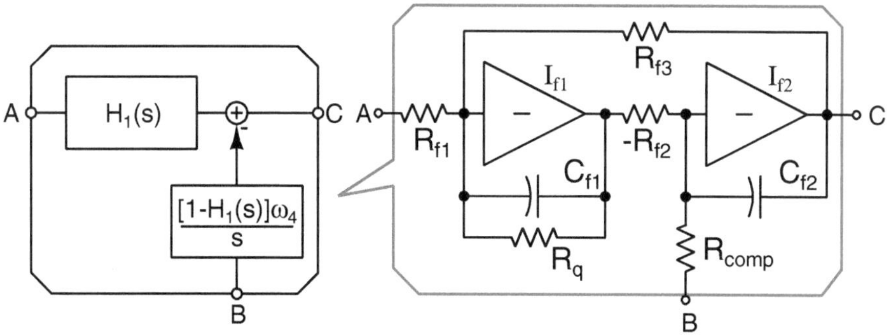

Fig. 4.8 Active RC realization of a second order filter and the compensation path

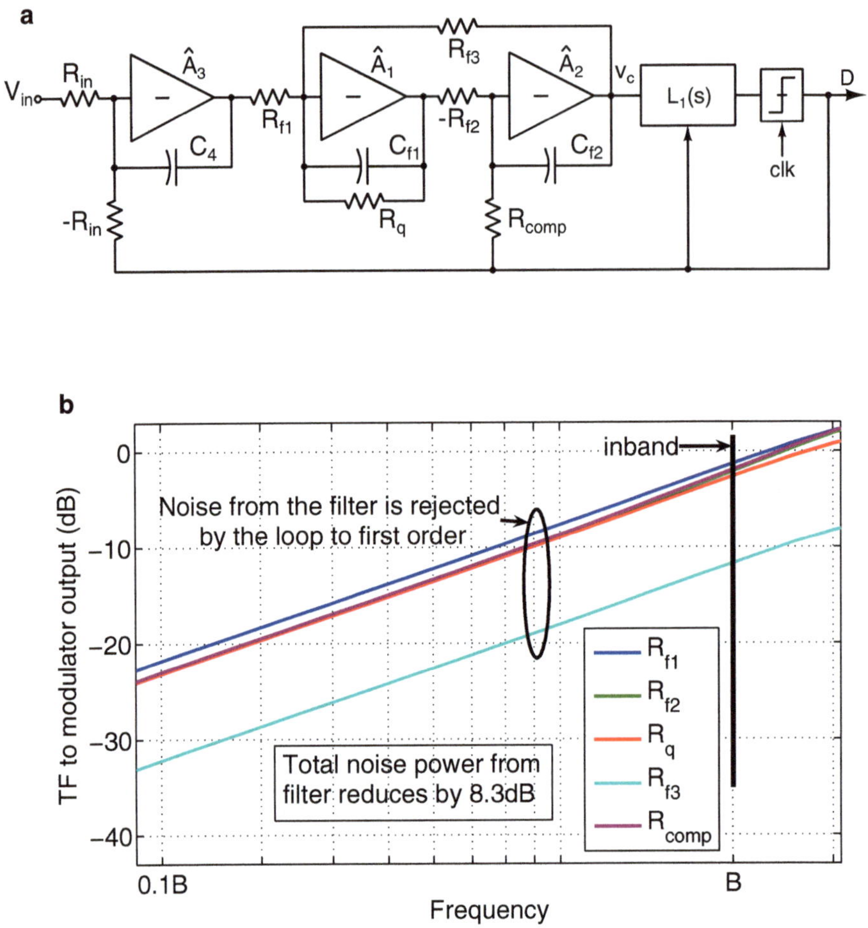

Fig. 4.9 (**a**) A second order filter embedded in a ΣΔ loop. (**b**) Magnitude of transfer functions from resistor noise sources in the embedded filter to the modulator output in our design

$$\frac{v_c(s)}{D(s)} = -\frac{1 + \frac{sR_{in}C_4R_{f3}R_{f2}}{R_qR_{comp}} + \frac{s^2R_{in}C_4R_{f3}R_{f2}C_{f1}}{R_{comp}}}{sR_{in}C_4\left[1 + \frac{sR_{f3}R_{f2}C_{f2}}{R_q} + s^2R_{f3}R_{f2}C_{f2}C_{f1}\right]} \tag{4.10}$$

It is apparent that if R_{comp} is chosen to be $R_{in}C_4/C_{f2}$, the numerator and the second order polynomial in the denominator cancel, yielding $\frac{v_c(s)}{D(s)} = \frac{-1}{sR_{in}C_4}$. This result makes intuitive sense due to the following. In the stand-alone CTΣΔM, the transfer function from D to v_c is $(\omega_4/s) = 1/(sR_{in}C_4)$. The same must hold for the CTΣΔM with embedded filtering. From Fig. 4.9a, we see that the path from D to v_c at high frequencies is through the second integrator of the filter, whose transfer function is $1/(sR_{comp}C_{f2})$. Since this must equal $1/(sR_{in}C_4)$, it follows that $R_{comp} = R_{in}C_4/C_{f2}$. As far as quantization noise is concerned, it is as if the filter does not exist—implying that the NTF is unchanged. A filter of arbitrary order can be embedded in a CT$\Sigma\Delta$M and compensated in a similar manner.

The technique discussed above can be used to move a filter into any CT$\Sigma\Delta$M. However, the benefits of doing so—namely, reduction of filter noise when referred to the input, accrue only when the gain of the first integrator (ω_4/s in Fig. 4.7a) over the signal band is larger than unity. To better understand this, consider our specific case of a CIFF-B CT$\Sigma\Delta$M with an embedded second order filter, shown in Fig. 4.9a. The low frequency gain from V_{in} to the output of $\hat{A}_3$ is $R_{f1}R_{f2}C_{f2}/(C_4R_{in}R_q)$. This can be deduced as follows. The low frequency current flowing through C_{f2} is ≈ 0. Since the STF is unity in the signal band, it must follow that the low frequency gain from V_{in} to the output of $\hat{A}_1$ is R_{f2}/R_{comp}. The voltage v_c at the input of $L_1(s)$ will be identical to that which would exist at the output of the first integrator of a CIFF-B modulator preceded by a second order filter. Further, due to the CIFF-B design, the gain from V_{in} to v_c in the signal band will be ≈ 0. Thus, the in-band current flowing through R_{f3} is ≈ 0, and the low frequency gain from V_{in} to the output of $\hat{A}_3$ is $R_{f1}R_{f2}/R_{comp}R_q$. Using $R_{comp} = R_{in}C_4/C_{f2}$ [from (4.10)], we see that the gain from V_{in} to the output of the first integrator is $R_{f1}R_{f2}C_{f2}/(C_4R_{in}R_q)$. As the filter bandwidth reduces (C_{f2} increases), C_4 has to increase to prevent $\hat{A}_3$ from saturating, thereby decreasing the gain of the first integrator in the signal band. Thus, choosing too low a filter bandwidth is counterproductive.

This is not restrictive, however, as the cutoff frequency of the filter should anyway be chosen so that the variation of the gain and group delay over the signal bandwidth is negligible. This holds whether the filter is embedded, or placed upfront. Denoting the signal bandwidth by B, calculations show that a second order Butterworth filter with a 3 dB bandwidth of $2B$ has only 0.25 dB gain droop across the signal band, whereas reducing the cutoff frequency to $1.5B$ increased this droop to 0.8 dB. Based on this, and the discussion in the previous paragraph, we chose a second order Butterworth filter with a bandwidth of $2B$. As shown in the next subsection, this choice still allows the first integrator to have a gain greater than unity at the edge of the signal band. In the discussion to follow, we compare the various performance metrics (noise, linearity, power consumption, active area) of a modulator with an embedded filter and its counterpart where the filter is placed up front.

4.3.1 Offset, Flicker and Thermal Noise

It is well known that offset and flicker noise are problematic in direct conversion receivers. Embedding the active filter into the $\Sigma\Delta$ loop eliminates its dc offset and greatly reduces the 1/f noise it contributes to the entire signal chain. Thus, the first opamp of the filtering CT$\Sigma\Delta$M remains, for all practical purposes, the only source of offset and 1/f noise. This makes it easy to manage. One way to do this is to chop the first OTA. Chopping renders the first integrator a linear periodically time varying (LPTV) system. Since the integrator processes shaped quantization noise whose out of band spectral density is several orders of magnitude higher than that in-band, down conversion of noise from multiples of the chopping frequency is a serious concern. Fortunately, this can be handled by the use of FIR feedback, which is also beneficial in other aspects as will be discussed in Sect. 4.4.

Figure 4.9b shows the transfer functions from the various resistor noise sources in the filter (Fig. 4.9a) to the modulator output. These transfer functions are calculated for our specific CIFF-B design where the oversampling ration (OSR) is 64. The signal bandwidth is denoted by B and the 3 dB bandwidth of the (Butterworth) filter is $2B$. It is seen that thermal noise from the filter is first order shaped out of the signal band. If the filter is placed up front, the transfer function from all its resistor noise sources to the modulator output will be ≈ 1 in the signal band. From Fig. 4.9b it is clear that the total output noise is reduced when the filter is embedded into the modulator. Δ-H_1-Σ can therefore be impedance scaled to have the same input referred noise as that in H_1-$\Sigma\Delta$. This reduces the power consumption and active area. In our design, the first integrator and filter in Δ-H_1-Σ were impedance scaled by 1.8 and 3 respectively, so as to have the same input referred noise as in H_1-$\Sigma\Delta$.

4.3.2 *Linearity*

To understand the effect of OTA nonlinearity in H_1-$\Sigma\Delta$ and Δ-H_1-Σ, we simulated normalized fourth order single bit designs (with CIFF-B loop filters). The sampling rate and OSR were assumed to be 1 Hz and 64 respectively. The filter's 3 dB bandwidth was set to twice the signal bandwidth. Keeping in view our final implementation, a 4-tap FIR feedback DAC was used. With the filter up front (H_1-$\Sigma\Delta$), the input resistance of the filter and CTDSM were chosen to be 1 Ω. The OTA is assumed to be a single stage design, with the current in each leg modeled as

$$i = g_m v_x - g_2 v_x^2 - g_3 v_x^3 \tag{4.11}$$

where v_x denotes the differential voltage at the input of the OTA. g_m, g_2, g_3 were chosen to be 100 S, 200 A/V^2 and 4000 A/V^3 respectively. Third order intermodulation distortion manifests due to odd nonlinearity. Second order distortion is deliberately introduced through g_2 in the equation above. This way, IM_2 can be simulated by introducing random mismatches in passive components. For the CT$\Sigma\Delta$M with embedded filter (Δ-H_1-Σ), R,C's in the first integrator and those in the filter were impedance (up)scaled by factors of 1.8 and 3 respectively. This choice resulted in an input referred noise spectral density which was equal to that of H_1-$\Sigma\Delta$. Regarding the OTAs, two cases were simulated, as described below.

Case A. The OTAs in the first integrator and the filter were also impedance scaled by 1.8 and 3 respectively. By "impedance scaling an OTA by a factor α", we mean the following—the output current of the OTA in (4.11) above is modified as $\hat{i} = i/\alpha = \frac{1}{\alpha}(g_m v_x - g_2 v_x^2 - g_3 v_x^3)$. In a real transistor level design, this would correspond to reducing all transistor widths (and their quiescent currents) by a factor α, thereby resulting in lower power dissipation. It is thus seen that by scaling OTAs, Δ-H_1-Σ has lower power dissipation compared to H_1-$\Sigma\Delta$.

Case B. The OTAs in the first integrator and the filter in Δ-H_1-Σ of Case A above were impedance scaled by 1/3,1/3 and 1 so that the power dissipation is the same as in H_1-$\Sigma\Delta$.

Since distortion depends on the product of the OTA's G_m and the integrating resistor [13], one should expect improved linearity in Case B when compared to Case A. This is confirmed by simulation results showing the in-band and out-of-band linearity of H_1-$\Sigma\Delta$ and Δ-H_1-Σ. First, both designs were excited by two closely spaced −10.5 dBFS in-band tones. Figure 4.10a compares the strength of the IM_3 tone. The distortion of H_1-$\Sigma\Delta$ and Δ-H_1-Σ-Case A are comparable, but when power saved due to impedance scaling is put back into the OTAs as in

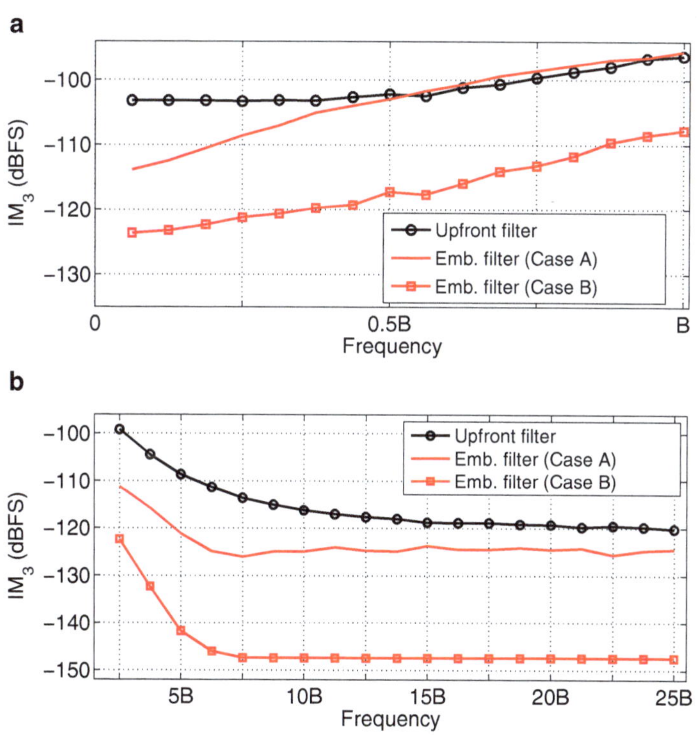

Fig. 4.10 Comparison of IM_3. (**a**) Tones are in-band. The x-axis is the average frequency of the two input tones. (**b**) Tones are out-of-band at frequencies f and $2f + \Delta f$

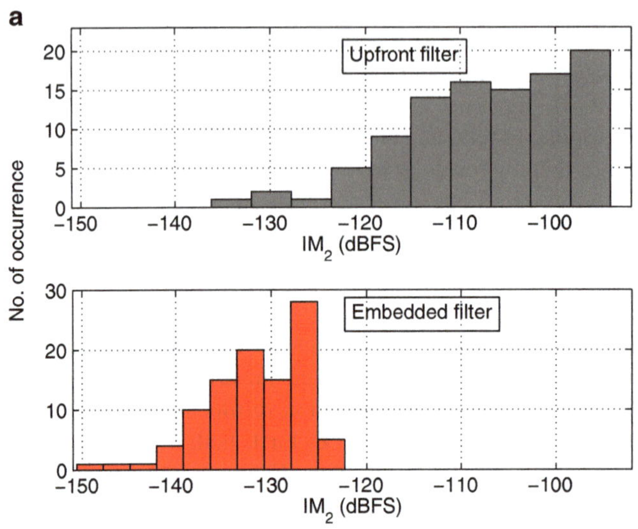

Fig. 4.11 Histogram of the power in inband IM_2 tone from 100 Monte Carlo runs. The input consists of two -10.5 dBFS tones around B

Case B, embedding the filter results in enhanced linearity. Figure 4.10b shows out-of-band IM_3 for both cases—the improvement is linearity is apparent. Similar results were obtained for IM_2, as shown in the histograms of Fig. 4.11. The input for these simulations consisted of two -10.5 dBFS tones around B, where B denotes the edge of the signal band. From the discussion above, it is apparent that other metrics (power and noise) remaining the same, embedding the filter into a CT$\Sigma\Delta$M improves linearity and reduces active area. Simulations (not given here due to space constraints) show embedding the filter does not influence the robustness of the CTDSM with respect to RC variations in any way.

4.4 Architectural Considerations

Having convinced ourselves that embedding an active filter into a $\Sigma\Delta$ modulator is beneficial on all counts when compared to having the filter up front, the next task at hand is to examine the various design choices regarding the modulator itself.

Number of quantizer levels: CT$\Sigma\Delta$Ms which achieve high levels of linearity have typically used multibit quantizers. Such a choice has several advantages. First, the increased number of levels enables a lower OSR to achieve a desired signal to quantization noise ratio (SQNR). The reduced step size in the feedback waveform reduces the sensitivity of the modulator to clock jitter, as well as the slew rates required of the opamps in the loop filter. However, the complexity of the ADC used in the quantizer increases exponentially with the number of bits. Even though the comparators in the quantizer lend themselves to low power operation, clock generation and distribution can consume significant current. This, unfortunately, is difficult to estimate during the architectural design phase. Moreover, mismatch in the unit elements of the feedback DAC degrade in the inband SNDR of the modulator, necessitating some form of mismatch correction, like calibration or dynamic element matching (DEM). This further increases the power dissipation and design time of the quantizer.

An alternative to using a flash design for the ADC is to use a successive approximation register (SAR) ADC. The latter's complexity increases linearly with the number of bits resolved. Comparator offset, which can be potentially problematic in a flash design, is benign. This approach has found favor in many recent designs, particularly those realized in nanometer CMOS technologies. The complexity of the DAC (which now needs a binary to thermometer decoder in the $\Sigma\Delta$ loop, apart from DEM) still remains.

In contrast, using a single bit quantizer, where the feedback DAC is inherently linear, dramatically simplifies the quantizer design. However, the full scale two level feedback waveform places increased demands on the linearity of the loop filter, as well as increases the sensitivity of the modulator to clock jitter. From the discussion above, it is seen that a multibit loop complicates the quantizer design at the expense of a simplified loop filter. The opposite is true in a single bit modulator. Recognizing this, several recent works have attempted to alleviate the linearity and clock jitter problems associated with a single bit design. Pavan and Sankar [14] and Nandi et al. [15] use integrators based on opamp assistance, which is a circuit technique that enhances the linearity of an opamp in a power efficient manner. Shettigar and Pavan [16] uses a feedback DAC which incorporates a finite impulse response filter (FIR-DAC).

The FIR DAC principle is not new [17, 18]—however, relatively few designs seem to have exploited this technique. The basic idea behind the FIR DAC is to feedback a filtered version of the single-bit quantizer output, as shown in Fig. 4.12a. Due to the high frequency attenuation of the FIR filter, the DAC output is a multi-level waveform, like in a multibit quantizer. Thanks to this, the FIR-DAC approach has low clock jitter sensitivity and relaxes the linearity requirements of the loop filter. In practice, the filter and DAC combination are implemented in a semi-digital fashion, as shown in Fig. 4.12b, which makes the FIR DAC inherently linear in spite of mismatch. Due to the single-bit quantizer, the ADC design is simple and consumes very little power. A modulator employing a single-bit quantizer and an FIR DAC, therefore, combines the best features of single-bit and multibit operation.

FIR feedback presents several design challenges. Due to the delayed nature of the feedback waveform, the modulator needs to be carefully stabilized to restore the noise transfer function. It can be shown that the NTF can be restored *exactly*. The design procedure and intuition behind this is given in [19].

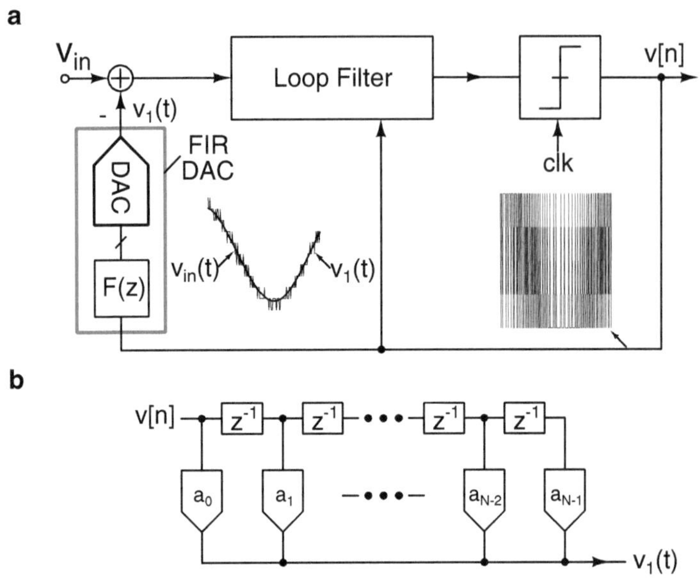

Fig. 4.12 (**a**) A single-bit CTΣΔM with FIR feedback and (**b**) a semi-digital implementation of the feedback DAC

4.4.1 Offset and Flicker Noise Reduction by Chopping

DC offset and flicker noise are very problematic in direct conversion receivers. While there are many mechanisms that result in offset and 1/f noise, the baseband section is a significant contributor. Using a filter up front to modify the modulator's STF results in increased offset and flicker noise. Embedding the filter, on the other hand, reduces its contribution to the low frequency noise of the signal chain since it is shaped by the response of the first integrator of the modulator. Since the source of offset is the first integrator, it can be more easily managed.

Chopping is a classical and effective technique to address 1/f noise problems in low frequency amplifiers. The block diagram of a fully differential CTΣΔM (with or without embedded filtering), employing chopping in the first integrator is shown in Fig. 4.13. A natural question that arises is how one should choose the chopping frequency f_{chp}. Chopping renders the first integrator a linear periodically time varying (LPTV) system with frequency f_{chp}, where frequency translation effects cause signals and noise from frequencies around multiples of f_{chp} to dc. It turns out that in a chopped integrator where f_{chp} has a 50 % duty cycle, shaped noise aliases only from *even* multiples of f_{chp}. The first integrator of the CTΣΔM processes shaped quantization noise, whose out of band spectral density can be 100 dB higher than the in-band density. Thus, even a small amount of aliasing can dramatically degrade in-band performance.

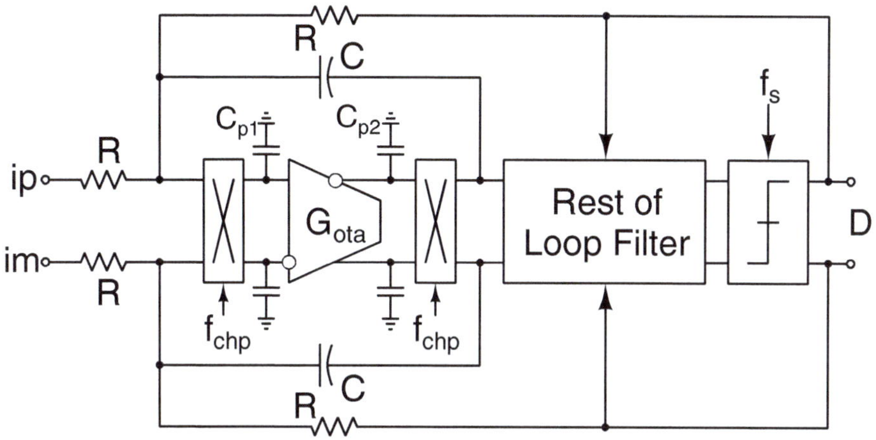

Fig. 4.13 Chopping the first integrator of a CTΣΔM to reduce flicker noise

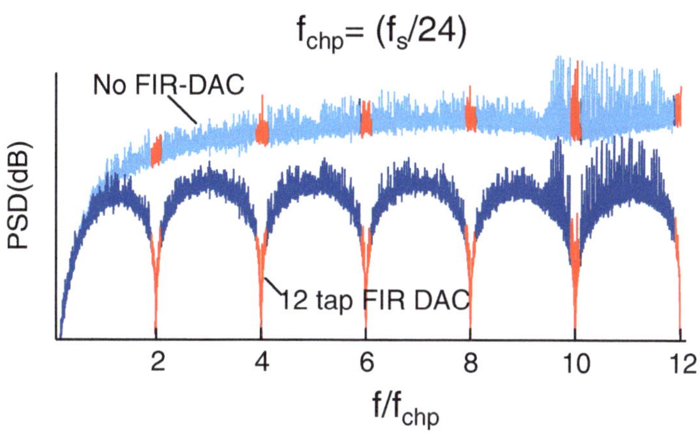

Fig. 4.14 PSD of the signal processed by the first integrator without and with FIR feedback. A 12 tap FIR DAC, with equal taps, is assumed

An FIR DAC solves the problem of shaped noise aliasing due to the following.The FIR filter introduces nulls in the spectrum of feedback waveform at multiples of f_s/N, as shown in Fig. 4.14 for the special case of a 12-tap FIR filter. If the chopping frequency is chosen such that $f_{chp} = f_s/2N$, noise that can potentially alias into the signal band is attenuated. Thus, frequency translation has minimal impact on the in-band spectral density.

Thus, the use of FIR feedback in a filtering CTΣΔM not only results in improved linearity of the loop filter and reduces jitter sensitivity, but also enables "chopping for free" if f_{chp} is appropriately chosen. A detailed analysis and experimental results can be found in [20].

4.5 Implementation Details and Measurements

Two modulators—one with up front filtering and another with embedded filtering were designed for the same nominal specifications in a 0.13 μm CMOS process. We attempted to achieve about 80 dB of instantaneous peak dynamic range in a signal bandwidth of 2 MHz and a 0–18 dB variable gain for both designs. In view of the trade off between filter bandwidth and performance (see Sect. 4.3), a Butterworth filter with 4 MHz bandwidth was chosen.

4.5.1 *Modulator with Upfront Filter (H_1-ΣΔ)*

Figure 4.15 shows a simplified single-ended schematic of H_1-ΣΔ. The filter was implemented as a Tow-Thomas biquad. Active-RC integrators were used for low noise and high linearity. Programmable gain was achieved by splitting it across the two stages of the filter (A_{v1} and A_{v2}). A variable gain A_{v1} was implemented by varying R_{f1} by a factor 1, 2 or 4. The current in A_1 was made variable depending on A_{v1}. A_{v2} can be programmed to be 1 or 2 by changing R_{f2} and R_{f3}, as shown in the figure. Thus, up to 18 dB of gain can be incorporated into the filter. Two stage feedforward compensated opamps were used with an AC coupled feedforward stage to increase the output swing (similar to those in [16]).

The modulator employed a single bit quantizer, chosen for low power operation. Stability considerations, therefore, restricted the out-of-band gain of the NTF to 1.5. This necessitated a fourth order NTF with OSR = 64 (f_s = 256 MHz) to achieve an in-band SQNR of about 96 dB—well below the desired signal to thermal noise ratio. A CIFF-B topology was used for the loop filter. R_{31} implements the feedforward path across the integrator I_3. R_{12} realizes complex zeros in the NTF, and is implemented using a T-network. All capacitors were implemented as digitally

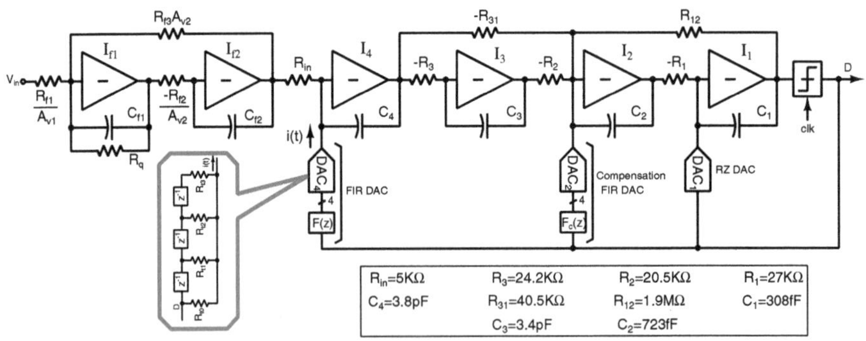

Fig. 4.15 Fourth order modulator with a CIFF-B loop filter with a second order filter up front

tunable banks to trim RC variations. Switches are placed across all the integrating capacitors to reset their states to recover from overload.

The main feedback element, DAC_4, is a 4-tap FIR DAC with an NRZ pulse shape. This reduces jitter sensitivity and relaxes linearity requirements of the first integrator [16–18]. Rather than use equal tap weights, they are optimized to minimize noise due to jitter. The weights [0.2, 0.3, 0.3, 0.2] reduce jitter noise by 15 dB when compared to a single bit CTDSM without an FIR DAC. The delay introduced by DAC_4 can destabilize the modulator. Another 4-tap FIR DAC, DAC_2, with transfer function $F_c(z)$ is used at the input of I_2 to compensate the loop. The FIR DACs are realized in a semi-digital fashion as shown in Fig. 4.15. Any mismatch in the DAC resistors manifests as a variation in the FIR tap coefficients and not as nonlinearity.

4.5.2 Modulator with Embedded Filter (Δ-H_1-Σ)

The simplified single-ended schematic of Δ-H_1-Σ is shown in Fig. 4.16. This is derived from the circuit of Fig. 4.15 by moving the filter beyond the first integrator of the modulator as discussed in Sect. 4.3. Since the main DAC (DAC_4) is of the FIR kind, DAC_3, which compensates for the delay of the embedded filter, is also an FIR DAC with the same coefficients. Variable gain is achieved by programming the input resistor. At high gains, the current in the first opamp is increased. Since the noise of the filter is shaped by the first integrator, it was impedance scaled. Power saved by doing this was used to improve linearity as described below.

Figure 4.17 compares the impedance levels of H_1-$\Sigma\Delta$ and Δ-H_1-Σ. In a well designed active-RC integrator, the input referred noise is dominated by the contribution from the resistors. Thus, the input referred noise spectral density of

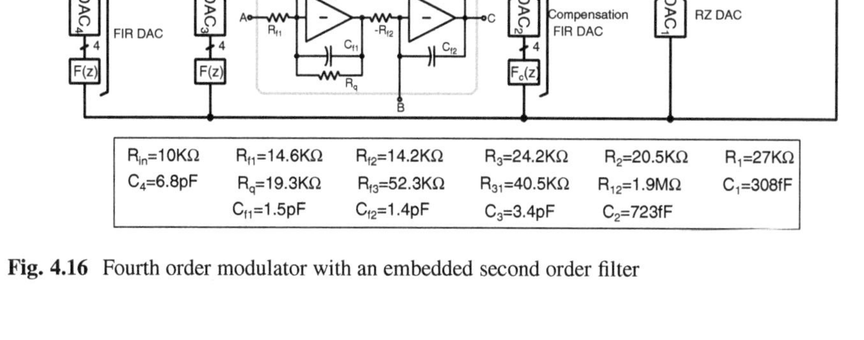

R_{in}=10KΩ	R_{f1}=14.6KΩ	R_{f2}=14.2KΩ	R_3=24.2KΩ	R_2=20.5KΩ	R_1=27KΩ
C_4=6.8pF	R_q=19.3KΩ	R_{f3}=52.3KΩ	R_{31}=40.5KΩ	R_{12}=1.9MΩ	C_1=308fF
	C_{f1}=1.5pF	C_{f2}=1.4pF	C_3=3.4pF	C_2=723fF	

Fig. 4.16 Fourth order modulator with an embedded second order filter

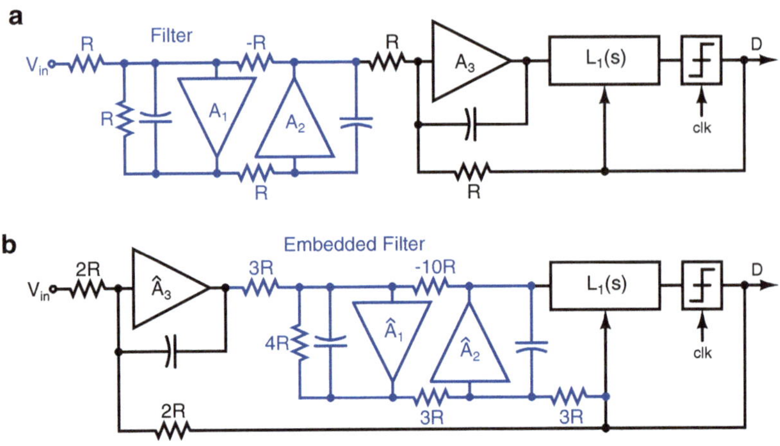

Fig. 4.17 Impedance levels in (**a**) CIFF-B CTDSM with up front filter (**b**) CTDSM with embedded filter

H_1-$\Sigma\Delta$ is $\approx$ 6R $\times$ 4kT. When the filter is embedded, the CTDSM input and DAC impedances are doubled and the filter impedance is scaled by $\approx$ 3$\times$. In spite of this, the input noise density is smaller, at $\approx$ 5.5R $\times$ 4kT. Thus, in principle, the currents in the OTAs $\hat{A}_3$, $\hat{A}_1$ and $\hat{A}_2$ can be reduced by factors of $\approx$ 2, $\approx$ 3 and $\approx$ 3 respectively. Some of the power saved in this process can be used to enhance linearity of the OTAs. In our work $\hat{A}_1$, $\hat{A}_2$ and $\hat{A}_3$ are scaled so that they consume 0.65, 0.33 and 0.75 times the currents of A_1, A_2 and A_3 respectively. Denoting the transconductance of an OTA by G_{OTA}, the G_{OTA}R products for $\hat{A}_1$, $\hat{A}_2$ and $\hat{A}_3$ are about 1.95, 1, 1.5 times the corresponding products for A_1, A_2 and A_3, indicating improved linearity of the integrators in Δ-H_1-Σ. Thus, not only is overall power dissipation lower when the filter is embedded into the CTDSM, linearity is also enhanced with respect to a filter-CTDSM cascade, while achieving similar noise.

In our design, the LPF and the loop filter in H_1-$\Sigma\Delta$ consume 3.4 mA to 4.1 mA (depending on the gain). The corresponding current for Δ-H_1-Σ is 2.3 mA to 2.7 mA. The clock generation block consumes 0.47 mA, while the quantizer and D-flip-flops draw about 0.3 mA.

4.5.3 Measurement Results

Both designs described in the previous section were fabricated in a 0.13 µm CMOS process through Europractice. H_1-$\Sigma\Delta$ and Δ-H_1-Σ have active areas of 0.42 mm^2 and 0.33 mm^2 respectively. For our choice of impedance scaling factors, embedding the filter results in approximately 25 % lesser active area. The modulator outputs were brought out using LVDS buffers. The single bit data streams were captured

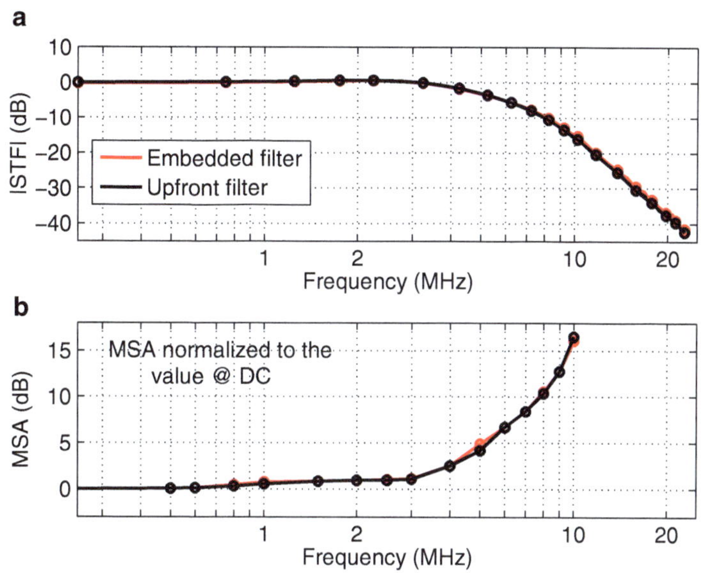

Fig. 4.18 Measured (**a**) |STF| and (**b**) MSA of the two modulators

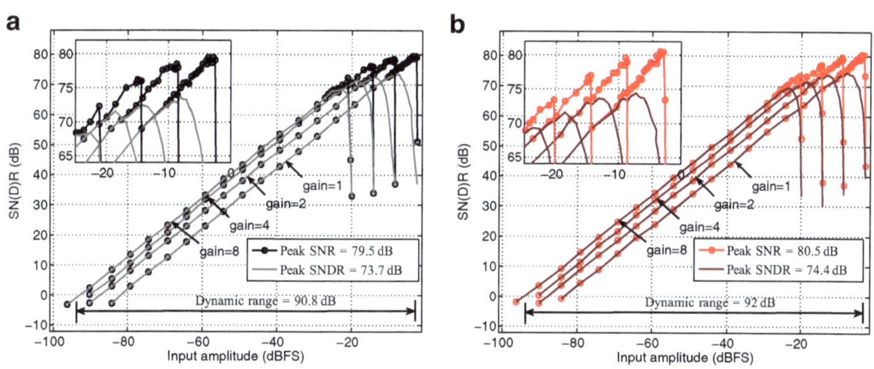

Fig. 4.19 SNR/SNDR vs. input amplitude with (**a**) up front filter and (**b**) embedded filter. (**a**) H_1-$\Sigma\Delta$. (**b**) Δ-H_1-Σ

using a digital storage oscilloscope and then processed offline. 64K data points were used for spectral estimation and a Blackman-Harris window was used to minimize spectral leakage effects.

Figure 4.18a b show the measured STF and maximum stable amplitude (MSA) for the two modulators, where the latter is normalized to the value at DC. It is seen that the STF and MSA of the two filtering modulators are virtually identical, indicating the same functionality. The SNR/SNDR vs input amplitude plots are shown in Fig. 4.19a, b. H_1-$\Sigma\Delta$ and Δ-H_1-Σ have an overall DR of 90.8 dB and 92 dB, and peak instantaneous DR/SNR/SNDR of 81/79.5/73.7 dB

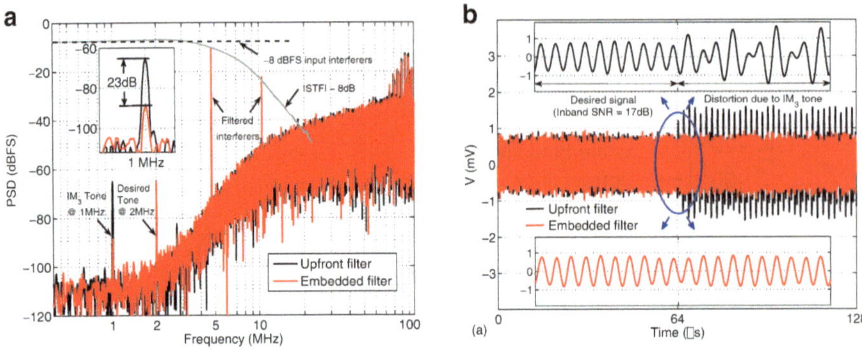

Fig. 4.20 Out-of-band linearity measurements in the presence of two −8 dBFS tones at 4.75 MHz and 10.5 MHz, gain = 1. (**a**) PSD at output of the modulators and (**b**) time domain demonstration of the linearity improvement

and 82/80.5/74.4 dB respectively. H_1-$\Sigma\Delta$ consumes 5.9–6.8 mW depending on the gain setting (including references) while Δ-H_1-Σ consumes 4.34–5 mW. From these results, we see that embedding the filter achieves slightly better performance with lower area and power dissipation.

Figure 4.20a compares the linearity of the two modulators in the presence of out-of-band interferers, which are placed such that one of their IM_3 products falls at 1 MHz. The input consists of a small (−65 dBFS) desired signal and two large interferers at 4.75 MHz and 10.5 MHz (each of amplitude −8 dBFS). A gain of 1 is used in the modulators. The filtering effect of the CTDSM is apparent. Further, the IM_3 product at 1 MHz is around 23 dB smaller with the embedded filter, representing an IIP_3 improvement of $\approx$ 11 dB. Again, it is seen that embedding the filter within the modulator enhances linearity for both in-band and out-of-band frequencies. To better appreciate this, we show time domain results, obtained by decimating the modulators' output sequence. The input consists of a small desired signal and one −8 dBFS blocker at 4.75 MHz. Another blocker (−8 dBFS at 10.5 MHz) is added to the input after 64 μs. Figure 4.20b shows the decimated outputs, with the insets zooming in around 64 μs. For the modulator with up front filter, the effect of distortion is clearly seen after the second interferer is applied, while the output of the modulator with embedded filter remains virtually unchanged, demonstrating improved linearity.

4.6 Conclusions

When interferers are present, a low pass filter is needed to reduce the in-band dynamic range requirements of a conventional continuous-time $\Sigma\Delta$ modulator. The filter can be embedded into the ADC. This achieves the same functionality as using a filter up front, but with better in band and out-of-band linearity, lower area and reduced power dissipation. Flicker noise and dc offset can be easily addressed by chopping if an FIR feedback is used. The latter also relaxes requirements on clock jitter and improves the linearity of the loop filter.

References

1. H. Shibata, R. Schreier, W. Yang, A. Shaikh, D. Paterson, T.C. Caldwell, D. Alldred, P.W. Lai, A DC-to-1 GHz tunable RF ADC achieving DR 74 dB and BW 150 MHz at 450 MHz using 550 mW. IEEE J. Solid State Circuits **47**(12), 2888–2897 (2012)
2. J. Kauffman, P. Witte, J. Becker, M. Ortmanns, An 8.5 mW continuous-time ΣΔ modulator with 25 MHz bandwidth using digital background DAC linearization to achieve 63.5 dB SNDR and 81 dB SFDR. IEEE J. Solid State Circuits **46**(12), 2869–2881 (2011)
3. G. Mitteregger, C. Ebner, S. Mechnig, T. Blon, C. Holuigue, E. Romani, A 20 mW 640 MHz CMOS continuous-time ΣΔ ADC with 20 MHz signal bandwidth, 80 dB dynamic range and 12 bit ENOB. IEEE J. Solid State Circuits **41**(12), 2641–2649 (2006)
4. F. Munoz, K. Philips, A. Torralba, A 4.7 mW 89.5 dB DR CT complex ΣΔ ADC with built-in LPF, in *IEEE ISSCC Dig. Tech. Papers* (2005), pp. 500–501
5. M. Sosio, A. Liscidini, R. Castello, F.D. Bernardinis, A complete DVB-T/ATSC tuner analog base-band implemented with a single filtering ADC, in *Proc. IEEE ESSCIRC* (2011), pp. 391–394
6. M. Andersson, M. Anderson, L. Sundstrom, S. Mattisson, P. Andreani, A 9 MHz filtering ADC with additional 2-nd-order ΣΔ modulator noise suppression, in *Proc. IEEE ESSCIRC* (2013), pp. 323–326
7. M. Anderson, M. Anderson, L. Sundstrom, L. Sundstrom, P. Andreani, A filtering ΣΔ ADC for LTE and beyond. IEEE J. Solid State Circuits **49**(7), 1535–1547 (2014)
8. A. Nejdel, X. Liu, M. Palm, L. Sundstrom, M. Tormanen, H. Sjoland, P. Andreani, A 0.6–3.0 GHz 65nm CMOS radio receiver with ΣΔ-based A/D-converting channel-select filters, in *Proceedings of the European Solid-State Circuits Conference (ESSCIRC)* (IEEE, Graz, 2015), pp. 299–302
9. R. Rajan, S. Pavan, A 5mW CT ΣΔ ADC with embedded 2nd-order active filter and VGA achieving 82dB DR in 2MHz BW, in *2014 IEEE International Solid-State Circuits Conference Digest of Technical Papers (ISSCC)* (IEEE, San Francisco, 2014), pp. 478–479
10. R.S. Rajan, S. Pavan, Design techniques for continuous-time ΣΔ modulators with embedded active filtering. IEEE J. Solid State Circuits **49**(10), 2187–2198 (2014)
11. J. De Maeyer, J. Raman, P. Rombouts, L. Weyten, Controlled behaviour of STF in CTΣΔMs. Electron. Lett. **41**(16), 896–897 (2005)
12. K. Philips, P. Nuijten, R. Roovers, A. van Roermund, F. Chavero, M. Pallares, A. Torralba, A continuous-Time ΣΔ ADC with increased immunity to interferers. IEEE J. Solid State Circuits **39**(12), 2170–2178 (2004)
13. S.V. Thyagarajan, S. Pavan, P. Sankar, Active-RC filters using the Gm-assisted OTA-RC technique. IEEE J. Solid State Circuits **46**(7), 1522–1533 (2011)
14. S. Pavan, P. Sankar, Power reduction in continuous time ΣΔ modulators using the assisted opamp technique. IEEE J. Solid State Circuits **45**(7), 1365–1379 (2010)
15. T. Nandi, K. Boominathan, S. Pavan, Continuous-time modulators with improved linearity and reduced clock jitter sensitivity using the switched-capacitor return-to-zero DAC. IEEE J. Solid State Circuits **48**(8), 1795–1805 (2013)
16. P. Shettigar, S. Pavan, Design techniques for wideband single-bit continuous-time ΣΔ modulators with FIR feedback DACs. IEEE J. Solid State Circuits **47**(12), 2865–2879 (2012)

17. B. Putter, $\Sigma\Delta$ ADC with finite impulse response feedback DAC, in *IEEE ISSCC Dig. Tech. Papers* (2004), pp. 76–77
18. O. Oliaei, Sigma-delta modulator with spectrally shaped feedback. IEEE Trans. Circuits Syst II Analog Digit. Signal Process. **50**(9), 518–530 (2003)
19. S. Pavan, Continuous-time delta-sigma modulator design using the method of moments. IEEE Trans. Circuits Syst. I Regul. Pap. **61**(6), 1629–1637 (2014)
20. S. Billa, A.Sukumaran, S.Pavan, A 280μW, 24 kHz BW, 98.5 dB SNDR, chopped single-bit CT$\Sigma\Delta$M achieving < 10 Hz 1/f noise corner without chopping artifacts, in *IEEE ISSCC Dig. Tech. Papers* (2016)

Chapter 5
Blocker and Clock-Jitter Performance in CT $\Sigma\Delta$ ADCs for Consumer Radio Receivers

Sebastián Loeda

5.1 Introduction

WLAN and Cellular communication standards have been the main drivers for low-area, low-power *and* high-performance wideband ADCs. This comes from a requirement for efficient, high-rate wireless data transmissions in a crowded and segmented radio spectrum using large-volume, low-cost, battery-powered handheld devices.

Zero-IF receivers are the most cost-effective radio architecture from a reduced number of high-Q external filter components and a smaller die size. However, a radio transmitting in the vicinity of a zero-IF receiver that is trying to decode a weak, distant signal on the adjacent channel will dominate the receiver's signal budget. ADC performance is a critical factor in a radio's selectivity, as it must convert the in-band received signal with high sensitivity while accommodating a poorly-filtered, large adjacent interferer (blocker) from a lack of high-Q external filters (Fig. 5.1). WLAN and Cellular support a range of channel bandwidths requiring the same sensitivity at similar efficiency across several MIMO antennas and/or aggregated carriers, in order to optimise the use of the spectrum and regulate radio power consumption. This has resulted in an order of magnitude increase in the number of ADCs required to support a radio receiver, and a heavier emphasis towards area-optimised ADC designs for lower-cost consumer radios.

CT $\Sigma\Delta$ ADCs dominated the ADC literature for two decades until SAR ADCs saw an increase in conversion efficiency at the beginning of this decade [1]. However, system-level implications from low-power zero-IF receivers have often been overlooked for either type of ADC in the literature. For instance, unlike CT $\Sigma\Delta$ ADCs, SAR ADCs create challenging anti-aliasing input filtering requirements with

S. Loeda (✉)
Broadcom Limited, Sydney, NSW, Australia
e-mail: ieee@loeda.com

© Springer International Publishing Switzerland 2017
A. Baschirotto et al. (eds.), *Wideband Continuous-time $\Sigma\Delta$ ADCs, Automotive Electronics, and Power Management*, DOI 10.1007/978-3-319-41670-0_5

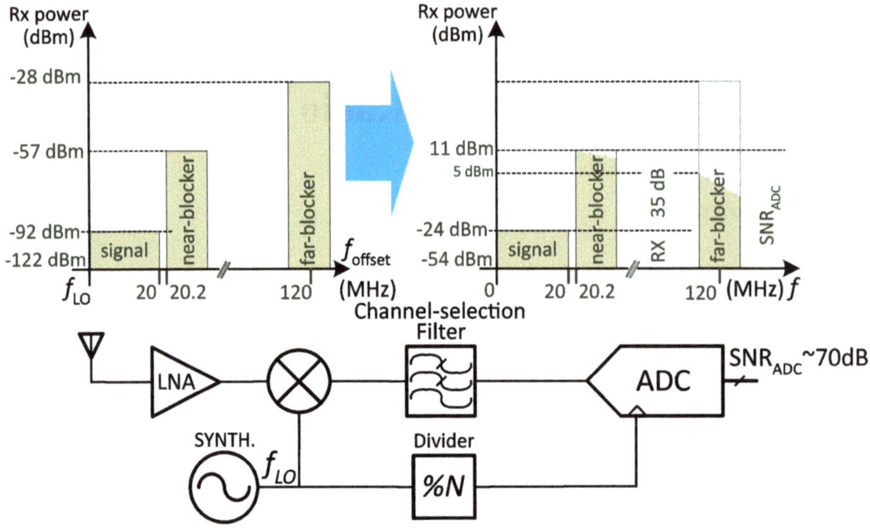

Fig. 5.1 A CT ΣΔ ADC in a zero-IF receiver with radio blockers

power-hungry sampling-input driving loads [2, 3]. Conversely, out-of-band (OOB) gain peaking in the signal transfer function (STF) of CT ΣΔ ADCs impairs zero-IF receiver performance [4]. STF OOB gain is typically seen in CT ΣΔ ADCs with efficient loop-filter architectures. It provides gain to (poorly-filtered) dominant blockers near the edge of band; thereby reducing the available ADC dynamic range for signal in-band.

Another limiting factor for CT ΣΔ ADCs is its higher sensitivity to clock-timing jitter, in the feedback DAC [5]. This is typically managed in CT ΣΔ ADCs by reducing the energy introduced in the feedback DAC from one output sample to the next, i.e. reducing the magnitude of the signal transitions modulating clock jitter. This is typically achieved with either: A larger number of bits in the feedback DAC [5, 6], or by spreading the energy of a single-bit DAC pulse across several time samples, e.g. an FIR DAC [7–10] (or in some cases, both [11]). Single-bit CT ΣΔ ADCs with an FIR DAC result in large area savings when compared to multi-bit DAC implementations with comparable linearity and jitter performance [10], from a simple and compact design. However, this also comes at the cost of (further) OOB STF degradation [12].

Wideband, high conversion-efficiency ADCs that occupy a small area in a zero-IF radio receiver are therefore a balancing act between conflicting requirements. We find that CT ΣΔ ADCs employing feedforward (FF) loop-filter compensation and a single-bit FIR DAC with a flat STF [13] most efficiently accommodate a zero-IF receiver when its requirements on the ADC are carefully reviewed. In this paper, we analyse the practical implications of the choice of ADC architecture in a wide-band, large-volume, low-cost, power-efficient consumer radio receiver to support this conclusion.

5.2 CT ΣΔ ADC STF and Loop Robustness to the FIR Feedback DAC

Loop compensation in high-order ΣΔ ADCs was elegantly resolved through the nesting of 1st order ΣΔ ADCs [14] (coined multiple feedback or FB-compensated), or later on with loop-filter compensating zeroes [15] (coined interpolative, feed-forward or FF-compensated). Lee's interpolative loop-filter results in a more (area and conversion) efficient architecture from a wider first integrator stage unity gain-bandwidth (GBW), fewer (input-referred) noise contributors, a single feedback DAC, and no internal memory. However, accommodating the delay from an FIR DAC in the feedback of a FF-compensated CT ΣΔ ADC to reduce clock jitter noise is challenging from a loop stability perspective. None of the examples of FIR CT ΣΔ ADCs found in the literature prior to [13] are FF-compensated. Instead, they either employ FB-compensation [7–9], or a hybrid FF-FB-compensation [10, 11]. Nesting inner ΣΔ modulators in a FF loop filter eases an FIR DAC CT ΣΔ loop implementation by splitting a noise and delay-sensitive single-feedback DAC FF CT ΣΔ loop (Fig. 5.2a) into a loop-delay sensitive inner CT ΣΔ loop and a noise-sensitive outer CT ΣΔ loop (Fig. 5.2b). A FF-FB-compensated CT ΣΔ ADCs benefits from a low-noise FF front-end and an easier to implement FIR DAC in the loop, but this comes at the area and power cost of a fast inner FB stage.

A FF loop-filter front-end in either FF or FF-FB-compensated CT ΣΔ ADCs, results in OOB STF peaking at near-blocker offset frequencies. The STF of a CT

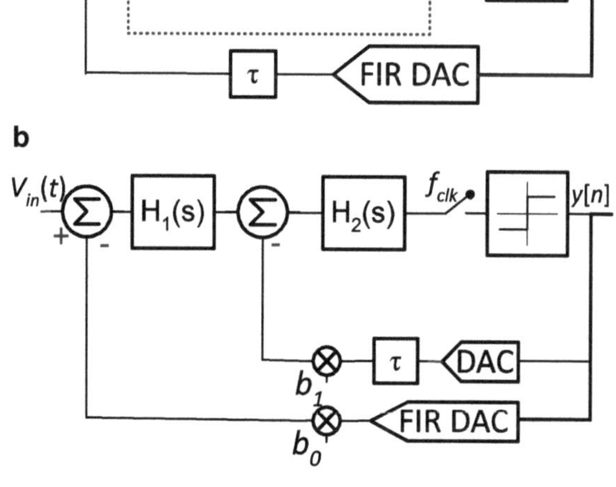

Fig. 5.2 (**a**) **FF CT ΣΔ ADC**; (**b**) **FB CT ΣΔ ADC** when $H_1(s)$ is a single integrator stage and $H_2(s)$ may be one integrator, or one integrator followed by nested FB CT ΣΔ stages; or a hybrid **FF-FB CT ΣΔ ADC** when $H_1(s)$ is a FF-filter and $H_2(s)$ may be one integrator, or one integrator followed by nested FB CT ΣΔ stages

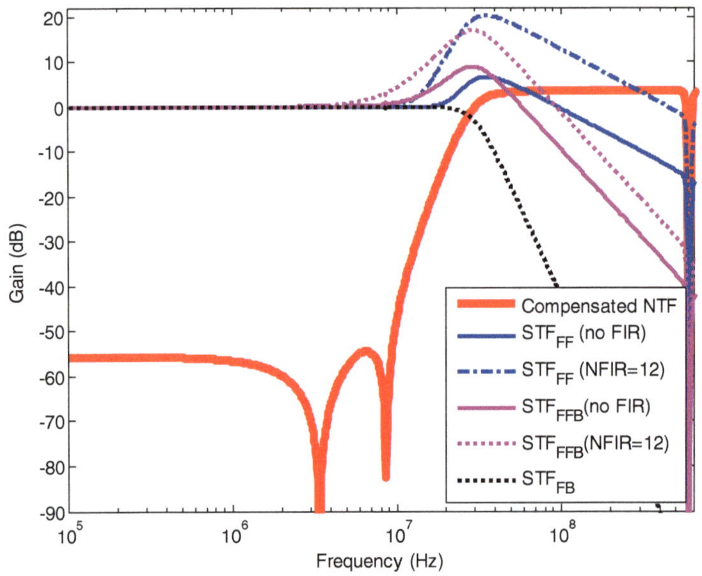

Fig. 5.3 STF in delay-compensated FF, FF-FB and FB CT ΣΔ ADCs with a NRZ DAC and a 12-tap FIR DAC

ΣΔ ADC is defined by [16]:

$$STF(\omega) = H_a(j\omega)\, NTF\left(e^{j\omega}\right), \tag{5.1}$$

where $H_a(s)$ is the transfer function from the input of the CT ΣΔ ADC to the input of its quantiser. The FF-compensation zeroes in $H_a(s)$ resulting from the FF part of the loop filter slow down the gain response roll-off near the 0 dB crossing point for loop stability. This crossing occurs around frequencies where the NTF rises rapidly, resulting in an OOB STF peak near the edge of band from the product of the two, and a degradation in receiver's blocker performance. A FF CT ΣΔ ADC is the most efficient CT ΣΔ to date [17] at 28 fJ/conv.-step but results in a larger OOB STF peak. A FF-FB CT ΣΔ ADC has a more moderate OOB STF peak from fewer stabilising zeroes, at the cost of the power and area impact from one or many additional fast FB loop-filter stages (e.g. 47 fJ/conv.-step in [18]). Finally, FB CT ΣΔ ADCs have a monotonic low-pass filter STF (Fig. 5.3), but are the least efficient.

However, FB CT ΣΔ ADCs may in fact result in an overall more efficient radio, if the OOB STF gain of a FF or FF-FB CT ΣΔ ADC exceeds its conversion efficiency gain over the FB CT ΣΔ ADC, e.g. [19]. This is further emphasised when the STF peak further deteriorates with FIR DAC length and the order of the FF part of the loop-filter (See Fig. 5.3). This effect arises because the low-pass filter response of an FIR DAC in the outer loop causes the FF loop-filter component

Table 5.1 FF CT ΣΔ ADC loop robustness to RC device spread (σ_{RC}) and mismatch (σ_{mis}), measured as a proportion of the deviation from the NTF norm-1(α 1/MSA) with an NRZ & FIR DAC, over 1000 normally distributed samples

	NRZ DAC		2-tap FIR DAC		12-tap FIR DAC	
$(3\sigma_{mis}, 3\sigma_{RC})$	$\mu_{norm\text{-}1}$	$\sigma_{norm\text{-}1}$	$\mu_{norm\text{-}1}$	$\sigma_{norm\text{-}1}$	$\mu_{norm\text{-}1}$	$\sigma_{norm\text{-}1}$
(0.5, 25 %)	1.0126	0.07	1.037	0.13	1.067	0.15
(0.5, 2 %)	1.0004	0.0098	1.0006	0.01015	1.0017	0.01
$(0.5, 2\,\%) + \sigma_{GBW}$	1.001	0.019	1.0019	0.0248	1.0025	0.0247

The $(+\sigma_{GBW})$ term in the last row includes an additional compensated 0.45/f_{clk} amplifier GBW-dependent delay, spread ±30 % (3σ) over its nominal value

(and $H_a(s)$) gain to increase at higher frequencies to compensate for the FIR loss, resulting in a higher OOB STF peak. This was reported in [12] for FF-FB CT ΣΔ ADCs and indicates a compromise between FIR DAC length and receiver blocker performance. Figure 5.3 also incorporates FF CT ΣΔ ADCs with an FIR DAC. Note that all CT ΣΔ ADCs inherently provide some amount of filtering above a certain frequency, particularly around Nyquist.

Another important factor of an FIR DAC in CT ΣΔ ADCs is its loop stability and robustness. Reference [12] provides an insight into the robustness of a FIR DAC FF-FB CT ΣΔ loop against RC spread. However, its observations exclude the impact of component mismatch on the NTF, or STF.

Table 5.1 reports the robustness of an FIR DAC FF-compensated CT ΣΔ ADC against RC process spread and component mismatch, measured in terms of the proportional impact on the nominal $norm_1$ of the (loop) noise transfer function (NTF), i.e. the sum of the absolute values of the samples of the impulse response of the NTF or $\sum_{n=0}^{\infty} |ntf\,[n]|$. $norm_1$ is a measure of the signal range quantisation noise occupies in a CT ΣΔ ADC loop, and by extension, the signal range left for the input signal. It is therefore a measure of the (inverse of the) maximum-stable amplitude (or MSA) of a ΣΔ converter. Reference [12] reports that a typical ±25 % (3σ) RC spread does not adversely affect the MSA of a 12-tap FIR DAC FF-FB-compensated CT ΣΔ ADC, any more than a NRZ DAC FF-FB-compensated CT ΣΔ ADC would. This is in line with the observed impact from RC spread on the mean value of the $norm_1$, $\mu_{norm\text{-}1}$, in row (0.5, 25 %) of Table 5.1, for a FF-compensated CT ΣΔ ADC with an FIR and an NRZ DAC. However, the standard deviation, $\sigma_{norm\text{-}1}$, doubles from 7 to 15 % from a NRZ DAC to a 12-tap FIR DAC, when a 25 % (3σ) RC spread and a 2 % (3σ) RC component mismatch are considered. A CT ΣΔ ADC with an un-calibrated ±25 % RC process spread would therefore have to operate within 55 % of its MSA such that it does not overload across a typical 3σ RC mismatch found in large-volume production in a relatively small area on the die.

Table 5.1 indicates that RC calibration (e.g. with a 2 % error) is required to keep ADC performance and yield degradation acceptably low in production. It also demonstrates in row $(0.5, 2\,\%) + \sigma_{GBW}$ that an FF CT ΣΔ ADC with an FIR DAC is robust in terms of loop stability (with RC calibration), even when a 30 % process spread over a power-optimised $0.45/f_{clk}$ amplifier GBW delay-compensation such as [13] is included.

5.3 Filtering CT ΣΔ ADCs

From Sect. 5.2, compromising CT ΣΔ ADC performance or conversion efficiency for a more desirable STF may in fact result in a more efficient radio receiver, albeit at a lower stand-alone ADC conversion efficiency.

Some publications aim to absorb the channel-selection filter in Fig. 5.1 into a (monotonic low-pass STF) FB CT ΣΔ ADC [19, 20]. Reference [21] also tackles FF-FB CT ΣΔ ADC STF OOB peaking with an embedded filter within the loop. Other publications attempt to reduce STF OOB peaking in FF CT ΣΔ ADC architectures by turning the peaky STF into a monotonic low-pass filter STF response through input feedforward coefficients [22]. However, there are a number of problems with the overall premise of filtering in a CT ΣΔ ADC STF.

5.3.1 In-Band Conversion Efficiency Degradation from Filtering in CT ΣΔs

Embedding the channel-selection filter within a FB or a FF-FB CT ΣΔ ADC results in a lower overall power than a channel-selection filter followed by a FB or a FF-FB CT ΣΔ ADC. From [21], we estimate that embedding the filter in a FF-FB CT ΣΔ ADC results in a ~50 % reduction in filter power, and a ~20 % reduction in ADC power. This is the result of the first stage of the FF-FB CT ΣΔ ADC loop-filter attenuating the embedded filter noise and linearity contributions, in addition to a lower ADC noise contribution to the overall noise budget from fewer stages in the receiver, e.g. a higher (filter-embedded) FF-FB CT ΣΔ ADC noise achieves the same noise overall than a channel-selection filter followed by an ADC. However, the state-of-the-art filtering (FB or) FB-FF CT ΣΔ ADC [20] is much less efficient *in-band* than a state-of-the-art SAR ADC [23] that includes a channel-selection filter [24], and an input buffer with anti-aliasing [25], e.g. 256 fJ/conv.-step for [20] versus 76 fJ/conv.-step for [23] and [24] with [25]. This may be surprising when SAR ADCs, albeit ~3× to ~7× more efficient standalone ADCs than CT ΣΔ ADCs, are known to have very demanding anti-alias and input driving requirements [2, 3, 25], i.e. requiring an additional stage of filtering in the radio for anti-aliasing and a relatively large multiplier of the SAR ADC power to drive it from a buffer at low-distortion and noise.

Signal filtering at the ADC requires heavy processing of input signal at every stage of a CT ΣΔ ADC loop filter; may that be by injecting input signal from the output of the ADC (e.g. with FB loop-filter stages), or by injecting input signal using feedforward coefficients directly from the ADC input (e.g., [22]). A stand-alone active-filter is more efficient at filtering blockers than a CT ΣΔ ADC when it does not have to spend dynamic range in each Op-Amp processing quantisation noise, while a CT ΣΔ ADC is most efficient *in-band* when it does not have to filter the input signal (monotonically). The power saving from embedding a filter within a FB or FF-FB CT ΣΔ ADC should therefore be weighed against what can be achieved with a state-of-the-art CT ΣΔ ADC and channel-selection filter.

One may compare the embedded-filter FB or FF-FB CT ΣΔ ADCs to the most efficient (FF) CT ΣΔ ADC [17] at 28 fJ/conv.-step with an input filter. However, the STF OOB in a FF CT ΣΔ ADC typically peaks near the edge of band and requires a sharp OOB filter frequency response [26]. The input-filter requirements for a FF CT ΣΔ ADC are comparable to those of a 2.5 times oversampled SAR ADC anti-aliasing filter, but at a much lower stand-alone in-band ADC conversion efficiency (sub-10 fJ/conv.-step [23] vs 28 fJ/conv.-step [17]) and an inability to reuse the existing channel-selection filter that already exists in the radio (Fig. 5.1) [13]. Reference [4] proposes inserting a low-pass filter and its inverse in a FF CT ΣΔ ADC to achieve a low-pass STF in a FF CT ΣΔ ADC and deal with the STF peak with no impact on the loop-filter, but this comes at the cost of an impaired loop stability at high clock frequencies.

Another important factor to consider with filtering ADCs is the sensitivity of the STF to process parameters [27], when an FIR DAC is included. [27, 28] propose the dual-feedback architecture to reduce sensitivity to process mismatch for a NRZ DAC. Figure 5.4 shows the sensitivity of a low-pass STF in a FF CT ΣΔ ADC to RC component mismatch with a short and a long FIR DAC. The STF reaches a large OOB gain peak near the edge of band with a long FIR DAC that will impact in-band DR in the radio when component mismatch is considered.

5.3.2 ADC Out-of-Band Performance

Filtering ADCs tend to provide better OOB linearity-intermodulation performance, e.g. OOB IIP3, when compared to a cascaded filter stage with an ADC stage. This is a result of a blocker filter with gain embedded within the CT ΣΔ ADC loop. However, OOB linearity is not equally critical across the radio receiver chain.

OOB non-linearity in the RF stages of a radio cause a loss in selectivity expressed as cross-modulation and/or desensitisation, from the intermodulation product of blockers falling in the channel of interest. Deterministic signals are not spectrally efficient, so modern radios transmit noise-like signals with a high peak-to-average ratio that is in the order of 12 dB for OFDM in WLAN or 10 dB for WCDMA in cellular. Unlike signals at RF stages of the receiver that are defined as *rms*, ADC performance is defined relative to a hard and deterministic limit: supply or digital full-scale at 0 dBFS. OFDM or WCDMA signals are backed-off by their peak-to-average ratio to accommodate their peak below 0 dBFS at the ADC output

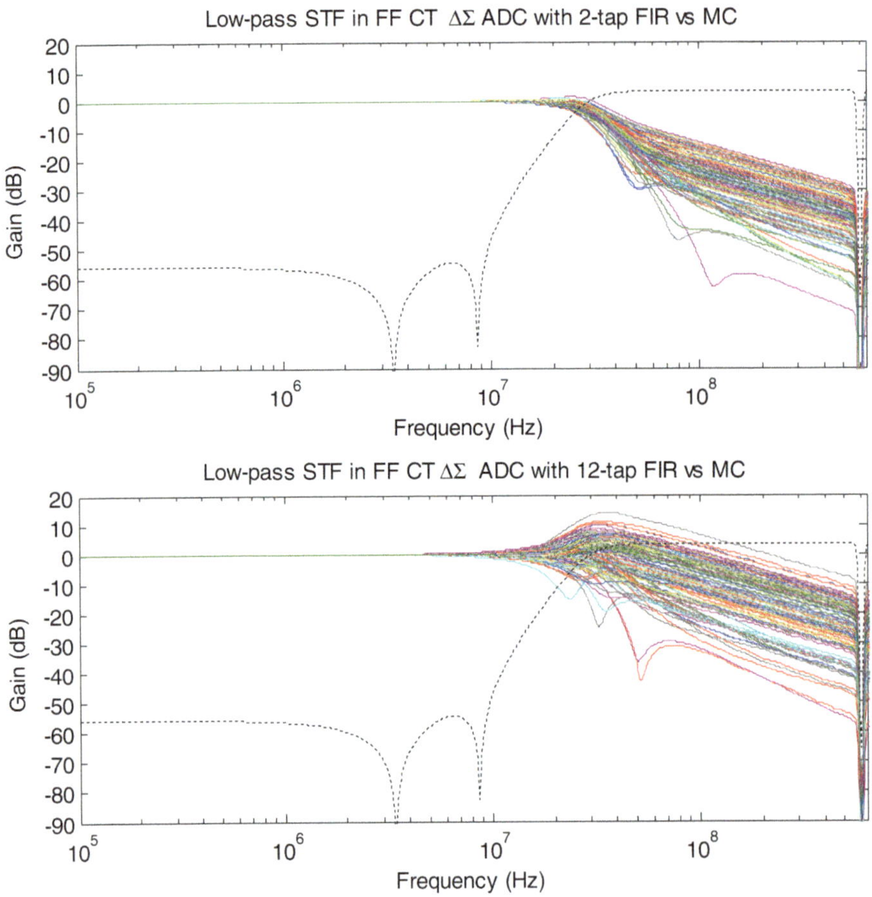

Fig. 5.4 STF robustness in a delay-compensated FF with a low-pass STF with 100 MC samples and RC calibration

and avoid clipping. OOB linearity therefore has a 10–12 dB less weight than in-band dynamic range in the ADC of a modern radio. An OOB-linearity figure-of-merit typically used to characterise filters is often used to compare between filtering ADC performance and a filter followed by an ADC, e.g. $FOM_{filt} = P/\left(N\ BW\ 10^{2/3(IIP3-P_{noise})/10}\right)$ in [19–21]. A more accurate comparison would weigh OOB linearity, e.g. IIP_3, lower than in-band noise, P_{noise}, by the peak-to-average ratio back-off used in the receiver. It's worth noting that FOM_{filt} is infinitely less efficient for a stand-alone non-filtering ADC, i.e. N = 0 in FOM_{filt}, than any filtering ADC when in fact OOB ADC linearity and ADC filtering play a secondary role in a radio receiver.

In summary, conversion efficiency and jitter-noise sensitivity (through FIR DAC length) in CT ΣΔ ADCs with a FF loop filter component are directly correlated with

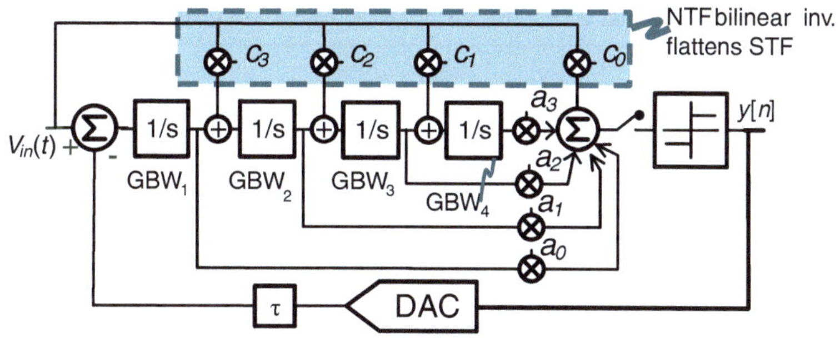

Fig. 5.5 A FF CT ΣΔ ADC with flat STF from input feedforward coefficients, C_x

radio-blocker performance degradation in a radio in terms of dynamic range in-band with an increasing OOB STF peak. However, the OOB IIP3 improvement gained from ADC OOB filtering rates much lower than in-band ADC performance due to the peak-to-average ratio back-off in OFDM or WCDMA signals. Fundamentally, a CT ΣΔ ADC STF should not provide OOB gain, while achieving monotonic filtering at the STF comes at a higher cost for a low benefit. A flat STF in an FF CT ΣΔ ADC has a lower impact on FF loop-filter scaling with a maximised ADC conversion efficiency and the least aggressive insertion possible of signal from the input, with the same blocker performance degradation of in-band noise than any other block driving the ADC.

5.4 Flat STF CT ΣΔ ADCs

A flat STF in a FF CT ΣΔ ADC may be achieved by turning $H_a(s)$ in (5.1) into the bilinear inverse of the NTF [29] using feedforward coefficients from the ADC input (C_x in Fig. 5.5).

Figure 5.6 indicates that a flat STF is more robust to component mismatch with an FIR DAC than a low-pass response in Fig. 5.4. This is due to the fact the feed-forward coefficients are two orders of magnitude smaller than, say, a low-pass STF in a FF CT ΣΔ ADC. One valid observation is the reduced anti-aliasing attenuation in the STF of a flat STF, when compared to the inherent STF of CT ΣΔ ADCs or a filtering CT ΣΔ ADCs (see Fig. 5.7).

However, a small RC at the ADC resistive input can provide significant additional signal attenuation around Nyquist at ∼30 times oversampling. The ADC internal noise (from a reference or through coupling) present around half the sampling clock will see as much attenuation as there is with quantisation noise before folding in-band. Finally, the spread in attenuation around Nyquist due to mismatch reduces its practical use for a low-pass STF when compared to a flat STF, with a more modest difference in anti-aliasing when considering the worse-case scenario.

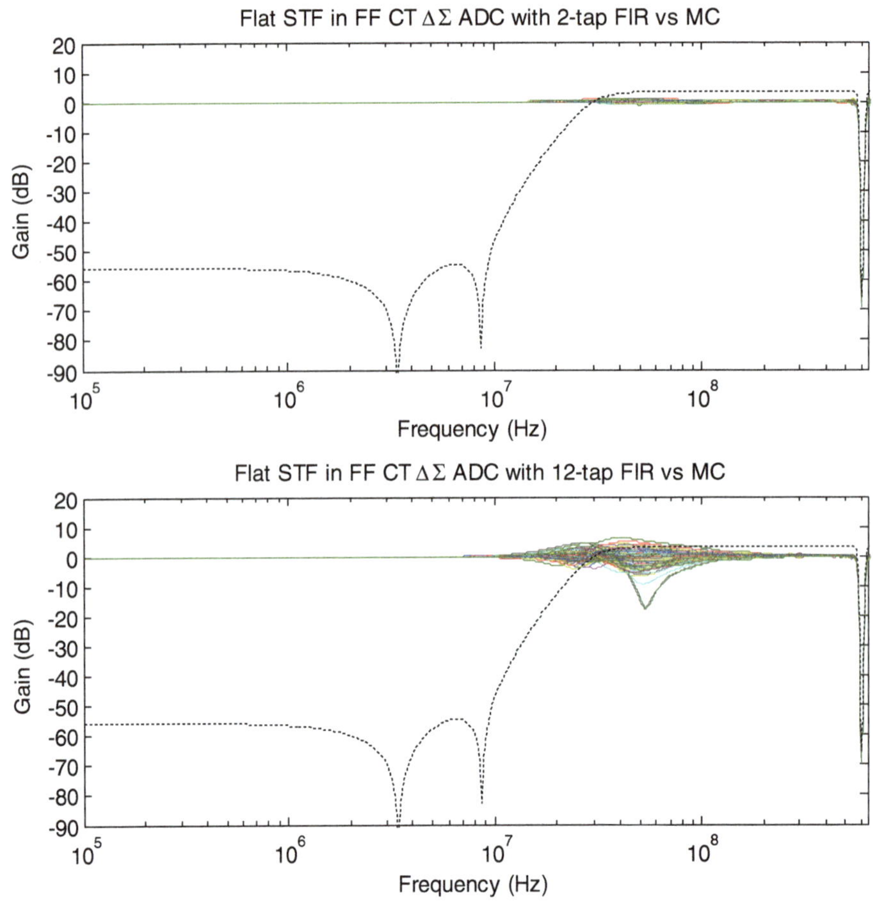

Fig. 5.6 STF robustness in a delay-compensated FF with a flat STF with 100 MC samples and RC calibration. Note that anti-aliasing is no worse than in-band quantisation noise with a low sensitivity

Table 5.2 captures the blocker degradation spread from OOB STF variation in the low-pass STF of Fig. 5.4 alongside the flat STF of Fig. 5.5 shown in Fig. 5.6, against the same MC RC-calibrated conditions than in Table 5.1 but over 100 samples. This quantifies how robust a flat STF is when compared to a low-pass STF in a FF CT ΣΔ ADC. The impact of component mismatch on radio performance is measured as the 3σ OOB STF peak the ADC may to achieve in large-volume production and is denoted as a peak OOB gain blocker margin. This in return defines the margin that should be allocated for in-band dynamic range loss from OOB blocker gain across process and mismatch. Table 5.1 shows that a 12-tap FIR FF CT ΣΔ ADC is robust

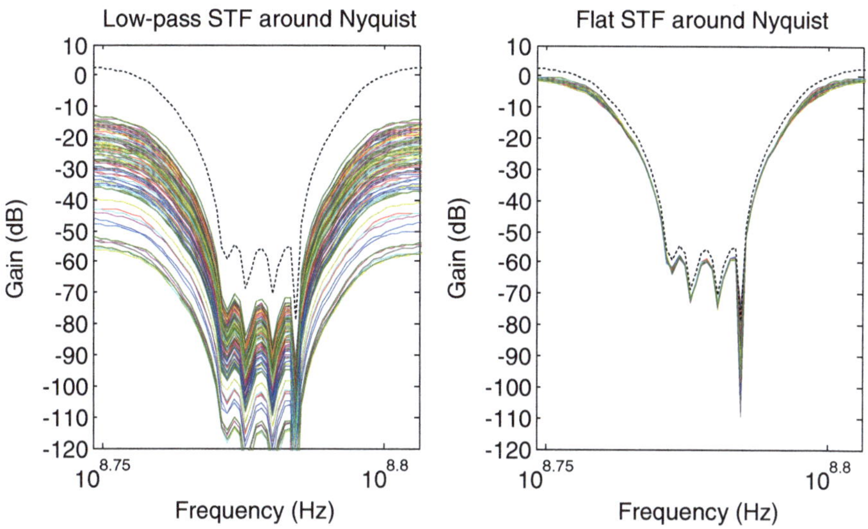

Fig. 5.7 Close-up around Nyquist, for a low-pass and flat STF with a 12 tap FIR

Table 5.2 STF OOB peak robustness MC over 100 normally distributed samples, for a FF CT ΣΔ ADC with a flat STF and a low-pass STF and GBW delay compensation

	2-tap FIR DAC		12-tap FIR DAC	
	μ_{peak} (dB)	σ_{peak} (dB)	μ_{peak} (dB)	σ_{peak} (dB)
Low-pass STF	**0.69 dB**		**12.28 dB**	
Peak OOB gain blocker margin	0.09	0.2	3.06	3.13
Flat STF	**1.17 dB**		**6.3 dB**	
Peak OOB gain blocker margin	0.45	0.24	2.16	1.38

in terms of loop performance and stability. However, a 12-tap FIR DAC results in a peak OOB over mismatch that is equivalent to a 6 dB loss in ADC conversion efficiency in a radio in production. This is a significant degradation.

It's therefore important to quantify how long an FIR has to be with a representative clock jitter spectrum with a flat STF FIR DAC FF CT ΣΔ ADC.

5.5 Clock Jitter in CT Σ Δ ADCs

The clock available in a WLAN receiver must have an integrated phase noise that is at least 0.8° for 64 QAM, i.e. 0.92 ps of rms jitter at 2.4 GHz (in a cellular radio, this is closer to 0.5 ps). If we apply a white 0.92 ps (rms) jitter to a single-bit FF CT ΣΔ ADC with a NRZ DAC, where jitter is subject to $NRZ(z) = \left(1 - z^{-1}\right)$ [5], the ADC noise degrades by 10 dB in Fig. 5.8.

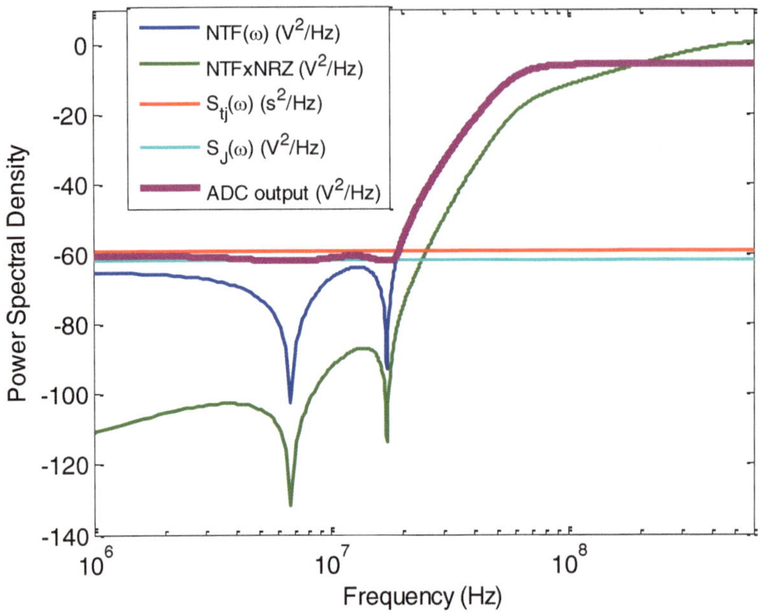

Fig. 5.8 Jitter degradation in a single-bit FF CT ΣΔ ADC with a NRZ DAC and 0.8% integrated phase noise

If we apply the same jitter noise to single-bit FF CT ΣΔ ADC with a 2-tap FIR DAC, jitter should be subject to $FIR(z) = \left(1 + z^{-1}\right)/2$ [8]. However, on should highlight the fact the response of jitter to the FIR is in fact: $FIR(z) = \left(1 - z^{-1}\right)/2$ [13], due to the strong correlation between DAC output noise transitions from a common clock noise source. Figure 5.9 shows a 3 dB degradation in the ADC noise floor from clock jitter when the FIR is included.

This would indicate either a longer FIR or a multi-bit ADC is required, at the cost of blocker performance degradation or area. However, clock jitter is seldom white. It is coloured. In the literature it is assumed a white clock-jitter or a white-noise equivalent of coloured jitter to simplify the analysis [13]. However, this results in a pessimistic noise jitter contribution.

Figure 5.10 shows a jitter response to a 0.92 ps (rms) coloured jitter with a 2-tap FIR DAC. The jitter spectrum is based on a simplified 200 kHz BW PLL mask for a WLAN radio that integrates to 0.92 ps (rms) of jitter.

A 2-tap FIR DAC results in a non-dominant jitter noise in a radio when a more realistic jitter spectrum shape is considered with the same integrated jitter.

Blockers will also modulate jitter. Figure 5.11 shows the effect of a blocker modulating jitter noise. The noise floor increases by 8 dB, but remains 18 dB below the noise floor, resulting in a robust loop with a flat STF response for a modern radio.

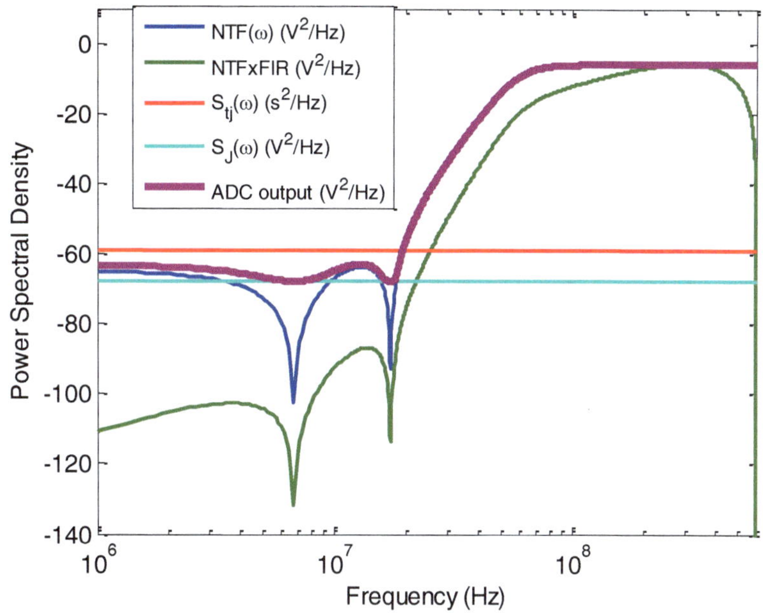

Fig. 5.9 Jitter degradation in a single-bit FF CT ΣΔ ADC with a 2-tap FIR DAC with 0.92 ps (rms) white noise

5.6 Conclusion

FF CT ΣΔ ADCs with an FIR DAC are the most area and conversion efficient CT ΣΔ ADC implementation but are more sensitive to FIR length in terms of STF OOB peaking and stability. It has been shown that a long FIR may be stabilised but that a short FIR is sufficient when the colour of typical radio clock jitter spectrum is considered. A flat STF is the optimal compromise between CT ΣΔ ADC conversion efficiency degradation and receiver blocker performance degradation, when OFDM back-off is taken into account in OOB ADC linearity requirements. On the other hand, filtering at the ADC results in a large area and FOM impact for a low return, especially when compared to a state-of-the-art filter driving a state-of-the-art CT ΣΔ ADC with a flat STF. SAR ADC converter efficiency in a radio suffers from anti-aliasing and sampling input buffer requirements that burn more power than the SAR ADC itself, and result in a poorer overall conversion efficiency in a radio, at 76 fJ/conv.-step vs 67 fJ/conv.-step in [13].

This paper confirms that a single-bit FF CT ΣΔ ADC with a short FIR DAC in the feedback and a flat STF is the most power and area efficient implementation in a modern receiver, when the wider system-level factors of a large-volume consumer radio device are considered.

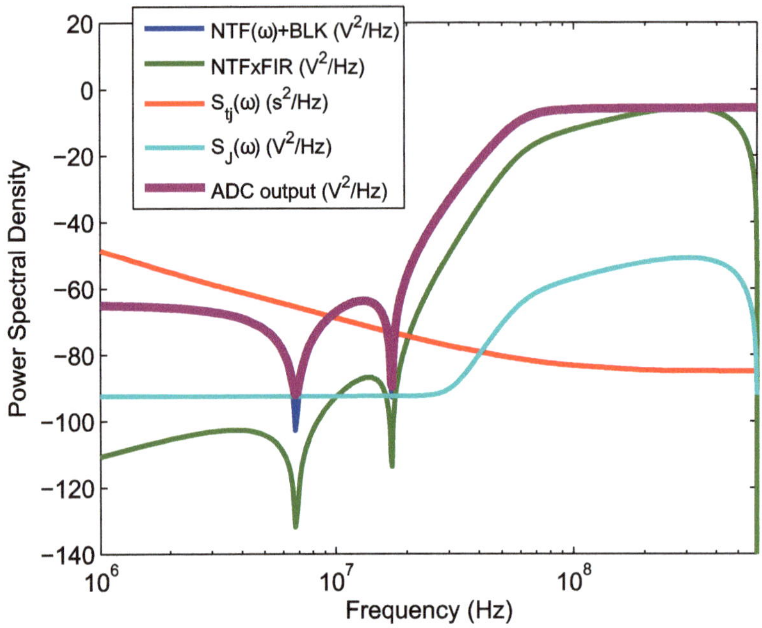

Fig. 5.10 Jitter degradation in a single-bit FF CT ΣΔ ADC with a 2-tap FIR DAC with shaped 0.92 ps (rms) jitter

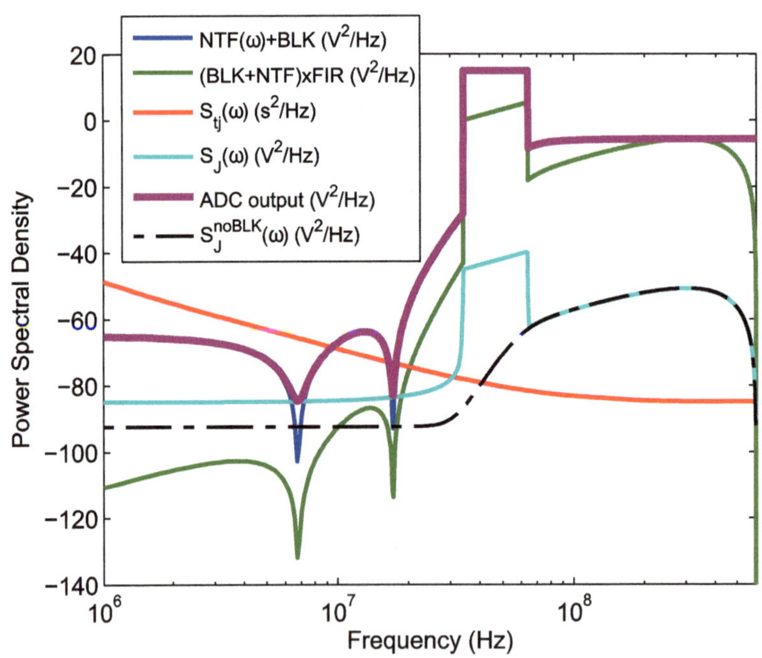

Fig. 5.11 Jitter degradation in a single-bit FF CT ΣΔ ADC with a 2-tap FIR DAC with shaped 0.92 ps (rms) jitter and a blocker

References

1. B. Murmann, ADC performance survey 1997–2015, [Online]. Available: http://www.stanford.edu/murmann/adcsurvey.html
2. M. Kramer, E. Janssen, K. Doris, B. Murmann, 14b 35MS/S SAR ADC achieving 75 dB SNDR and 99 dB SFDR with loop-embedded input buffer in 40 nm CMOS. in *IEEE International Solid-State Circuits Conference (ISSCC) Digest of Technical Papers*, 2015, pp. 1–3
3. B. Razavi, Design considerations for interleaved ADCs. IEEE J. Solid-State Circuits **48**(8), 1806–1817 (2013)
4. K. Philips, P.A.C.M. Nuijten, R.L.J. Roovers, A.H.M. van Roermund, F.M. Chavero, M.T. Pallares, A. Torralba, A continuous-time ΣΔ ADC with increased immunity to interferers. IEEE J. Solid-State Circuits **39**(12), 2170–2178 (2004)
5. J.A. Cherry, W.M. Snelgrove, Clock jitter and quantizer metastability in continuous-time delta-sigma modulators. IEEE Trans. Circuits Syst. II Analog Digit. Process. **46**(6), 661–676 (1999)
6. K. Reddy, S. Pavan, Fundamental limitations of continuous-time delta sigma modulators due to clock jitter. IEEE Trans. Circuits Syst. I Regul. Pap. **54**(10), 2184–2194 (2007)
7. B.M. Putter, ΣΔ ADC with finite impulse response feedback DAC. in *IEEE International Solid-State Circuits Conference (ISSCC) Digest of Technical Papers*, 2004, pp. 76–77
8. O. Oliaei, Sigma-delta modulator with spectrally shaped feedback. IEEE Trans. Circuits Syst. II Analog Digit. Process **50**(9), 518–530 (2003)
9. T. Okamoto, Y. Maruyama, A. Yukawa, A stable high-order delta-sigma modulator with an FIR spectrum distributor. IEEE J. Solid-State Circuits **28**(7), 730–735 (1993)
10. P. Shettigar, S. Pavan, Design techniques for wideband single-bit continuous-time ΣΔ modulators with FIR feedback DACs. IEEE J. Solid-State Circuits **47**(12), 2865–2879 (2012)
11. J. Gealow, M. Ashburn, L. Chih-Hong, S. Ho, P. Riehl, A. Shabra, J. Silva, Y. Qicheng, A 2.8 mW ΣΔ ADC with 83 dB DR and 1.92 MHz BW using FIR outer feedback and TIA-nased integrator. in *Proceedings of IEEE Symposium on VLSI Circuits*, 2011, pp. 42–43
12. A. Sukumaran, S. Pavan, Low power design techniques for single-bit audio continuous-time delta sigma ADCs using FIR feedback. IEEE J. Solid-State Circuits **49**(11), 2515–2525 (2014)
13. S. Loeda, J. Harrison, F. Pourchet, A. Adams, A 10/20/30/40 MHz feedforward FIR DAC continuous-time ΣΔ ADC with robust blocker performance for radio receivers. IEEE J. Solid-State Circuits **51**(4), 860–870 (2016)
14. J. Candy, A use of double integration in sigma delta modulation. IEEE Trans. Commun. **33**(3), 249–258 (1985)
15. W.L. Lee, A novel higher order interpolative modulator topology for high resolution oversampling A/D converters. Master's Thesis, MIT, Cambridge, 1987
16. F. Gerfers, M. Ortmanns, *Continuous-Time Sigma-Delta A/D Conversion: Fundamentals, Performance Limits and Robust Implementations*, vol. 21 (Springer, New York, 2006)
17. Y.-S. Shu, J.-Y. Tsai, P. Chen, T.-Y. Lo, P.-C. Chiu, A 28fJ/conv-step CT ΣΔ modulator with 78 dB DR and 18 MHz BW in 28 nm CMOS using a highly digital multibit quantizer. in IEEE *International Solid-State Circuits Conference (ISSCC) Digest of Technical Papers*, 2013, pp. 268–269
18. C.-L. Lo, C.-Y. Ho, H.-C. Tsai, Y.-H. Lin, A 75.1 dB SNDR 840MS/s CT ΣΔ modulator with 30 MHz bandwidth and 46.4fJ/conv FOM in 55 nm CMOS. in *Proc. IEEE Symp. VLSI Circuits*, 2013, 589 pp. C60–C61
19. M. Andersson, M. Anderson, L.S. Sundstrom, S. Mattisson, P. Andreani, A filtering ΣΔ ADC for LTE and beyond. IEEE J. Solid-State Circuits **49**(7), 1535–1547 (2014)
20. R. Ritter, J.G. Kauffman, J. Becker, M. Ortmanns, A 10 MHz bandwidth, 70 dB SNDR continuous time delta-sigma modulator with digitally improved reconfigurable blocker rejection. IEEE J. Solid-State Circuits **51**(3), 660–670 (1993)
21. R.S. Rajan, S. Pavan, Design techniques for continuous-time ΣΔ modulators with embedded active filtering. IEEE J. Solid-State Circuits **49**(10), 2187–2198 (2014)

22. T. Brueckner, C. Zorn, W. Mathis, M. Ortmanns, A single DAC CT sigma-delta modulator with butterworth STF characteristic. in *IEEE 54th International Midwest Circuits and Systems (MWSCAS)*, 2011, pp. 1–4
23. F. van der Goes, C. Ward, S. Astgimath, H. Yan, J. Riley, J. Mulder, S. Wang, K. Bult, A 1.5 mW 68 dB SNDR 80MS/s 2× interleaved SAR-assisted pipelined ADC in 28 nm CMOS. in *IEEE International Solid-State Circuits Conference (ISSCC) Digest of Technical Papers*, 2014, pp. 200–201
24. R.K. Palani, M. Sturm, R. Harjani, A 1.56mW 50MHz 3rd-order filter with current-mode active-RC biquad and 33dBm IIP3 in 65nm CMOS. in *IEEE Asian Solid-State Circuits Conference (A-ISSCC) Digest of Technical Papers*, 2013, pp. 373–376
25. R.K. Palani, R. Harjani, High linearity PVT tolerant 100MS/s rail-to-rail ADC driver with built-in sampler in 65nm CMOS. in *Proceedings IEEE Custom Integrated Circuits Conference (CICC)*, 2014, pp. 1–4
26. H.M. Geddada, C.-J. Park, H.-J. Jeon, J. Silva-Martinez, A.I. Karsilayan, G. Garrity, Design techniques to improve blocker tolerance of continuous-time ΣΔ ADCs. IEEE Trans. VLSI Syst. **23**(1), 54–67 (2015)
27. R. Ranjbar, A. Mehrabi, O. Oliaei, Continuous-time feed-forward ΣΔ—modulators with robust signal transfer function. in *2008 IEEE International Symposium on Circuits and Systems*, 2008, pp. 1878–1881
28. R. Ranjbar, O. Oliaei, A multibit dual-feedback CT ΣΔ modulator with lowpass signal transfer function. IEEE Trans. Circuits Syst. I Regul. Pap. **58**(9), 2083–2095 (2011)
29. J. De Maeyer, J. Raman, P. Rombouts, L. Weyten, Controlled behaviour of STF in CT ΣΔ modulators. IET Electron. Lett. **41**(16), 19–20 (2005)

Chapter 6
Continuous-Time MASH Architectures for Wideband DSMs

Hajime Shibata, Yunzhi Dong, Wenhua Yang, and Richard Schreier

6.1 Introduction: Wideband $\Sigma\Delta$ ADC for Communication Applications

Continuous-time $\Sigma\Delta$ ADCs are widely used in the radio signal chain in various modern wireless communication systems. Continuous-time $\Sigma\Delta$ ADCs not only digitize the input signal but also provide additional benefits such as oversampling and inherent anti-aliasing. These additional benefits relax the filtering requirements and simplify the radio signal chain. Since optimized circuit blocks in continuous-time converters consume lower power for a given noise requirement than their discrete-time counterparts, radio systems using continuous-time $\Sigma\Delta$ ADCs also enjoy reduced power consumption.

In high-performance radio applications such as cellular base-stations, an IF (intermediate frequency) sampling radio architecture shown in (Fig. 6.1) is commonly used to avoid the quadrature matching problem. The IF of the system is generally located higher than $1.5 \times$ RF bandwidth to avoid the interference associated with the second harmonic of the down-converted IF signal appearing in-band. For LTE applications, a bandwidth up to 200 MHz is required. Therefore the IF is located above 300 MHz and the signal spans ± 100 MHz around the IF. If a low-pass $\Sigma\Delta$ ADC is employed for this application, the ADC is required to achieve more than 400 MHz of bandwidth.

H. Shibata (✉) • R. Schreier
Analog Devices, Toronto, ON, Canada
e-mail: hajime.shibata@analog.com

Y. Dong
Analog Devices, Somerset, NJ, USA

W. Yang
Analog Devices, Wilmington, MA, USA

© Springer International Publishing Switzerland 2017

A. Baschirotto et al. (eds.), *Wideband Continuous-time $\Sigma\Delta$ ADCs, Automotive Electronics, and Power Management*, DOI 10.1007/978-3-319-41670-0_6

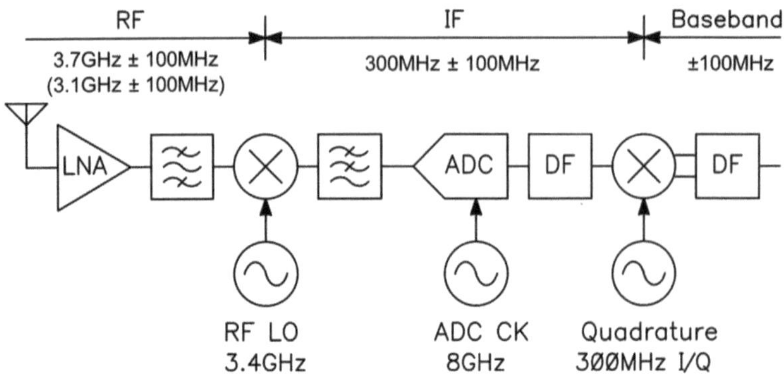

Fig. 6.1 An application of a wideband continuous-time ΣΔ ADC in a wireless radio system

In base-station radio applications, power consumption of the ADC is primarily limited by thermal considerations since the transmitter consumes much more power than the receiver. Since the trend in base-station design is to include several receivers (for diversity), an IC power budget of a few Watts leads to an ADC power target of less than 500 mW. The job of the ADC designer is therefore to design a low-noise, low-distortion and high dynamic range ADC which digitizes more than 400 MHz of bandwidth while consuming no more than 500 mW.

The bandwidth of an oversampled ADC is given by

$$BW = \frac{f_S}{2\,OSR}$$

where the OSR is the oversampling ratio, f_S is the sampling frequency, and BW is the raw bandwidth of the ΣΔ ADC [1]. The maximum clock frequency is limited by the combination of the regeneration time constant of the comparator in the quantizer and the logic delay from the quantizer to the feedback DAC [2]. Nanometer CMOS process technologies have enabled continuous-time ΣΔ ADCs operating at clock frequencies in the GHz range [3–6]. In a 28-nm CMOS process, a 4-ps regeneration time constant can be achieved, which supports a maximum clock frequency of approximately 8 GHz. Therefore to achieve more than 400 MHz bandwidth with an 8-GHz clock frequency, OSR = 8 is required for wideband wireless radio applications [6].

6.2 Background: Power-Efficient ADC

Typical high-resolution ADC systems consist of series-connected gain stages followed by a coarse ADC (quantizer) as shown in Fig. 6.2. The final coarse ADC has a quantization noise power that depends on the full-scale, number of

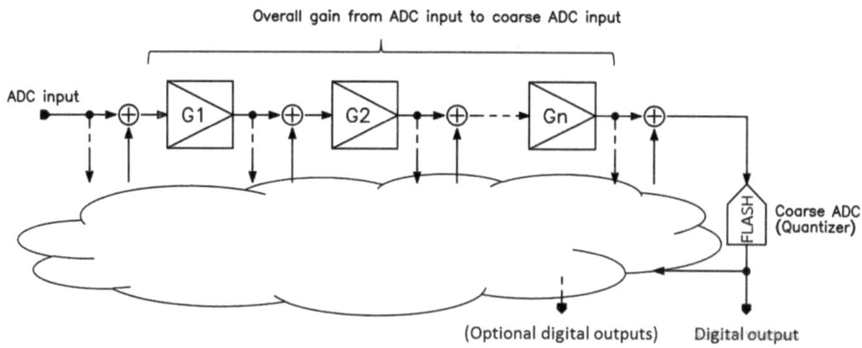

Fig. 6.2 Typical high-resolution ADC system

bits and OSR. The overall gain from the ADC input to the final coarse ADC, $G_1G_2\cdots G_n = \prod G_n$, and the quantization noise of the coarse ADC determine the input referred quantization noise of the overall ADC. In typical high-resolution ADCs, more than 40 dB of overall gain is needed. To make the following arguments simple, we assume the allowed maximum input and output signal level of all the gain stages are equal. Each gain stage may have additional feedback or feedforward paths to prevent the gain stages from being saturated.

One example of such a high-resolution ADC is a pipeline ADC. Let's assume that five gain stages, each having 4× gain, are connected in series and that the final ADC has 4-bit resolution. In this case, the overall gain is $4^5 = 60$ dB, which suppresses the quantization noise by 60 dB, providing 10 extra bits when the quantization noise is input-referred. Therefore, the overall ADC achieves $10 + 4 = 14$ bit resolution.

The same argument can be applied to ΣΔ ADCs. In the case of ΣΔ ADCs, the overall gain is frequency dependent and represented as a transfer function L_0 while a transfer function L_1 is used to represent the frequency-dependent gain of the feedback path [1]. For the sake of simplicity let's assume $L_0 = L_1 = L$. L must reduce its gain below 0 dB toward $f_S/2$ to keep the ΣΔ loop stable. However, since the bandwidth of interest is $BW = f_s/(2\ \mathrm{OSR})$, the most important parameter is $|L_1(BW)|$, the loop gain at the edge of the in-band region.

Now let us discuss the power efficiency of such ADC systems. The power consumption of each gain stage is proportional to the conductance level at that stage. The thermal noise power spectral density of a stage is proportional to the impedance level. As a result, the optimal power allocation for the stages is to scale the impedance level by $\sqrt{T}$ when the gain of the each stage is T. (This argument assumes the proportionality constant is the same for all stages.) This also suggests that a low-power and low-noise system needs higher stage gain T, especially in the first stage. The higher gain also helps to reduce any backend non-ideality such as distortion.

In the case of the ΣΔ ADC, the first-stage gain to be considered is the first-stage gain at $f = BW$. This gain can be used as an index to discuss the power efficiency of ADC architecture for a given OSR. We discuss the power efficiency of ΣΔ ADC systems based on this argument in the following sections.

6.3 Single-Loop ΣΔ ADCs for Low-OSR Applications

As discussed in Sect. 6.1, continuous-time ΣΔ ADCs are well-suited to wireless radio applications. However, ΣΔ ADC systems are oversampling systems and inherently narrow-band because the loop gain has to be decreased below 0 dB at $f_s/2$. Therefore the challenge for ΣΔ systems in wideband radio applications is to keep the ADC power efficiency at the passband edge and drive the modulator with a high-frequency clock. We review the traditional ΣΔ ADCs with feedback and feedforward loop filter configurations in this respect and discuss the limitation of those traditional approaches.

Figure 6.3 shows traditional 3rd-order ΣΔ ADCs; Fig. 6.3a is a feedforward ΣΔ ADC, and Fig. 6.3b is a feedback ΣΔ ADC. In these ΣΔ ADCs, the three-cascade-connected integrators provide the overall gain from the ADC input to the last quantizer as discussed in Sect. 6.2. The overall shape of this loop gain $L_1(z)$ can be calculated by the Delta Sigma Toolbox [7] and is shown in Fig. 6.4 assuming $BW = 500$ MHz, OSR = 8, $f_s = 8$ GHz and $||H||_\infty = 2.5$, without the NTF zero optimization. As shown in this figure, $L_1(z)$ has 16 dB gain at BW.

Table 6.1 shows the unity-gain frequency of the each integrator in the loop filter for both the feedback and the feedforward configurations after dynamic-range and other scaling. The gain at BW is also shown in the table. As indicated in the table, the feedback configuration has a first-stage gain below 0 dB while the feedforward has 11 dB. This difference can be also intuitively understood by the shape of L_1 in Fig. 6.4. In the feedback configuration, the frontend stage forms the 3rd-order path. Therefore the first-stage integrator is designed to have the lowest unity-gain frequency among the three integrators in the loop filter. Whereas in the feedforward configuration, the 1st-stage integrator forms the 1st-order path and thus the first-stage integrator has the highest unity-gain frequency among the three integrators.

Table 6.1 Stage gain comparison between the feedback modulator and the feedforward modulator shown in Fig. 6.3

	3rd-order Feedback	3rd-order Feedforward
1st-stage unity-gain frequency	169 MHz	1.73 GHz
1st-stage gain at the band edge (OSR = 8)	−9 dB	11 dB
STF peaking	0 dB	13 dB

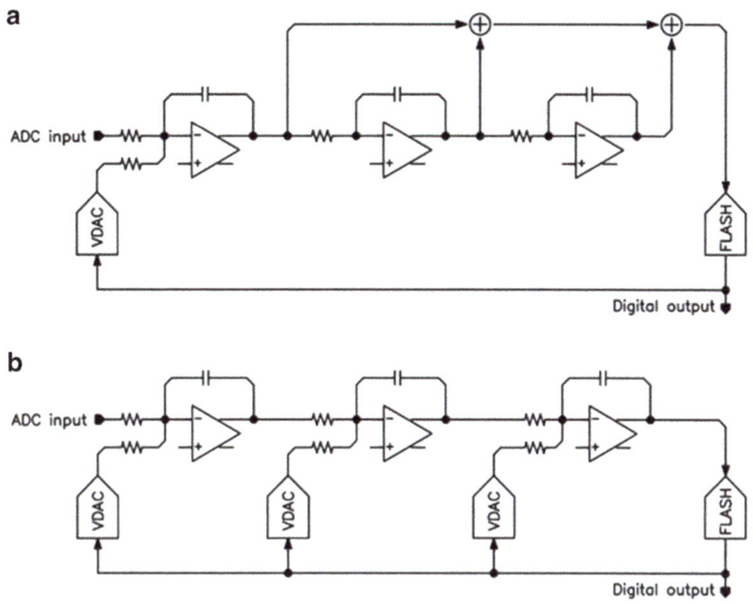

Fig. 6.3 Traditional ΣΔ ADCs: (**a**) 3rd-order feedforward configuration and (**b**) 3rd-order feedback configuration

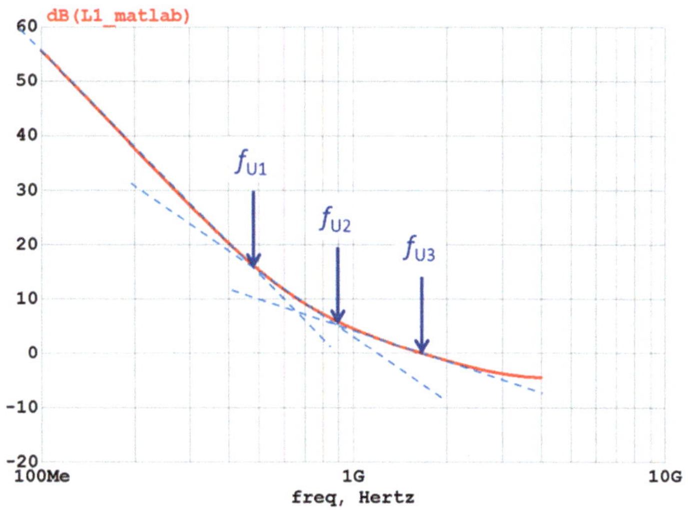

Fig. 6.4 $L_1(s)$ of the 3rd-order ΣΔ ADCs calculated by the Delta Sigma Toolbox. f_{U1}, f_{U2}, and f_{U3} are the unity gain frequencies of the internal integrators

Based on the efficiency argument in Sect. 6.2, each stage needs to have a gain $T > 0$ dB, especially in the first stage. The gain allocation of the traditional feedback ΣΔ ADC contradicts this requirement. Consequently, traditional feedback ΣΔ ADCs are not appropriate for an application requiring OSR ≤ 8.

The feedforward configuration is power efficient based on the 1st-stage gain argument. However, as discussed in the literature [4, 8], the feedforward architecture inherently comes with STF peaking. In wireless communication applications, one of the most difficult usage scenarios is an out-of-band blocker scenario. In that scenario, the STF peaking requires additional filtering in the radio chain or full-scale relaxation which limits the dynamic range of the inband signal. Both modifications diminish the benefits of the simplified radio architecture enabled by continuous-time ΣΔ ADCs. Based on these arguments, we conclude that the traditional single-loop ΣΔ ADCs are not suitable for wideband radio systems with OSR ≤ 8.

6.4 Continuous-Time MASH Σ Δ ADCs for Low-OSR Applications

We argued that the traditional single-loop ΣΔ ADCs with OSR = 8 are inefficient for wideband wireless communication systems. A multi-loop, or MASH, ADC is a ΣΔ ADC in which the internal sub-ADCs or sub-ΣΔ ADCs are connected with residue generation stages in series [5, 6, 8–11].

A simplified block diagram of the residue generation stage is shown in Fig. 6.5. The input signal is split into two paths: one goes directly to the summer while the other path goes though the FLASH ADC, the coarse DAC and then connects to the summer. With this configuration, the summer subtracts the quantized input signal from the original input signal and the continuous-time difference between the input signal and the quantized signal is generated as the residue. When the two paths are designed to match each other, the main component of the residue is the quantization noise of the coarse ADC. In the ideal case when the frequency of the input signal is much smaller than the sampling frequency, the residue amplitude is bounded by $(1/M)\ V_{FS}$ where the M is the quantization levels of the coarse ADC and V_{FS} is the full scale of the coarse ADC. However in practical cases, the residue amplitude can be larger than this bound. Thus we define the cancellation factor $k < M$ such that the bound on the residue is (V_{FS}/k).

Since the signal level is reduced by a factor of k, a gain block having a gain of k can be added without saturating the processed signal as shown in Fig. 6.5. Therefore the overall gain from the input of the ADC to the final coarse ADC shown in Fig. 6.2 can be increased by the cancellation gain k.

The residue generation circuit can be implemented using regular circuit blocks used in traditional ΣΔ ADC systems, a coarse ADC, a coarse DAC, and summation nodes provided by an opamp in the active-RC filter as shown in Fig. 6.6.

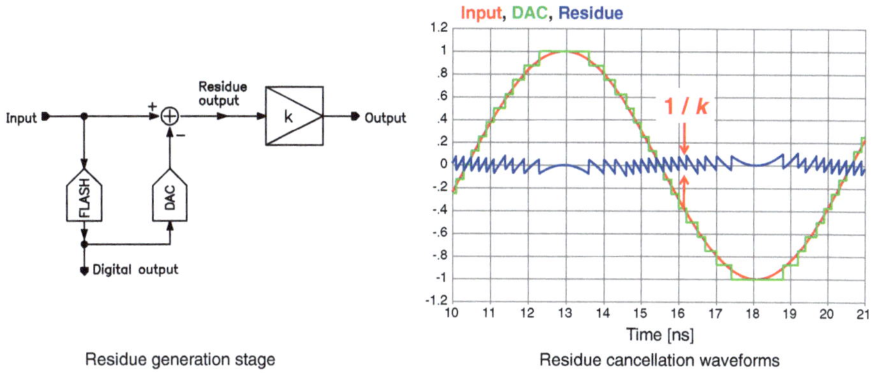

Fig. 6.5 Residue cancellation operation: (**a**) residue cancellation stage and (**b**) residue cancellation waveforms

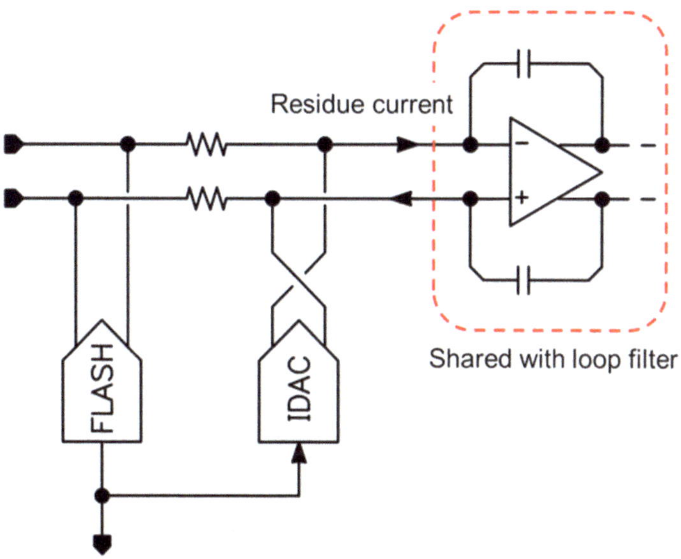

Fig. 6.6 Active-RC implementation of the residue generation circuit

Example MASH ADC systems are shown in Fig. 6.7. Figure 6.7a is a traditional MASH ADC where a frontend ΣΔ ADC and a backend ΣΔ ADC are connected through a residue generation circuit. The overall gain from the ADC input to the final coarse ADC can be increased for two reasons. The first reason is due to the residue cancellation gain k, discussed above. This gain can be a flat gain over frequency. The second reason is due to the additional low-order path gain(s) provided by the MASH configuration. In a single-loop ΣΔ ADC, gain stages can be inserted as needed but the bandwidth of the additional stages has to be decreased to satisfy stability requirements. In the MASH configuration, the additional gain stages can

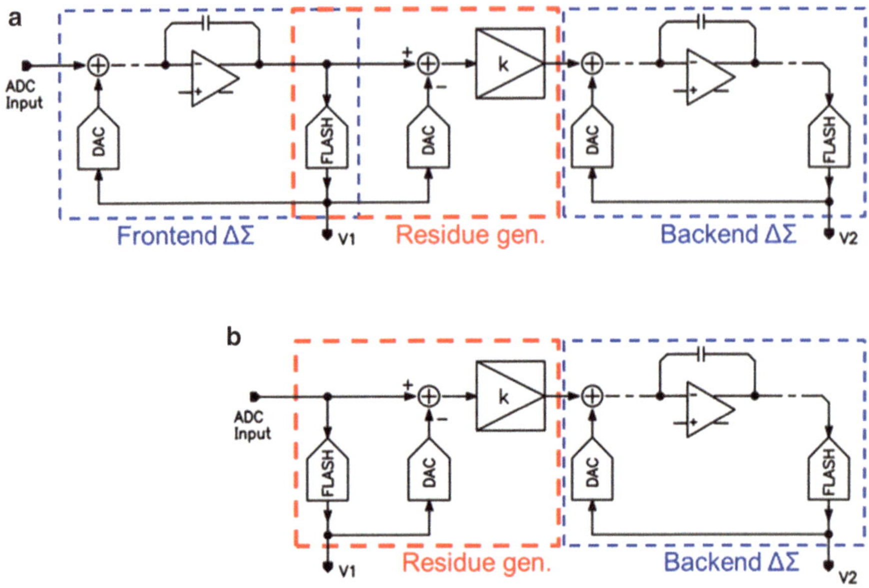

Fig. 6.7 MASH ADCs. (**a**) Traditional MASH. (**b**) ∅–x MASH

be in a different loop so that low-order wide-bandwidth gain stages can be added in the gain chain from the ADC input to the last quantizer. Therefore the additional stages increase the gain at the band edge.

The MASH ADC structure shown in Fig. 6.7b is a 0-*x* MASH ADC, which consists of the residue generation stage followed by a traditional ΣΔ ADC [5, 12–15]. In this configuration, the signal amplitude provided to the ΣΔ ADC is reduced by the cancellation factor *k* so that the first-stage gain of the backend ΣΔ ADC can be increased by *k*. Therefore this configuration also increases the overall gain from the ADC input to the last quantizer.

In the MASH examples in Fig. 6.7, the diagrams are drawn in a redundant fashion for explanation purposes. In an actual implementation, the gain block *k* can be merged into the backend ΣΔ ADC by scaling the first DAC and the first integration capacitor in the backend ΣΔ ADC.

In Sect. 6.2, we discussed how the ADC power efficiency can be estimated based on the first-stage gain at *BW*. Next we investigate the power efficiency of the MASH ADC in the same style. The first-stage gain of the traditional MASH is the same as the traditional ΣΔ ADC because the frontend ADC is identical to a traditional single-loop ΣΔ ADC. However in a MASH configuration, low-order ΣΔ ADCs can be cascaded together to achieve the required overall gain where each gain stage still provides a gain at *BW* higher than 0 dB. Therefore a MASH ADC with cascade-connected low-order ΣΔ ADCs can be a candidate for a wideband ADC achieving OSR = 8. Similarly, the first integrator gain of the 0-*x* MASH ADC can be higher than in a traditional ΣΔ ADC because the signal amplitude provided by the backend ADC is reduced by the cancellation factor *k*.

Other interesting properties of the 0-x MASH are the implementation cost and the thermal noise impact. The 0-x MASH requires four additional blocks: a coarse ADC, a coarse DAC, a summer and a gain block. However the summer and the gain block k can merged into the backend $\Sigma\Delta$ ADC. Also the DAC can be identical to the first DAC in the traditional $\Sigma\Delta$ ADC. Therefore the remaining additional blocks are the FLASH in the residue generation stage and the DAC in the backend $\Sigma\Delta$ ADC. However, since the DAC in the backend ADC is scaled, it consumes a fraction of the power consumed by the first DAC. Also, the additional FLASH ADC can be replaced with a more efficient SAR or VCO-based ADC, and thus the power penalty associated with the 0-x architecture can be small. Consequently, the traditional MASH and the 0-x MASH can be used for OSR = 8 applications and we expect the 0-x MASH will be more power efficient.

6.5 Residue Generation in Continuous-Time MASH ADCs

As discussed in Sect. 6.4, a continuous-time MASH $\Sigma\Delta$ ADC is an ADC consisting of two or more sub-ADCs (often $\Sigma\Delta$ ADCs) connected together with continuous-time residue generation circuits. Since the ADC performance is essentially limited by the circuits in the gain chain, we now discuss the properties of the residue generation circuit in detail.

Discrete-time residue generation circuits have been used in pipeline ADCs or discrete-time MASH $\Sigma\Delta$ ADCs for many decades [10]. In contrast, the residue generation circuit used in continuous-time MASH ADCs has not been well discussed in the literature. For a discrete-time residue-generation circuit, the amplitude of the residue is limited within ± 0.5 LSB of the coarse DAC when the components are all ideal. However in continuous-time residue generation, the amplitude of the residue can be higher than ± 0.5 LSB of the coarse DAC for two reasons. The first reason is transfer function mismatch between the main path and the cancelation path. The second reason is due to the DAC image signals.

A primitive residue generation circuit is shown in Fig. 6.8 along with the FLASH-DAC timing diagram. The transfer function $G_0(s)$ represents the main path and $G_{\mathrm{FD}}(s)$ represents the cancellation path. As shown in the timing diagram, the DAC outputs the signal based on a quantized signal one clock (T_{CK}) earlier. Also, assuming non-return-to-zero waveform, the DAC outputs a constant current for a whole clock period. Therefore the two transfer functions can be written as

$$G_0(s) = \frac{1}{2R}$$

and

$$G_{FD(s)} = \frac{I_{FS}}{V_{FS}} \frac{1 - \exp(-T_{CK}s)}{s} \exp(-T_{CK}s)$$

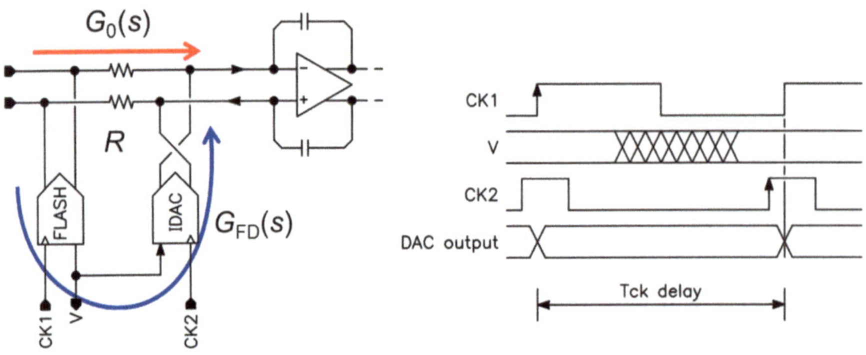

Fig. 6.8 Transfer functions in the residue generation circuits

where $G_0(s)$ and $G_{FD}(s)$ are the transconductances through the resistor R and the FLASH-DAC path, respectively. I_{FS} and V_{FS} are the full-scales of the current-output DAC and FLASH ADC. The full-scales are adjusted to match the transfer functions at DC:

$$\frac{I_{FS}}{V_{FS}} = \frac{1}{2R}$$

Using these two transfer functions, the residue error current can be written as

$$I_{RES}(s) = \{G_0(s) - G_{FD}(s)\}\, V_{IN}(s)$$

As shown in this equation, even if the magnitude of the two transfer functions match at DC, the residue error current increases as frequency increases. Figure 6.9a plots the normalized residue error current vs. frequency. As shown in the figure, the error reaches 10 % even at OSR = 50. At OSR = 8, the error is 58 % and the cancellation gain is thus limited to $k < 1/0.58 = 1.7$ even if a fine-resolution FLASH and DAC are used in the residue-cancellation stage. Clearly, transfer function matching is essential in a continuous-time MASH ADC.

This matching error can be reduced by introducing additional components in either or both of the transfer functions [5, 6, 13, 16, 17]. Since the DAC's sinc response only changes its magnitude by less than 1 % at OSR = 8 and 2.5 % at OSR = 4, matching the phase or the delay is more important. Since the middle of the DAC pulse is at 0.5 T_{CK}, the effective delay through the FLASH-DAC path is 1.5 T_{CK}. The equivalent phase rotation can be introduced by replacing the transconductance $1/2R$ in the main path with an all-pass network such as: (a) a terminated delay line, (b) a terminated LC lattice delay line, or (c) an RC lattice delay line, as shown in Fig. 6.10. All the networks have a constant transconductance but rotates its phase. The matching improvement can also be implemented in the FLASH-DAC path by adding an inverse-delay transfer function. However the frequency range is limited due to the need to satisfy causality [16].

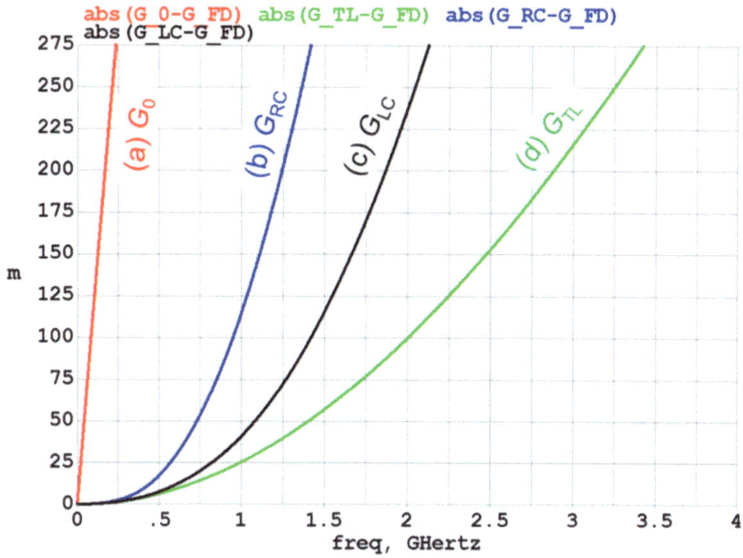

Fig. 6.9 Normalized residue error current. $f_S = 8$ GHz

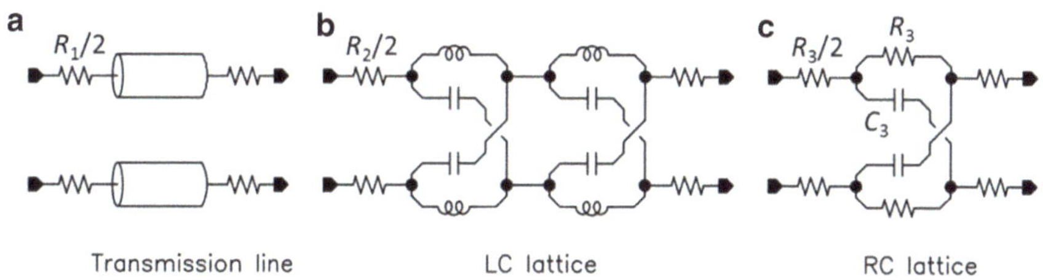

$$G_{TL}(s) = \frac{1}{2R_1}\exp\left(\frac{-s}{1.5T_{CK}}\right) \quad G_{LC}(s) = \frac{1}{2R_2}\left(\frac{1-\frac{s}{\omega_0}}{1+\frac{s}{\omega_0}}\right)^2 \quad G_{RC}(s) = \frac{1}{4R_3}\left(\frac{1-sR_3C_3}{1+sR_3C_3}\right)$$

Fig. 6.10 Phase matching networks in the continuous-time residue generation circuit

The residue error of each case is depicted in Fig. 6.9 as well. As shown in Fig. 6.9, the transmission lines provide the best cancellation because the phase response of the two transfer functions are identically matched and the only difference is the sinc magnitude difference. Unfortunately, the transmission line trace length for 1.5 T_{CK} delay on silicon is $l_{TL} = \frac{1.5\,T_{CK}\,c}{\sqrt{\varepsilon_r}} = 28mm$, where c is the speed of light and ε_r is relative dielectric constant of silicon. This length is not practical for integration even if the trace is folded. The cascaded LC lattice dclay line is the second best, supporting $k > 20$ even at OSR = 4. However, since on-chip inductors can be susceptible to external magnetic flux, introducing such components in the

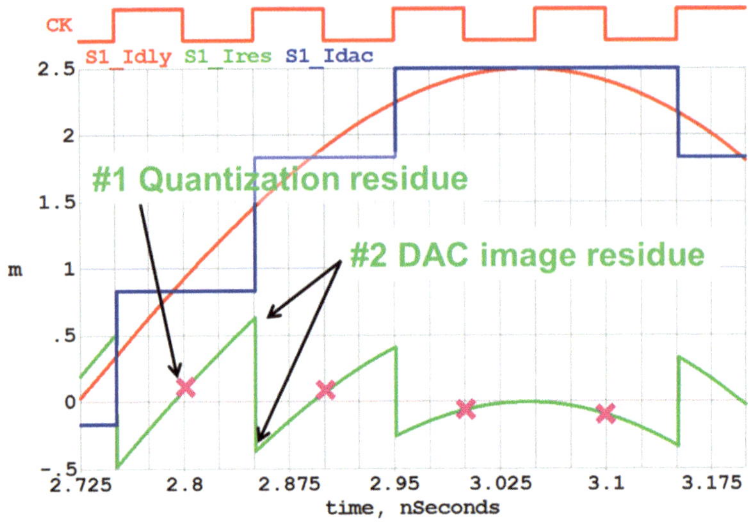

Fig. 6.11 Residue generation in time domain. $f_S = 10$ GHz and $f_{signal} = 1$ GHz

frontend of the ADC is a concern. Also the inductors realizing the required values occupy significant silicon area. Therefore the most attractive choice is an RC-lattice implementation. With the RC lattice arrangement, the error magnitude is 1.7 % of the full scale at OSR = 8, which is still smaller than a 4-bit cancellation residue of 1/16 = 6 %. However the error quickly increases at high frequencies. At OSR = 4, the cancellation error reaches 12 % and thus the cancellation of out-of-band signals is eventually limited.

In the case of continuous-time residue generation, the cancellation DAC injects image signals also at 2nd, 3rd and higher Nyquist frequencies following the sinc(s) envelope. Figures 6.11 and 6.12 show the residue generation waveforms in the time domain and in the frequency domain, respectively. In the frequency domain plot, it is obvious that only the 1st Nyquist component is used for cancellation and the other terms remain. These residual image signals also potentially limit the achievable cancellation gain k. Since the 2nd and 3rd Nyquist image signals follow the sinc(s) envelope, the magnitude increases when the fundamental signal frequency increases. For example, a 500 MHz signal at OSR = 8 and $f_s = 8$ GHz creates image signals at 7.5 and 8.5 GHz, each having 6 % of the fundamental magnitude. With a 1 GHz input signal, the image appears at 7 and 9 GHz with 14 % of the fundamental magnitude. These image signal levels are higher than the 4-bit cancellation residue level in the 1st Nyquist zone as shown in Fig. 6.12. Therefore a sub-ADC driven by the residue generation circuit needs to handle such high-frequency image signals.

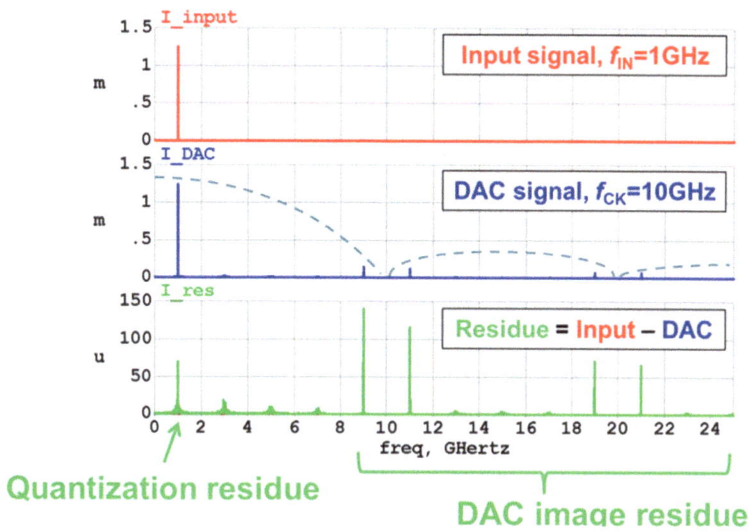

Fig. 6.12 Residue generation in frequency domain. $f_S = 10$ GHz and $f_{\text{signal}} = 1$ GHz

6.6 MASH Σ Δ ADCs Implementations in 28 nm CMOS

We reported two wideband continuous-time MASH ADCs implemented in a 28 nm CMOS technology [5, 6]. In this section we discuss the benefits and drawbacks of those two ADC architectures in the context of wideband wireless communication applications. Simplified block diagrams of the two ADC systems are shown in Figs. 6.13 and 6.14, and the architectural comparison is summarized in Table 6.2. The first design shown in Fig. 6.13 is a 0-x MASH with a 4-bit FLASH ADC as a frontend ADC and 3rd-order feedforward ΣΔ ADC as a backend ADC (0-3FF MASH). Those frontend and backend sub-ADCs are connected by a residue generation circuit. The phase matching network in the residue generation circuit consists of an off-chip transmission line. The second design is a continuous-time MASH employing a 1st-order ΣΔ ADC as the frontend and a 2nd-order feedback ΣΔ ADC as the backend (1-2FB MASH). The two sub-ADCs are connected though the RC delay line discussed in Sect. 6.5. The OSR target of the second design is OSR = 8. At this OSR = 8 target, a 1st-order RC all-pass delay network is sufficient. The FLASH ADCs in the 1-2FB MASH have 17 levels.

Both MASH ADCs increase the cancellation gain by virtue of the residue-generation circuits. The 0-3FF and 1-2FB enjoy $16/6 \approx 2.7\times$ and $6\times$ cancellation gains, respectively. These cancellation gains are determined by considering the cancellation DAC resolution, transfer function mismatches, and DAC image signal magnitude in addition to the circuit non-idealities such as the offset of the comparators in the FLASH ADC.

Table 6.2 ADC architectural comparisons

	3FB ΣΔ	3FF ΣΔ	0-3FF MASH ΣΔ	1-2FB MASH ΣΔ
1st-stage unity-gain frequency	169 MHz ($f_s = 8$ GHZ)	1.73 GHz ($f_s = 8$ GHZ)	1.2 GHz[a] ($f_s = 3.2$ GHZ)	1.1 GHz[a]($f_s = 8$ GHZ)
1st-stage gain at the band edge ($OSR = 8$)	−9 dB	11 dB	15 dB	6 dB
Energy efficiency ($OSR = 8$)	Poor	Good[b]	Very good (FOM = 171 dB)	Good (FOM = 159 dB)
Anti-aliasing	Very good	Good	Moderate	Good
STF peaking	0 dB	13 dB	Moderate	0 dB
Out-of-band blocker robustness	Very good	Poor	Moderate	Very good

[a]After normalizing the full-scale voltage
[b]Not good if the inband DR has to be reduced by out-of-band blockers

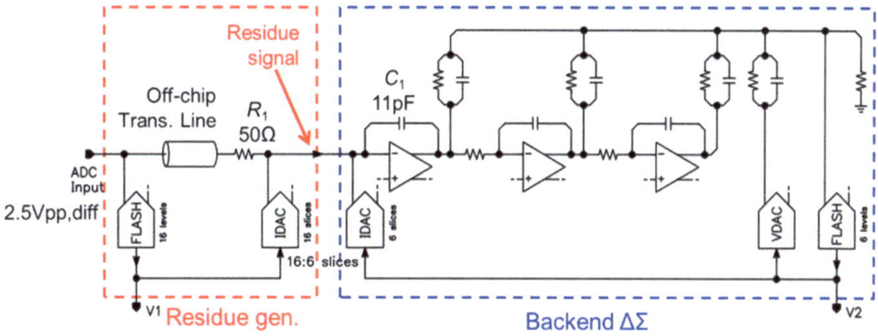

Fig. 6.13 0-3FF MASH ADC

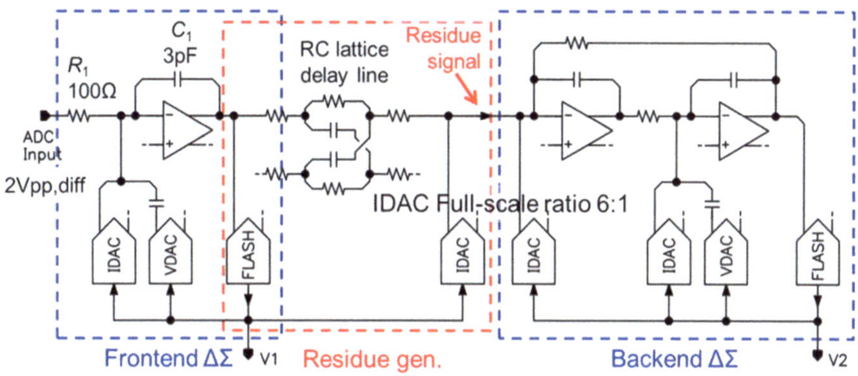

Fig. 6.14 1-2FB MASH ADC

The immunity to out-of-band blocker signals is an important metric for ΣΔ ADCs in wireless communication applications. The 1-2FB design has better immunity than the 0-3FF design because the frontend ΣΔ ADC also provides a low-pass filtering function whereas the frontend ADC in the 0-3FF architecture does not provide any filtering but also injects image signals.

All ΣΔ ADCs consist of three main circuit blocks: loop filter, quantizer and DAC. In the active-RC loop filter, the amplifier is the main active circuit. In the 1-2FB MASH, a 5th-order feedforward amplifier, shown in Fig. 6.15, is employed in the integrators [4, 18]. The feedforward opamp design is used to satisfy two conflicting requirements, namely moderate phase shift at f_s /2, and high gain at *BW*. The low-gain, high-bandwidth 1st-order path in the amplifier satisfies the first requirement. The high-gain, low-bandwidth path through the gm4a-gm5b-gm5c-gm3c-gm2c path satisfies the second requirement and thereby provides low-distortion operation for in-band signals.

The DACs used in both the 0-3FF MASH and the 1-2FB MASH are push-pull current steering DACs [19]. One of the DAC slices is shown in Fig. 6.16. PMOS and NMOS sub-DACs are driven by the same digital signal to minimize the thermal

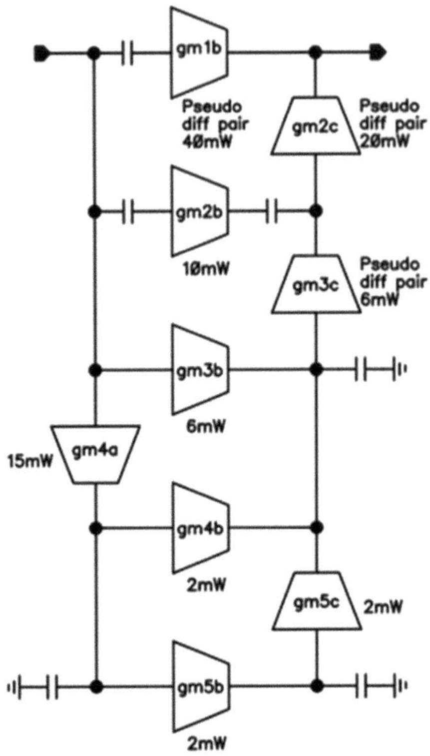

Fig. 6.15 AMP1 in Fig. 6.14

noise injected into the summing node for a given full-scale current. The majority of the block is powered with 1.0 V core voltage while extra supply voltages of 1.8 and −1.0 V are provided to the DAC current sources to reduce their thermal noise.

The performance of the ADCs is summarized in Table 6.3. The 0-3FF MASH design and 1-2FB MASH have OSR = 30 and 8 with 3.2 and 8 GHz clock frequencies, respectively. The single-tone spectra in Figs. 6.17 and 6.18 show bandwidths of 53 and 465 MHz with −166 and −160 dBFS/Hz noise densities. In both designs, the digital output streams V_1 and V_2 from the frontend and the backend ADCs are combined via a digital filter to minimize the in-band noise. The performance of the 1-2FB design satisfies the wideband high-performance IF radio applications discussed in Sect. 6.1.

Die photos are shown in Fig. 6.19. The power efficiency of each ADC can be quantified via the thermal noise figure-of-merit $FOM = DR + 10\log_{10}(BW/P)$ [1]. A FOM plot using published ADC data points is shown in Fig. 6.20 [20]. The 0-3FF MASH design achieves FOM = 171 dB, which is competitive among converters having $BW = 50$ MHz. The 1-2FB MASH achieved the widest BW among the ΣΔ ADCs published as of this writing. However, the FOM of this ADC drops to 159 dB because of the high clock frequency and the lower-efficiency ADC architecture.

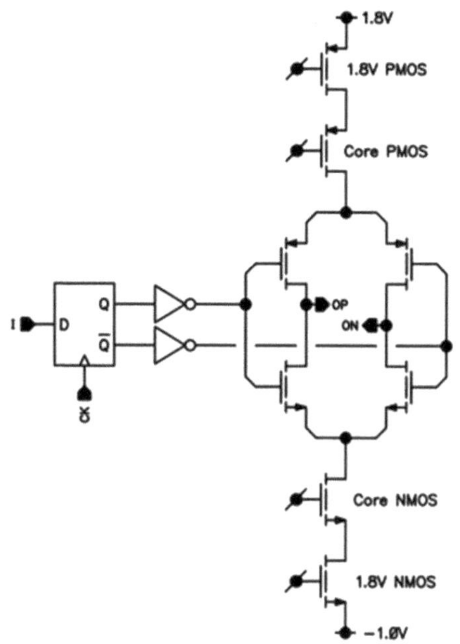

Fig. 6.16 DAC slice in MASH ADCs

Table 6.3 ADC performance summary

	0-3FF MASH [5]	1-2FB MASH [6]
Process technology	28 nm CMOS	28 nm CMOS
Active area	0.9 mm^2	1.4 mm^2
Supply Voltages	0.9 V/1.8/−1.0 V	±1 V/1.8 V
Power (*P*)	235 mW	890 mW
Sampling rate (f_s)	3.2 GHz	8 GHz
Bandwidth (*BW*)	53 MHz (OSR = 30)	465 MHz (OSR = 8.6)
Input full scale	2.5 Vp-p, diff	2.0 Vp-p, diff
Average NSD (small signal)	4.3 nV/$\sqrt{}$Hz (−166 dBFS/Hz)	7.1 nV/$\sqrt{}$Hz (−160 dBFS/Hz)
Dynamic range (*DR*)	87 dB	72 dB
Peak SNR	83 dB	68 dB
Peak SNDR	70 dB	67 dB
FOM = $DR + 10\log_{10}(BW/P)$	171 dB	159 dB

6.7 Conclusions

The reasons for employing continuous-time MASH ΣΔ ADCs in wideband wireless communication application was discussed. In wideband applications with an OSR less than ten, the MASH ADC structure is inevitable from an efficiency perspective. MASH architectures require continuous-time residue generation

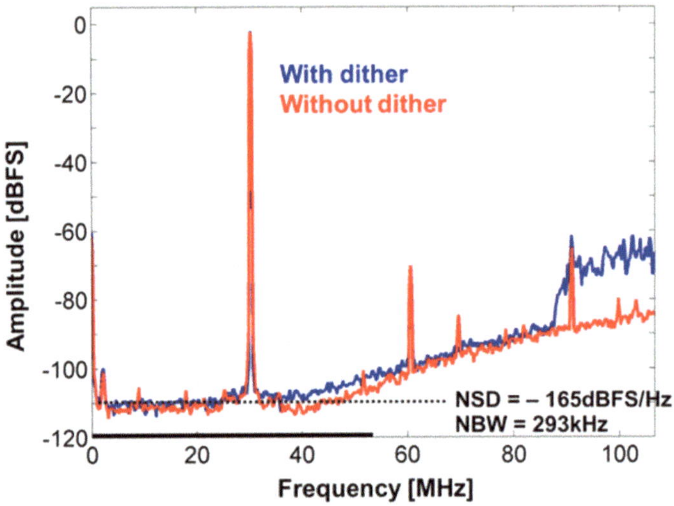

Fig. 6.17 Single-tone output spectrum of the 0-3FF MASH ADC with and without an external high-pass dither signal

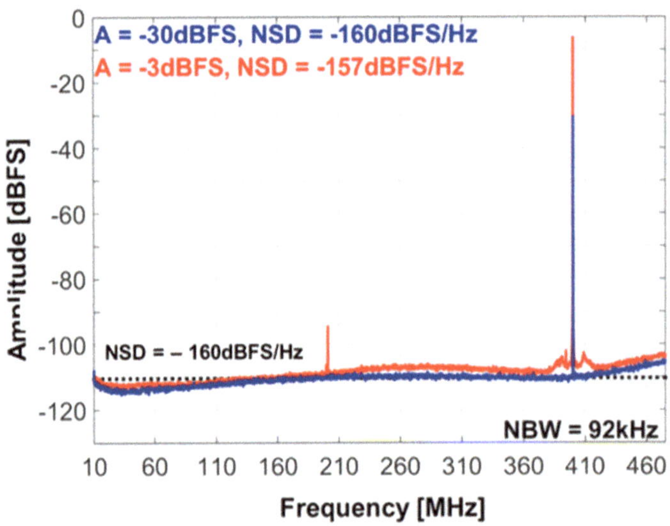

Fig. 6.18 Single-tone output spectrum of the 1-2FB MASH ADC

circuits, and the important properties of this circuit were discussed in detail. Two wideband MASH ADC implementations were compared in the context of a wideband radio receiver and it was shown that a MASH ADC can be designed to satisfy the requirements of a high-performance radio having IF = 300 MHz and $BW = 200$ MHz while achieving a respectable power efficiency.

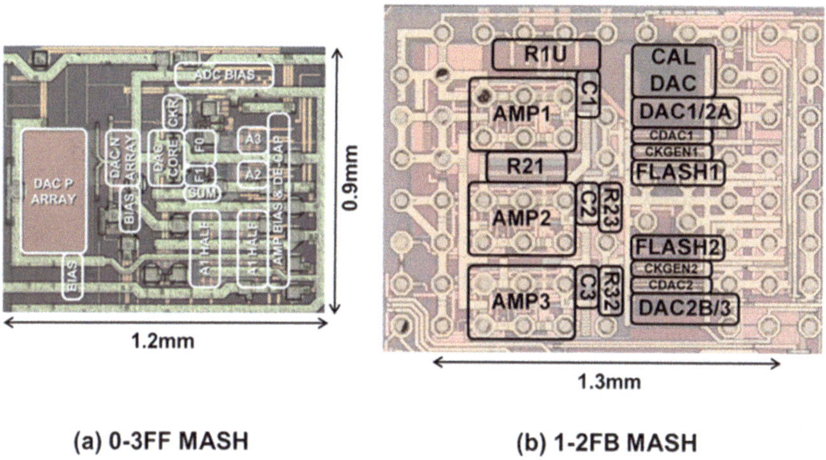

Fig. 6.19 Die photo of (**a**) 0-3FF MASH ADC and (**b**) 1-2FB MASH ADC

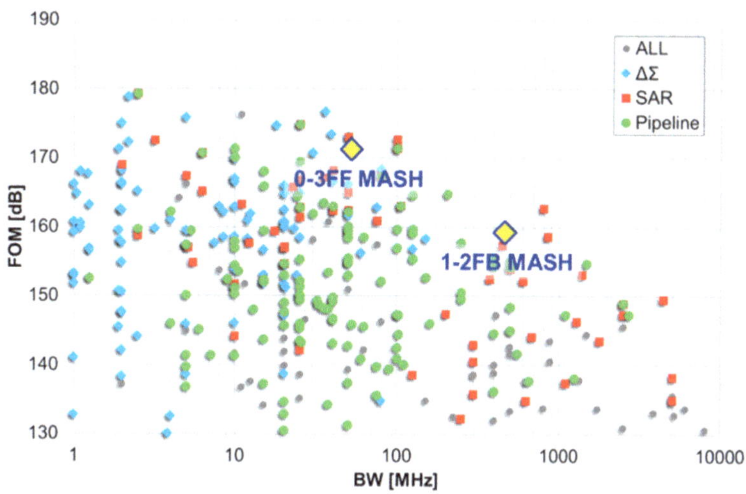

Fig. 6.20 *FOM* vs. *BW* plot

References

1. R. Schreier, G.C. Temes, *Understanding Delta-Sigma Data Converters* (Wiley-IEEE Press, New York, 2005)
2. T. Caldwell, D. Alldred, R. Schreier, H. Shibata, Y. Dong, Advances in high-speed continuous-time ΣΔ modulators, in *2014 IEEE Proceedings of the Custom Integrated Circuits Conference*, pp. 1–8, 2014
3. M. Bolatkale, L.J. Breems, R. Rutten, K.A.A. Makinwa, A 4GHz continuous-time ADC with 70dB DR and 74dBFS THD in 125MHz BW. IEEE J. Solid State Circuits **46**(12), 2857–2868 (2011)

4. H. Shibata, R. Schreier, W. Yang, A. Shaikh, D. Paterson, T. Caldwell, D. Alldred, P.W. Lai, A DC-to-1GHz tunable RF ADC achieving DR=74dB and BW=150MHz at f_0=450MHz using 550mW. IEEE J. Solid State Circuits **47**(12), 2888–2897 (2012)
5. Y. Dong, W. Yang, R. Schreier, A. Sheikholeslami, S. Korrapati, A continuous-time 0-3 MASH ADC achieving 88dB DR with 53MHz BW in 28nm CMOS. IEEE J. Solid State Circuits **49**(12), 2868–2877 (2014)
6. Y. Dong, J. Zhao, W. Yang, T. Caldwell, H. Shibata, R. Schreier, Q. Meng, J. Silva, D. Paterson, J. Gealow, A 930mW 69dB DR 465MHz BW CT 1-2 MASH ADC in 28nm CMOS, in *IEEE International Solid-State Circuits Conference Digest of Technical Papers*, pp. 278–279, Feb 2016
7. R. Schreier, Delta Sigma Toolbox, Matlab Central, http://www.mathworks.com/matlabcentral/fileexchange/19-delta-sigma-toolbox, 2016
8. K. Philips, P. Nuijten, R. Roovers, A. Roermund, F. Chavero, M. Pallarés, A. Torralba, A continuous-time ΣΔ ADC with increased immunity to interferers. IEEE J. Solid State Circuits **39**, 2170–2177 (2005)
9. L.J. Breems, R. Rutten, G. Wetzker, A cascaded continuous-time modulator with 67-dB dynamic range in 10MHz bandwidth. IEEE J. Solid State Circuits **39**(12), 2152–2160 (2004)
10. L.J. Breems, R. Rutten, R. Veldhoven, G. Weide, A 56mW continuous-time quadrature cascaded ΣΔ modulator with 77dB DR in a near Zero-IF 20MHz band. IEEE J. Solid State Circuits **42**(12), 2696–2705 (2007)
11. Y. Matsuya, K. Uchimura, A. Iwata, T. Kobayashi, M. Ishikawa, T. Yoshitome, A 16-bit oversampling A-to-D conversion technology using triple integration noise shaping. IEEE J. Solid State Circuits **SC-22**(6), 921–979 (1987)
12. J. Steensgaard, *High-Performance Data Converters*, Chapter 9, Ph.D. Thesis, The Technical University of Denmark, Lyngby, Denmark, 1999
13. A. Gharbiya, D.A. Johns, A 12-bit 3.125MHz bandwidth 0-3 MASH Delta-Sigma modulator. IEEE J. Solid State Circuits **44**(7), 2010–2018 (2009)
14. Y. Chae, K. Souri, K.A.A. Makinwa, A 6.3μW 20bit incremental zoom-ADC with 6ppm INL and 1μV offset. IEEE J. Solid State Circuits **48**(12), 3019–3027 (2013)
15. B. Gonen, F. Sebastiano, R. Veldhoven, K.A.A. Makinwa, A 1.65mW 0.16mm^2 dynamic zoom-ADC with 107.6dB DR in 20kHz BW, in *IEEE International Solid-State Circuits Conference Digest Technical Papers*, pp. 282–283, Feb 2016
16. D. Gubbins, B. Lee, P.K. Hanumolu, U. Moon, Continuous-time input pipeline ADC. IEEE J. Solid State Circuits **45**(8), 1456–1468 (2010)
17. D. Yoon, S. Ho, H. Lee, A continuous-time sturdy-MASH ΣΔ modulator in 28nm CMOS. IEEE J. Solid State Circuits **50**(12), 2880–2890 (2015)
18. A. Thomsen, D. Kasha, W. Lee, A five stage chopper stabilized instrumentation amplifier using feedforward compensation, in *Symposium on VLSI Circuits Digest Technical Papers*, pp. 220–223, 1998
19. P. DeWit, G. Gielen, Complementary DAC topology for reduced output impedance dependency and improved dynamic performance. Electron. Lett. **48**(17), 1039–1041 (2012)
20. B. Murmann, ADC Performance Survey 1997-2015. [Online]. Available: http://www.stanford.edu/~murmann/adcsurvey.html

Part II
Automotive Electronics

The second part of this book is dedicated to recent developments in automotive electronics. The first chapter introduces the current status and outlines future directions. The second and the third chapter deal with power distribution in car environment. The last three chapters address the development of signal processing devices. The fourth chapter discusses robust ADCs, the fifth chapter deals with capacitive sensor front-end for automotive application. Finally, the sixth chapter discusses the design of an advanced accelerometer for smart tyres.

The first chapter, by Herman Casier, describes how electronics are rapidly becoming the major driver in new car developments. Since their humble beginnings in entertainment and driver comfort systems, they have gradually been used to optimize car behavior and are now beginning to take over more and more of the tasks of the human driver.

In the second chapter, Andreas Kucher discusses how power distribution is the overall term that describes how electrical power is transferred from its source to the different loads in a car. To overcome current limitations on the power net, relays and fuses should be replaced by semiconductor equivalents to reduce cost, power losses and vehicle weight. Moreover to gain more system benefits, the wiring harness, switches and loads have to be co-optimized. The main challenges for the required semiconductor devices are increased power capability, high reliability, self-diagnosis capability and low idle/quiescent currents.

The third chapter, by Bernhard Wicht, covers the design of highly integrated gate drivers and level shifters for high-speed, high power efficiency and dV/dt robustness with a focus on automotive applications. Such need for such circuits will be driven by the introduction of a 48 V power net in addition to the existing 12 V net.

In the fourth chapter, Carmelo Burgio and Mauro Giacomini present the development of a low-complexity digital self-calibration technique for a 12-bit SAR ADC for automotive applications in 40 nm CMOS. A key feature is that the ADC's DAC is split into an un-calibrated, binary, LSB part and a calibrated, thermometric, MSB part.

Capacitive sensor developments for automotive are discussed in the fifth chapter, by Paolo D'Abramo. In automotive electronics, capacitive sensors play a significant role because they can indirectly monitor many different physical quantities with the advantage of a contactless implementation. The design challenge is then how to implement high performance capacitive-sensing solutions, with relatively easy and low-cost construction, which can tolerate the harsh automotive electrical and environmental conditions.

Finally, Piero Malcovati presents the design of a low-power analog front-end for a 3-axis MEMS capacitive accelerometer. It consists of an analog preamplifier to boost the sensor's output, followed by a SAR ADC. Power minimization is achieved by using a continuous-time preamplifier to realize a constant-charge capacitance-to-voltage conversion and a SAR ADC with a split capacitive array.

Chapter 7
Trends and Characteristics of Automotive Electronics

Herman Casier

7.1 Introduction

After some difficult years, the market for light vehicles is predicted to grow steadily in the coming years by around 4 % per year to 110 m vehicles in 2020 [1]. The growth rate of electronics within the car is much higher and—despite the continuously decreasing cost of semiconductors—their share of the total car cost is predicted to reach 35 % in 2020 and 50 % in 2030 [2]. Where cars started as pure mechanical machines, fully controlled by the human driver, they are now complex electronic systems with optimized mechanical performances and increased comfort and safety for the driver. At this moment, the powertrain, chassis and body electronics are already well developed and new electronics developments are focusing more on safety, driver comfort and convenience and on advanced driver assistance systems (ADAS). The ultimate goal is the autonomous, driverless car, which is now advertised by different manufacturers and generates a lot of attention in the media. In the first paragraph the history and the future trends of automotive electronics and the disruptive nature of the driverless car will be discussed in more detail.

In parallel with this trend towards autonomous, driverless cars, a lot of effort in the automotive industry is also devoted to more energy efficient and cleaner vehicles e.g. mild and full hybrid vehicles, battery and fuel cell electric vehicles... Although this evolution requires a lot of electronics, it is mainly driven by the improvements of the energy sources, their energy content, weight and volume, lifetime, safety, refueling time and infrastructure... This trend is not driven by automotive electronics and is not further discussed here.

H. Casier (✉)
Avondster 6, B-8520 Kuurne, Belgium
e-mail: herman_casier@ieee.org

© Springer International Publishing Switzerland 2017

A. Baschirotto et al. (eds.), *Wideband Continuous-time ΣΔ ADCs, Automotive Electronics, and Power Management*, DOI 10.1007/978-3-319-41670-0_7

Semiconductors for automotive electronics are often perceived as not state-of-the-art and lagging mainstream chip design and technologies. This perception is now disappearing for the advanced micro control units (MCU), used in the advanced driver assistance systems [3], but remains for the large number of mixed signal ASICs used in the car. Indeed, many ASICs in the car use mature mainstream or special high voltage technologies with less advanced lithography and the signal processing specifications of the circuits are typically lower than for commercial systems. But, this is due to the very high reliability and safety requirements and to the "hostile" car environment in which these systems have to remain fully functional [4]. The high and spiky supply voltage, the extreme temperatures and the harsh interferences are very tough boundary conditions for automotive ASICs. Furthermore, low total system cost prohibits extensive environment shielding and protection of the circuits. These specific automotive characteristics and their effects on technology and circuit design are discussed in the second paragraph.

7.2 Trends in Automotive Electronics

The auto-mobile started as a pure mechanical system and remained a high performance mechanical system during more than half a century. Apart from the ignition system and the radio, introduced by the Galvin corporation in 1930 (Motorola) and by Ideal (Blaupunkt) shortly after, no electronics were used in the car. It was indeed difficult to compete with very cost efficient mechanical and hydraulic systems.

Electrical and electronic systems were introduced first as special and expensive features in high-end cars, mainly focusing on engine improvements, driver convenience and safety e.g. electronic injection, air-conditioning, airbags … While most of these applications were easily adopted by the driver community, some safety measures, with a direct impact on the driver and the passengers, raised controversy e.g. the still ongoing safety belt discussions. Airbags are another example of the human reluctance to trust new developments. Airbag patents for automotive use were already registered in 1951 in Germany and the US and were offered in the early 1970s as a high-end car feature [5]. Despite their proven safety improvement, the introduction failed commercially. They were reintroduced in the 1980s by Mercedes and deployed as standard equipment in addition to seatbelts in several high-end cars by the end of the decade.

Seatbelts were also introduced in the 1950s but were not generally installed in all cars, despite their high efficiency in saving thousands of deaths on the road. Laws had to be voted requiring fitting and wearing seatbelts in all cars, including mid-range and low-range cars. Legislation is indeed an important driver for electronics in the car by mandating better pollution, safety and traffic control systems in line with technological possibilities.

Car electronics are now also embraced by car manufacturers for their higher flexibility in manufacturing, easier model and upgrade differentiation and better servicing. Furthermore, new advanced features have become an important commercial differentiator between car manufacturers.

7.2.1 Automotive Electronics Evolution

The pioneering stage of horseless carriage ended around 1890 with the first regular series production of gasoline cars on a small scale in France. Soon other manufacturers in France and the U.S. followed with ever larger scale production facilities in moving assembly lines. These first cars were purely mechanical systems and required high technical skills to drive them and keep them going. Roads at that time were not adapted to car traffic but traffic was very low anyway.

After World War I, car prices had plummeted down and "Automobiling" was no longer a privilege for the rich but became accessible for everyday people. Traffic increased considerably and the number of motor-vehicle fatalities increased sharply. As a result, nation-wide traffic laws for cars were issued. Driver-licenses became mandatory and because the required technical skills to drive a car remained high, these licenses also included a technical exam. To cope with the increased traffic, roads were enlarged and the first highways were built in Germany in the 1930s. Cars were increasingly used for long trips and to make these journeys more comfortable, the first electrical appliances, head- and tail-lamps, radio, wiper/washers... were installed in the cars. These first electrical systems were only introduced in non-critical areas: driver comfort, information and entertainment. The low intelligence electrical systems had no impact on engine performance nor on driver or driver skills. They were typically pushbutton controlled on/off systems and were stand-alone systems, not requiring any communication to other electrical systems in the car (Fig. 7.1).

After World War II, traffic became even more dense and the focus of driver-licenses shifted to driver skills and traffic rules. The technical exams disappeared since the cars became more reliable and easy to use. More highways were built in Europe and the U.S. and attracted even more traffic. More electronic systems entered the car for engine optimization (e.g. fuel injection), comfort and convenience (e.g. air-conditioning, cruise control), active and passive safety (e.g. airbags, ABS) and driver information and entertainment (e.g. GPS, service warnings). These increasingly complex and intelligent electronic systems optimized and controlled the engine performance and supported, but not controlled, the driver through reactive correction of driver errors (e.g. traction control, ESP, brake assist system). The advanced Electronic Control Units (ECU) require accurate sensors, powerful signal processing and communication with other ECUs in the car. Their communication and impact however remained within the car shell.

In the last 20 years, a third phase of automotive electronics has started. Electronic systems now gradually control more critical applications (e.g. adaptive cruise

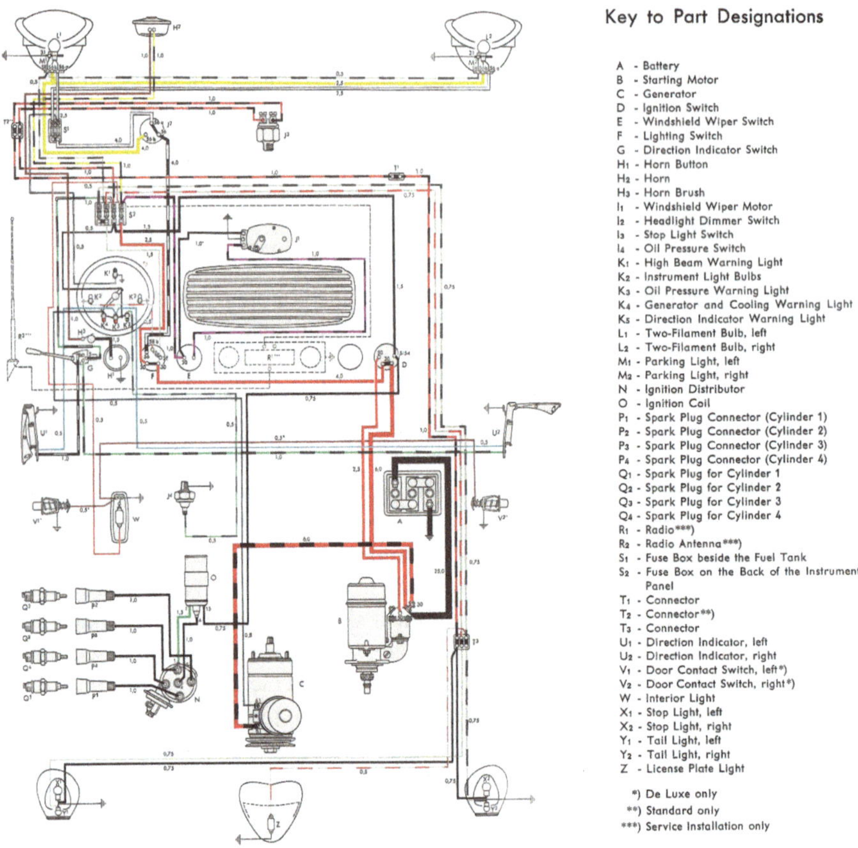

Fig. 7.1 VW Beetle 1954, complete electrical schematic

control, anti-collision radar, automatic car parking...) Since these intelligent systems can be authorized by the driver to take over full car control in well-defined situations and can also proactively prevent dangerous situations inside and around the car without driver control, they have to be very robust and safe. They are also increasingly complex. Nowadays, a typical car contains around 90 ECUs, 170 sensors, more than 15 million lines of code, and a few kilometer network with more than 1000 connection points [6].

7.2.2 Steps Towards Autonomous Driving

It is not realistic to rethink human mobility from scratch and devise a whole new traffic system for autonomous cars since this would require the replacement of all current cars and road infrastructure at once by a new system. In reality, autonomous cars and human driven cars will share the road for a long time. Hence, the electronic system that replaces the driver in an autonomous car must start from the human behavior of the driver [7]. Human senses, vision and hearing, can be replaced by sensors and communications. Human decision making however is much more difficult and requires huge amounts of software and continuous improvement by learning. The human decision making also depends on a large memory of human behavior, geographical situations and environment conditions (e.g. human body language, weather, road pavement . . .). Human behavior is also culture specific and not easy to reduce to an algorithm (e.g. a speed limit is interpreted differently by human drivers in different countries and an autonomous car which obeys every traffic law and speed limit will impede traffic flow and make other drivers nervous in many countries). Sensors, memory and software are key requirements to replace the human driver in the autonomous car.

Ultimately, the human mobility system could then become a traffic network, which takes control of the macro movements (upper layers) of the autonomous cars. Each autonomous car executes full control of the car and the immediate surroundings and communicates with neighboring cars and with local traffic signs. The driver is now a passenger for the complete or at least most of the journey. In such system, the car functions as a personal public transport with the highest flexibility. This is a disruptive vision of future mobility and opens new perspectives for the society on top of the mobility aspects. Although this landmark may be reached technically within 10–15 years for high-end cars, the adoption by the society and the penetration in lower-class vehicles is expected to be much slower [8]. It will also be an evolutionary process in which three phases can be distinguished [9].

In a first phase, driving is partially automatic. In a number of well-defined, not complex driving situations, the human driver can hand over control to the autonomous driver but he must be prepared to retake control instantly if unforeseen conditions occur. In this phase, the full responsibility and liability remains with the human driver. Examples of such driving situations are cruising at high speed on a highway where no crossing traffic can occur or driving at low speed in a traffic jam where no fast movements from other drivers are expected. These situations are monotonous and boring for the driver and his concentration can easily fade away. This is dangerous since handover to a sleepy driver can take more than 18 seconds, while a vigilant driver can react within one second [10]. The car must either detect fatigue, drowsiness or distraction of the driver and keep him alert or the car must be already intelligent enough to drive itself into a safe condition.

The next phase is highly automated driving where the autonomous driver handles very complex driving situations and recognizes, well in time, unknown driving situations. The autonomous driver then alerts the human driver to take over control and turns to safe operation until the driver reacts. Car to car and traffic sign to car communications and a developing traffic network control on major roads will augment the performance. The autonomous driver will be able to handle all car to car interactions on highways, rural roads and city streets. Interactions between the car and pedestrians though will be much more unpredictable due to difficult interpretation of human reactions and body language. In this highly automated driving phase, the human driver cedes full control to the electronic system and this will inevitably lead to legal issues regarding responsibility and liability of the driver or the car/ADAS manufacturer in case of accidents. Confidentiality of the data exchanged with other cars, with the traffic network and with the car manufacturers is also an important legal issue.

As described above, the last phase is the fully autonomous car, which completely replaces the human driver. The car functions as a personal public transport with high comfort and flexibility and can also move without a human driver inside. This opens up new possibilities. If public transport becomes so comfortable and flexible, there will be less incentive to buy and own your personal car, which is sitting idle 90 % of the time. This is a trend, which is already seen with young adults in Europe, who live in a city and extensively use public transport, despite its current inflexibility and lack of comfort. It is tempting to imagine mobility-on-demand with a pool of cars, small, medium and large size for short, medium and long distances. Customers would call-up a small or medium car for shopping and work commuting and a long distance car for business trips and leisure. Because they can rapidly move without driver, these cars do not stay idle at the customer premises and they can be concentrated in a few parking buildings in the city or constantly drive around to lower pick-up time of customers. Such shared mobility system would drastically lower the total number of cars by lowering their idle time and reclaim living area in the cities by eliminating their parking spaces. A smart traffic network with smart cars will also smooth urban traffic but traffic congestion is not expected to decrease since travelling by car will become more convenient and more people, including people who are currently unable to drive, will drive longer distances. Such mobility-on-demand is however only a viable approach in cities and not in rural areas. Also the car is still very much an individual status symbol and many drivers enjoy driving, which is not expected to change rapidly.

It will be a formidable technical task to build the autonomous car, the car to car and traffic sign to car communication system and the traffic network control system. Safety, predictability and protection against malicious attacks are key issues in this development. The system will also coexist with non-autonomous cars, with pedestrians, with unpredictable kids, with animals, with unforeseen road obstacles... Pedestrians and human drivers are likely to exploit known weaknesses in the system to their advantage (e.g. not allowed street crossing or priority taking since the autonomous car will obey the law and yield anyway). Unless standardized, manufacturers of the ADAS systems are likely to compete on human behavior algorithms. This could give some car types an, considered unfair, advantage over other cars. Eventually the current ranking of high-end, mid-end and low-end cars would shift to a first-class, business-class and economy-class ADAS system ranking.

7.3 Characteristics of Automotive Electronics

The main differences between automotive and typical commercial electronic systems are due to the supply by the car battery, the distribution of the system ECU's over the whole car-body and the stringent safety requirements for automotive applications.

Although the mainstream lead-acid battery is nominally 14 V, the supply voltage to the ECU's shows a large variation: typically 5 to 40 ... 60 V and it is polluted by various interfering signals. Due to new, high-power applications in the car and stringent emission requirements there is an evolution to implement a second 48 V Lithium battery for high-end and for mild-hybrid cars. This battery voltage is less polluted and the maximum ECU supply voltage is in both cases quite comparable. Internal in the ECU, standard low voltages are used for the high complexity signal processing MCU's and memories but the ASICs at the borders of the ECU must cope with the high voltages and disturbances on their supply pins, communication pins, sensor inputs and actuator outputs.

The distribution of the ECU's and the communication networks over the whole car-body make them vulnerable to EMC, ESD and transients. Shielding cannot be used for its extra cost and weight. As a consequence, EMC and ESD are very tough constraints in automotive design in contrast with most commercial electronic systems, which operate in much better shielded environments.

A major reason for the distribution of ECU's over the car-body is to bring them closer to sensors and actuators. This improves speed and accuracy of the applications but exposes the ECU's to the high temperatures in some car-locations. These high temperatures exceed the industrial and also the military (125 °C ambient) temperature range and there is a strong push for further increase. This puts chip design, chip reliability, package materials, PCB materials and manufacturing under pressure. In extreme temperature cases special technologies and packages have to be used.

Finally, since more and more critical car-functions are controlled by electronics, safety, reliability and predictable error behavior of the applications are vital and the communication between the ECU's must remain reliable and dependable under all conditions.

In conclusion, automotive electronics must fulfil all classical design challenges: low cost, fast time-to-market, complete functionality, low power consumption, testability ... On top of this they must also be designed according to the highest safety and reliability standards, cope with high voltages and large supply variations, function up to high temperatures, sustain large ESD pulses, not disturb neighboring applications and remain functional in the presence of large EMC and transient interferences.

7.3.1 Quality and Safety

The automotive industry has always been known for its very high reliability requirements. All classical reliability techniques are used in design and test: six sigma design, worst case and degradation simulation, high test coverage, IDDQ, Vstress and other advanced tests... On top of this other reliability engineering methods are used such as FMEA (Failure Mode and Effect Analysis) and FTA (Fault Tree Analysis). The former is a bottom-up, inductive evaluation of failure effects and becomes tedious for complex systems. The latter is a top-down, deductive chain-of-events tree model leading to the accident and assumes independence of failures, which is not always true.

Safety however is not the same as reliability [11, 12]. Where reliability focuses on component failures and their MTTF (Mean Time To Failure), safety of a complex system can be impacted even when no components fail their specification due to e.g. incomplete concepts and specifications or human or procedural interactions. The contrary can also occur. Therefore ISO 26262 [13] is now used to make safety part of all phases of the product lifecycle: management, concept, system, hardware, software, development, production, operation, service ...

From the discussion above on autonomous driving, it is clear that the requirements for safety of the complete ADAS will not decrease but become more stringent in the future.

7.3.2 High-Voltage

The mainstream lead-acid battery is a fairly reliable and production proven supply but it shows large fluctuations. While the nominal voltage is 14 V, the minimum can drop to less than 7 V and the maximum can rise to more than 19 V when charged by the alternator. On top of this large tolerance, the distribution of the battery voltage over the car-body adds another spread and makes it vulnerable to interferences from high-power actuators and EMC. These interferences create all sorts of spikes on the supply voltage but do not drastically change the DC value of the battery. A special case occurs when the battery is disconnected or defective and the alternator charges the electronic system of the car without the stabilizing effect of the battery. This is the load-dump condition. Since the low voltage alternator is very inefficient, it will react slowly on the generated overvoltage and, if not clamped, the voltage will rise to more than 100 V peak during half a second. A clamped load-dump limits this rise below 40–60 V. Finally, the battery connections can be reversed and a series diode is used to prevent this. This results in a battery voltage between 5 V and 40 ... 60 V with high voltage (150 ... 200 V) short impulses (ISO 7637 [14, 15]) and EMC interferences.

Due to the ever increasing peak power demand in high-end cars, new electronic applications which are difficult and inefficient to build on a 14 V supply and the

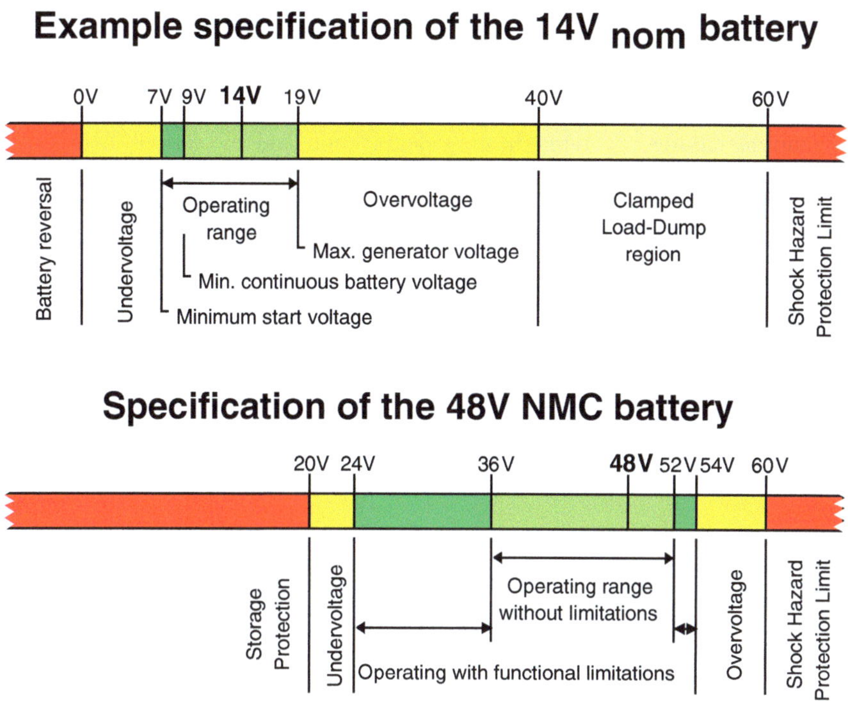

Fig. 7.2 Voltage ranges for the 14 V lead-acid and the 48 V NMC car-batteries

low efficiency of the 14 V alternator, a study to define a future automotive electrical system was conducted in MIT [16]. This resulted in the "42 Volt Powernet" draft specification in 1997, based on a 42 V lead-acid secondary battery (three times the 14 V battery). For different technical and commercial reasons, this was never implemented in large series production cars. But the underlying necessity for a higher power battery and alternator remained. With an extra push from the lower emission requirements for 2020, work on a high voltage battery has resumed and a new 48 V standard has been proposed in 2011 by five German carmakers. This battery is based on state-of-the-art battery technology (lithium nickel manganese cobalt oxide: NMC). Luxury and mild-hybrid cars are the first ones to use this new secondary battery. This battery voltage is less polluted by the load-dump than the 14 V battery and the maximum ECU supply voltage is in both cases quite comparable and remains below the shock hazard protection limit for DC voltages: 60 V (Fig. 7.2).

Various technology families with many different types of DMOS transistors have been developed for these high voltage applications. These technology families follow their own roadmap but they lag the mainstream CMOS roadmap by roughly three generations. As a result, the high voltage chips have a limited digital content and are usually at the border of the ECU, interfacing with the supply, the sensors, the actuators and the communication channels. The complex signal processing in MCU's and DSP's remains within the ECU and uses low voltage state-of-the-art technologies. The high voltage DMOS transistors have a quite complicated Safe Operating Area (SOA). The SOA is confined by instant overvoltage and overcurrent failures, by rapid heating degradation and by various field-related degradation mechanisms. The high voltage design requires different layout and circuit techniques while the high power induces self-heating and large temperature gradients on the chip [17].

With the advent of the new 48 V Powernet secondary supply and a 48 V alternator with much higher efficiency than the 14 V alternator, the power bottleneck for high-end cars and for power-hungry applications is relieved. There will be no push for higher voltages in the near future especially since this would exceed the 60 V shock hazard protection limit for DC voltages and would require costly and heavy protections and shieldings throughout the car.

7.3.3 High Temperature

High temperatures in the car typically rise to 150 ... 170 °C ambient in the engine compartment or in the chassis near the breaks. Close to the exhaust the maximum temperature can even exceed 200 °C ambient [18]. Junction temperatures are typically 20–50 °C higher than the ambient and can rise to a maximum value of 170...200...250 °C, far beyond even military specifications. Shielding the electronics from the high temperature is expensive and there is a strong demand for higher temperature electronics in the car.

With appropriate components and design techniques, bulk silicon can handle temperatures up to 200...300 °C. SOI technologies are even better and go up to 250... >300 °C. If higher temperatures are needed, GaAs, GaN and SiC technologies have to be used. The main circuit design issue is the exponential increase of junction leakage with temperature up to the point that the material becomes intrinsic and the junction becomes a short-circuit. This intrinsic temperature limit drops with lower majority concentrations in the junction. Hence high-voltage junctions cannot go as high in temperature as low voltage junctions. Here again, SOI has a large advantage over bulk silicon [19] (Fig. 7.3).

Silicon, bulk or SOI, can handle the required high temperatures up to e.g. 200 °C or even higher but the packaging is now the limiting factor [20]. Especially wire bonding degradation and depolymerization of classic plastic packages limit the maximum temperatures to much lower values. More expensive or newer plastic packages and chip connection techniques can be used to extend this limit.

The demand for higher ambient temperature electronics in automotive will continue. The chips also contain increasingly more high power devices and complex signal processing blocks. Both give rise to an increase of the maximum junction temperature, the former directly, the latter tempered by technology improvements. High temperature capability will continue to be an asset for silicon manufacturers.

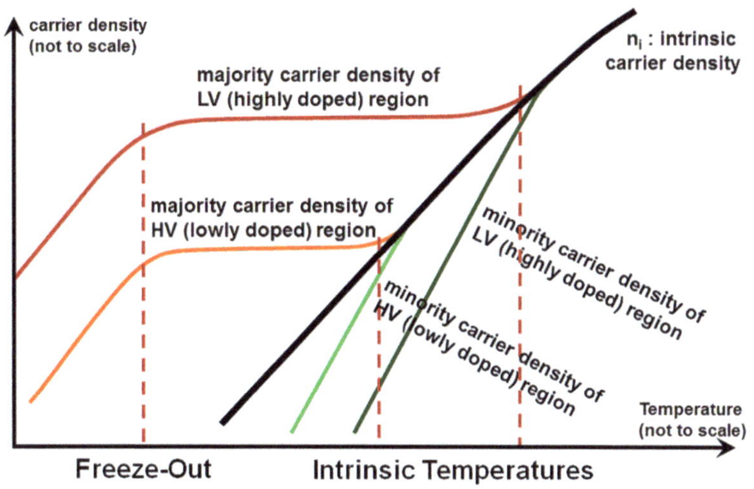

Fig. 7.3 Freeze-out and intrinsic temperatures for high-voltage (lowly doped) and low-voltage (highly doped) semiconductor junctions

An increase of the maximum temperature requires advancements on all levels: technology, characterization, chip layout and design, package materials and package to chip connections, PCB and ECU materials and connections . . .

7.3.4 EMC Interferences

The U.K. Defense Standard 59-41 describes EMC as: "Electro Magnetic Compatibility is the ability of electrical and electronic equipments, subsystems and systems to share the electromagnetic spectrum and perform their desired functions without unacceptable degradation from or to the specified electromagnetic environment".

EMC deals with electromagnetic fields, which are completely described by Maxwell's laws at the device and vehicle scale. These laws describe the fields in time and three-dimensional space for all signals, linear and non-linear, in all materials, homogeneous or not. They are virtually impossible to solve analytically except for very simplified structures but they can be solved by various EM simulation techniques for low complexity two and three-dimensional structures [21].

Small dipole and loop antennas with dimensions much smaller than the wavelength of the EM-field can be solved analytically. They radiate very low energy into the far field and most of their energy is kept in the near field, where it behaves as capacitive, respectively inductive coupling. The field behavior of these antennas can then be neglected and only the coupling behavior has to be taken into account. This can be modeled with classic capacitors and inductors in SPICE-based simulators.

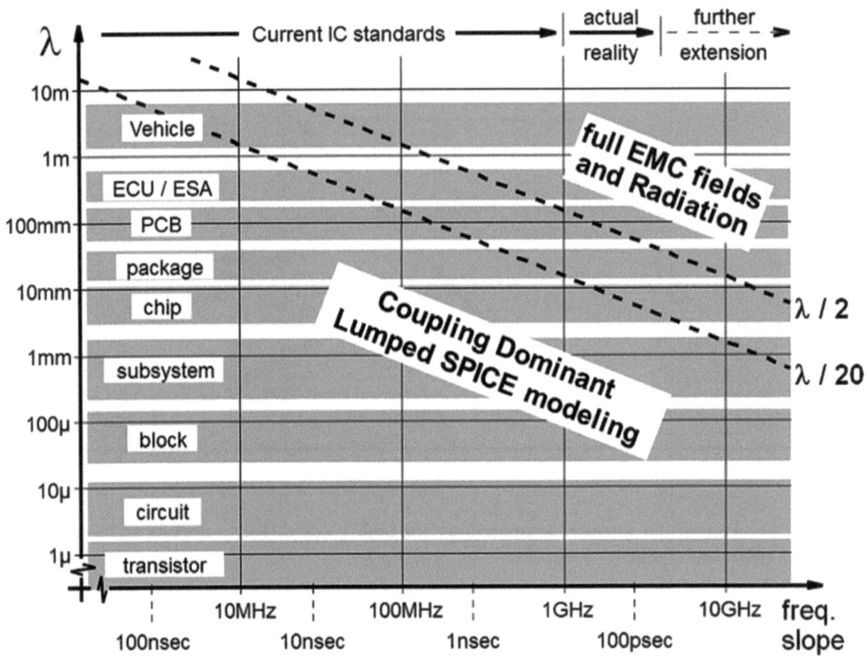

Fig. 7.4 EMC can be modeled by coupling at chip level. EMC has to be treated as a radiating field at ECU and vehicle level

Dimensions up to 2–3 mm can be treated as small antennas for frequencies below 3 GHz and slopes slower than 100 psec, Chip circuits, blocks and subsystems fall in this category and only for large chips and packages some on-chip antennas can start to radiate at the highest frequencies and fastest slopes. PCB's, ECU's and the wiring harness of the vehicle are much larger and must be treated as radiating antennas. Radiated emission and immunity is hence a major problem for ECU's and the wiring harness but not for the chip (Fig. 7.4).

The chip however generates fast and large differential and common-mode current and voltage changes at its outputs and power pins, which are transformed into a radiating field by the efficient antennas in the ECU and the wiring harness. Conversely, the efficient ECU and wiring harness antennas transform an incident EMC radiation field into conducted interference signals to the chip.

In short, EMC is treated at chip level as a non-radiating, conductive coupling while it is treated at ECU and vehicle level as a radiation. The translation between the EM radiating field and the conducted currents and voltages at the chip boundaries is very complex since it is dependent on the three dimensional structure of the ECU and on the complex impedances at the chip boundary. This results in a difficult situation where the car manufacturer and the Tier one suppliers think in EM-fields and are very uncertain (and conservative) in translating this into conductive specifications for the chip manufacturer. Only with a lot of experience and a close cooperation between the chip and module manufacturers this division can be bridged and adequate conduction limits for the chip designers can be set.

This is also reflected in the component radiation standards IEC 61967 and IEC 62132, which contain different measurement methods for radiated and for conducted emission and immunity measurements [22, 23].

Even with good conductive specifications, the simulation of EMC at chip level remains a very difficult task. With EMC frequencies into the GHz range, very small parasitic components and couplings on the chip and in the package have to be included in the simulation deck. Large EMC interferences can cause non-linearities, breakdown and substrate currents. Resonances between parasitics and components are difficult to detect in simulation. Charge pumping due to large signal distortion and charge accumulation requires large signal transient simulations with many parasitics and couplings over a long simulation period.

EMC interferences will further increase in the future, mainly in frequency and to a lesser extent in power. There is however also an ongoing push to limit the discrete protection devices in the ECU's. This increases the EMC power requirements with respect to immunity and requires higher accuracy power and communication drivers for emission control.

7.4 Conclusions

Electronics are a major driver in the evolution of the car and the human mobility. Due to the relentless advancements of sensors and software, the electronic system in a car is more and more taking over the operation of the human driver itself. This is a disruptive evolution which ultimately can transform the car in a personal public transport with high comfort and flexibility. This will not only transform traffic and human mobility but can also have a large impact on life organization and city urbanization. This will however not happen overnight due to the human reluctance to accept and adopt these new possibilities.

The evolution of automotive electronics towards autonomous cars requires very high levels of safety. The increasing power demand for advanced driver assistance systems can only be satisfied by high voltage supplies. It is also not expected that high temperature spots in the car will disappear and the interferences from EMC radiations will increase in power and extend into higher frequencies. The specific automotive characteristics of automotive electronics will hence remain important design and technology challenges in the future.

References

1. KPMG's Global Automotive Executive Survey 2015
2. PwC Semiconductor Report, Spotlight on Automotive. September 26, 2013

3. C. Takahashi, S. Shibahara, K. Fukuoka, J. Matsushima, Y. Kitaji, Y. Shimazaki, H. Hara, T. Irita, A 16 nm FinFET Heterogeneous Nona-Core SoC Complying with ISO26262 ASIL-B: Achieving 10^{-7} Random Hardware Failures per Hour Reliability. Paper 4.5, ISSCC 2016, San Francisco, Februari 1, 2016
4. H. Casier, Electronic circuits in an automotive environment. Tutorial Session "Circuit Design for Automotive", ESSCIRC 2013, Bucharest, 16 Sep 2013
5. "Airbag" from Wikipedia, retrieved January 2016
6. K. Paesler, M. Saemann, Continental, growth opportunities in the automotive group. in *Kepler Cheuvreux One Stop Shop Conference*, Milan, 29 June 2015
7. Think Act, Autonomous Driving, Roland Berger Strategy Consultants GMBH, November 2014
8. McKinsey & Company, Automotive Revolution—Perspective towards 2030. Advanced Industries, January 2016
9. NHTSA, Preliminary Statement on Policy Concerning Automated Vehicles. May 2013
10. ITS World Congress 2015, Connected and Automated Driving, Panel discussion at ITS World Congress 2015, Bordeaux, France, 6 Oct. 2015
11. N.G. Leveson, *Engineering a Safer World*. (MIT Press, 2011), ISBN 978-0-262-29911-4 (ebook)
12. Q. Van Eikema Hommes, Assessment of the ISO 26262 standard, road vehicles—functional safety. in *SAE 2012 Government/Industry Meeting*, 25 Jan 2012
13. ISO 26262, Road Vehicles—Functional Safety, Parts 1–10 version 1 released in 2011. Version 2 in preparation (with two extra parts)
14. ISO 7637-2:2011, Road vehicles—Electrical disturbances from conduction and coupling, Part 2: Electrical transient conduction along supply lines only
15. ISO 7637-3:2007, Road vehicles—Electrical disturbances from conduction and coupling, Part 3: Electrical transient transmission by capacitive and inductive coupling via lines other than supply lines. New Publication, March 2016
16. J. Kassakian, Challenges of the new 42 volt architecture and progress on its International acceptance. VDI Berichte Nr. 1415, 21–35 (October 1998)
17. H. Casier, "High-Voltage/High-Temperature Challenges and Limitations", Mead Education, Advanced Engineering Course on "High-Voltage and Automotive IC Design", Lausanne, 25 Aug 2015
18. W. Wondrak, A. Debhi, G. Umbach, A. Blessing, R. Getto, F.P. Pesi, W. Unger, Passive components for high temperatures: application potential and technological challenges. in *AEC Annual Reliability Workshop*, Nashville, 20 April 2004
19. F. Patrick McCluskey, R. Grzybowski, T. Podlesak (ed.), *High Temperature Electronics*, (CRC Press, Boca Raton, 1997), ISBN 0-8493-9623-9
20. P.C. De Jong, Smart sensor systems for high-temperature applications, PhD Dissertation, T.U. Delft, The Netherlands (1998)
21. H. Casier, Automotive Interferences: EMC and Transients. IDESA Course funded by the EC 7th Framework Program, 19 June 2008
22. IEC 61967, Integrated Circuits—Measurement of electromagnetic emissions, 150 kHz to 1 GHz
23. IEC 62132, Integrated Circuits—Measurement of electromagnetic immunity, 150 kHz to 1 GHz

Chapter 8
Next Generation of Semiconductors for Advanced Power Distribution in Automotive Applications

Andreas Kucher and Alfons Graf

8.1 Introduction

Power distribution is a quite complex topic in vehicles. One of the major legal requirements is the regulation for reduction of CO_2 given by the European Union. In Fig. 8.1, the development of the average CO_2 of major car makers fleets from 2007 to 2011 is shown.

The CO_2 target for 2015 was 130 g/km, for 2021 this target is decreased to 95 g/km. Please refer to [1]. Exceeding the targets given by European Union leads to a penalties for the car maker. This penalty is up to 95 €/g. Most of the car makers met the 2015 limits, nevertheless the 2021 limit will be much more difficult to attain. All the low hanging fruits (e.g. start/stop functionality) for reduction of emission are already implemented. To achieve lower values significant effort must be spent.

One gram of CO_2 is equivalent to 40 W electrical power dissipation. A similar consideration can be done with mass. Reduction of one gram CO_2 is equivalent to 20 kg mass reduction. Thus any reduction of power dissipation or mass helps to reduce CO_2.

For electrical system major trends can be seen in the market:

- Mechanical components are replaced by electro-mechanical.
  - Examples for this are water pump, oil pump, fan, turbo boost or power steering. The advantage of electro-mechanical components is the possibility of efficient and demand orientated control.

A. Kucher (✉)
Infineon Technologies, Siemensstraße 2, 9500 Villach, Austria
e-mail: Andreas.Kucher@infineon.com

A. Graf
Infineon Technologies, Am Campeon 1-12, 85579 Neubiberg, Germany
e-mail: Alfons.Graf@infineon.com

© Springer International Publishing Switzerland 2017

A. Baschirotto et al. (eds.), *Wideband Continuous-time ΣΔ ADCs, Automotive Electronics, and Power Management*, DOI 10.1007/978-3-319-41670-0_8

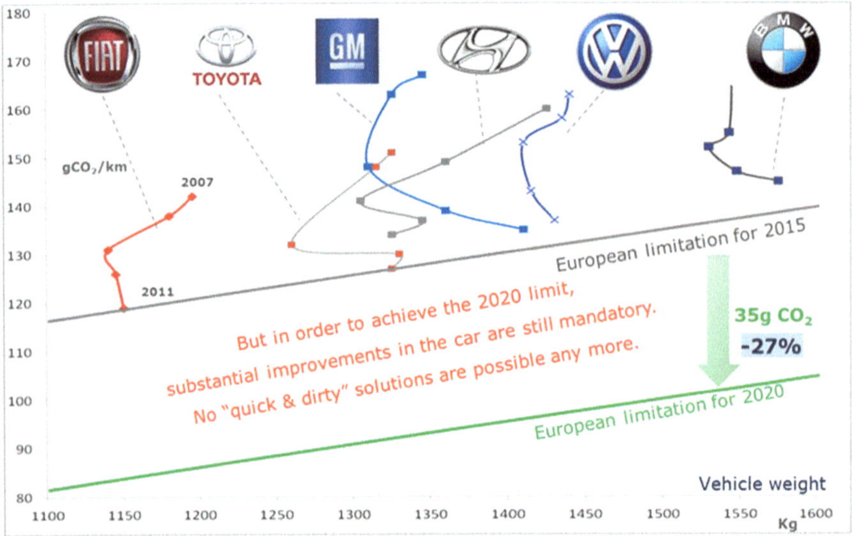

Fig. 8.1 Development of CO_2 emission for major carmakers from 2007 to 2011

- Reduction of electrical idle losses.
  - All operating currents of devices result in power dissipation. Semiconductors have here a clear advantage compared to standard relays.
- High reliability
  - The acceptance of a failure in the car is decreasing from year to year with the clear target: zero defects.
  - Due to increased number of loads and increased load activations high robustness is mandatory.
- Lower system cost

In parallel to the classical 12 V power network several carmakers have introduced the 48 V power net. The driver for this new voltage class is the supply of high power loads, energy recuperation and the use of the starter as motor for torque boosting in hybrid vehicles. Beside 48 V specific challenges the 48 V power network face the same challenges than the 12 V power net.

8.2 Power Network

8.2.1 Classical Centralized Power Network

In Fig. 8.2 a typical 12 V power network is shown. The system contains three fuse boxes.

Usually the pre-fuse box is located directly at the battery terminal thus the line from battery terminal to the pre-fuse box is a non-protected line. A broken pre-fuse is very rare, thus the pre-fuse box need not be accessible by driver or mechanics. The other two fuse boxes are located in the front (e.g. under the hood) and/or in the rear of the vehicle (e.g. trunk). Those two fuse boxes must be accessible by driver and must have the possibility to replace a fuse without going to the garage. Those requirements lead to a centralized power net.

The wiring harness of such a typical power net has approximately a mass of 30 kg, contains 90 fuses and 20 relays. The cable length of all wires is approximately of 2 km.

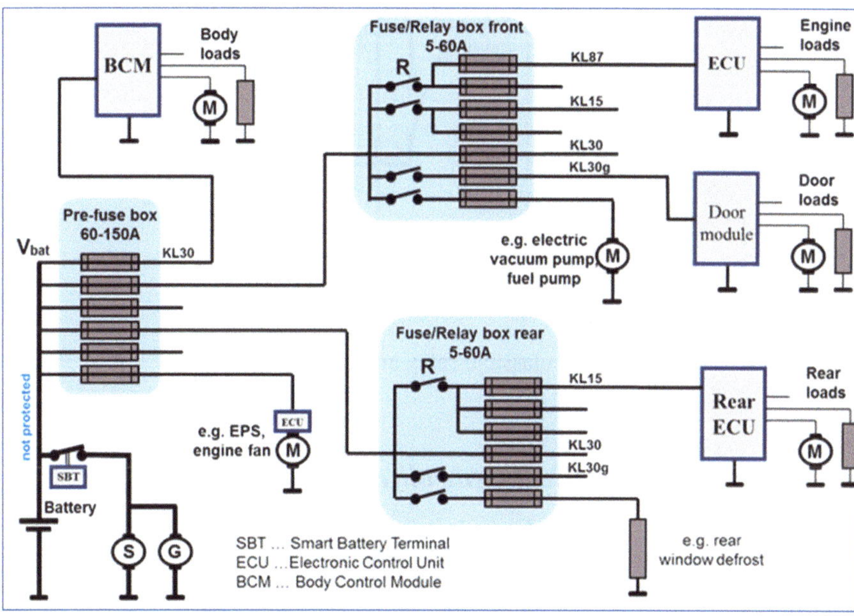

Fig. 8.2 Typical 12 V centralized power network

Fig. 8.3 Typical relay and fuse box

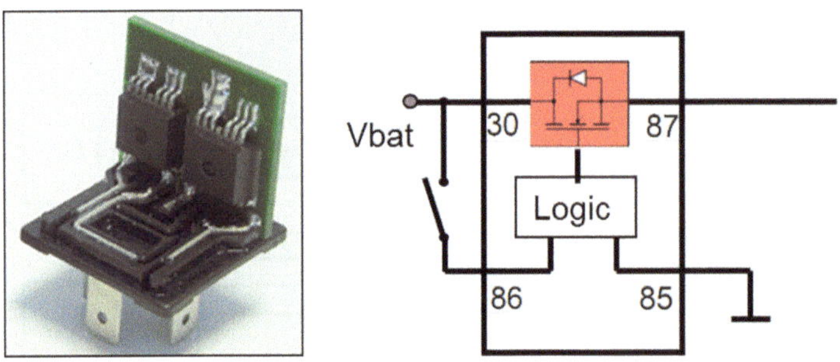

Fig. 8.4 Semiconductor switch inside relay housing

Figure 8.3 shows a typical classical relay and fuse box:

8.2.2 *Enhanced Classical Centralized Power Network*

Some vehicle functions (e.g. start/stop) require switching cycles larger than 200,000. Today those functions are driven by semiconductor switches since a relay has a limited number of switching cycles and does not fulfill the requirements.

In this system the relay box is a mix of relays, fuses and semiconductors. The change from relays to a semiconductor switch can be done even without having to make a change in the fuse box itself. Figure 8.4 shows the simplest way to replace a relay. This is done by placing semiconductors as a power switch inside the housing of a relay. This type of relay is called **S**olid-**S**tate **R**elay (SSR).

This simple relay replacement enables a high number of switching cycles, high reliability and low idle current consumption. A mechanical relay has power

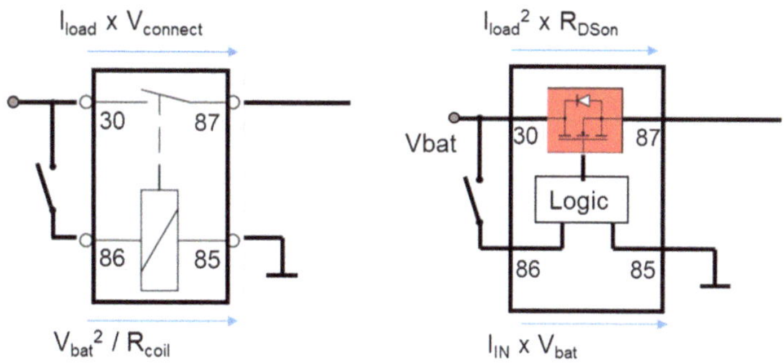

Fig. 8.5 Equivalent circuit diagram of relay and semiconductor switch

dissipation (P_{diss}) of (see Fig. 8.5):

$$P_{diss} = I_{load} * V_{connect} + {V_{bat}}^2 / R_{coil} \tag{8.1}$$

While I_{load} is the load current, $V_{connect}$ the voltage drop across mechanical switch, V_{bat} the battery voltage and R_{coil} the resistance of the coil driving the relay. For a semiconductor device I_{in} is the operating current of the switch itself and R_{dson} is the on resistance of the power switch.

The semiconductor switch (see Fig. 8.4) has a power consumption of:

$$P_{diss} = {I_{load}}^2 * R_{dson} + I_{in} x V_{bat} \tag{8.2}$$

A high current relay can have a power dissipation of 5 W typical. A semiconductor switch has below 1 W typical. In the case of replacing five relays in one vehicle this is a power dissipation reduction of five times 4 W = 20 W. 20 W is equivalent to a CO_2 saving of half a gram. Beside the standard relays there also exists a more expensive bi-stable relay where current through the coil is needed only in case of switching transition.

Some applications require reverse blocking. This means no current through the switch in case of reverse battery. Unfortunately the semiconductor power transistor has an intrinsic body diode which is conducting in reverse battery condition. In case of reverse blocking requirement a second power transistor is required to be used in serial connection as shown in Fig. 8.6. The mechanical switch of a relay has inherent reverse battery blocking.

To have the same R_{dson} compared to a single switch the R_{dson} of the two switches in serial have to be reduced by half of the original value. This leads to significant higher semiconductor cost.

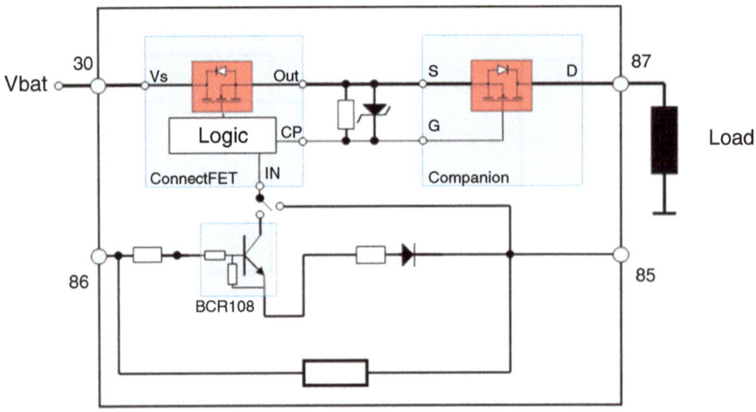

Fig. 8.6 Possible concept for reverse blocking semiconductor switch

Fig. 8.7 Classical vs. semiconductor based electronic power distributing box with semiconductor relay and fuse function

8.2.3 Semiconductor Based Centralized Power Network

To completely remove accessible fuses from a power distribution box a "semiconductor only system" approach is required. Figure 8.7 shows how a replacement of the fuse box could look like without changing the power network. Please refer to [2].

In addition to not having any accessibility requirements, the advantage of such a system is optimal usage of wires, small volume, enhanced diagnosis, low idle current and high reliability. The weight of this system is reduced by 50 % compared to a classical fuse and relay box. The three relays in this system implement the reverse blocking requirement.

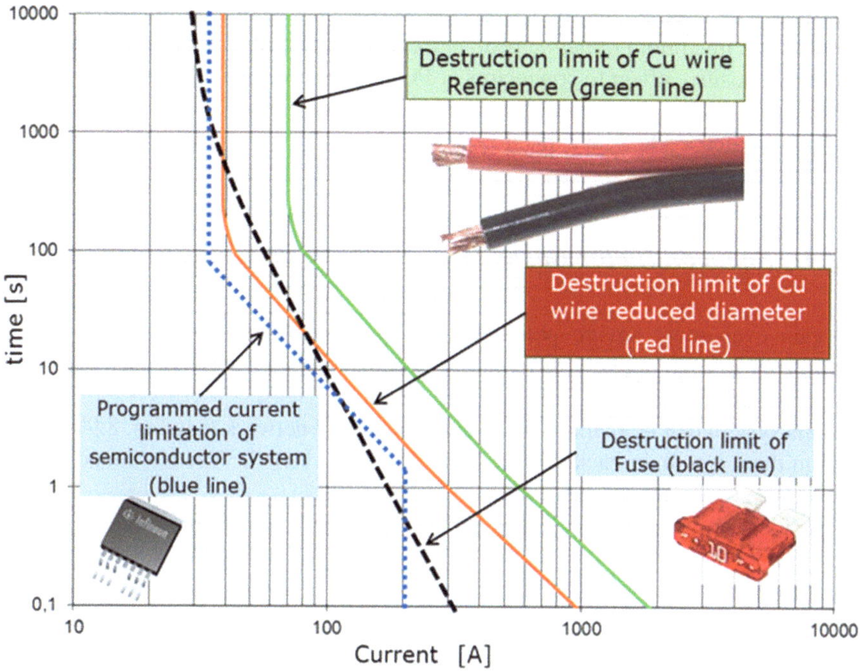

Graph 8.1 Current over time capability of cable, fuse and semiconductor

Let's have a more in-depth look on the optimal usage of wires.

The need of a certain current at the load determines what wire diameter is required and the wire determines what fuse is required. Aging, tolerance and temperature behavior of the fuse must be considered as well. To ensure safe operation the wire with its dedicated diameter must be able to handle more current over time than the fuse. With all the fuse tolerances and a wire more robust than the fuse, the safety margin of the wire compared to the load is usually quite significant.

A semiconductor power switch provides the possibility of a real time current measurement and the wire diameter can be optimized to the load profile. The safety margin of current over time of load versus wire depends on the current measurement accuracy of the semiconductor only.

Graph 8.1 shows the current over time diagram with two different wire diameters.

The black line is the current over time characteristic of the fuse. In this example the reference wire (green line characteristic) is needed to ensure a more robust wire compared to fuse. The use of a reduced wire diameter with the fuse would cause the risk of a damaged cable in case of overload at 40 up to 80 A. In this region the fuse has more current capability than the reduced wire. Using a semiconductor controlled by a microcontroller the overload protection can be programmed slightly below the maximum allowed value of the reduced wire diameter. This is an optimal use of the wire. Please refer to [3]. This saves cost and weight. For reliability reasons the absolute maximum dynamic peak current of semiconductors has to be limited. This has a positive effect to connectors or PCB traces, but also on voltage dips at the power net. On the other hand it has to be considered carefully in case high inrush currents are needed to load capacitors.

8.2.4 Semiconductor Based Decentralized Power Network

To gain even more saving potential there is the possibility to change from a centralized to a completely different decentralized power network (see Fig. 8.8).

This system can be realized as a backbone power network (see Fig. 8.9). In this system there is a main backbonefeed line across the vehicle. The backbone system provides a significant saving of wires.

For locations where power is needed there are small subsystems connected to the feed line with a low number of switches which supply the loads close to point of load. To avoid power losses the switches have to be in the low one digit mΩ area or even below. The subsystem (see Fig. 8.10) has a microcontroller and communication unit (e.g. Lin/CAN) on board.

A further new challenge is the current consumption in the on state. Some of the switches must be in on state even during parking/key off. Here the big challenge is to

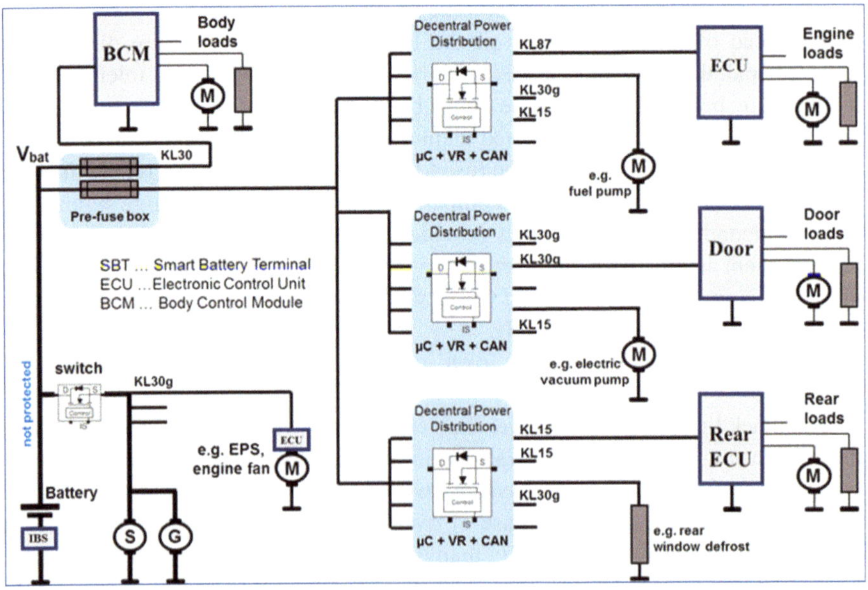

Fig. 8.8 Example of decentralized power network

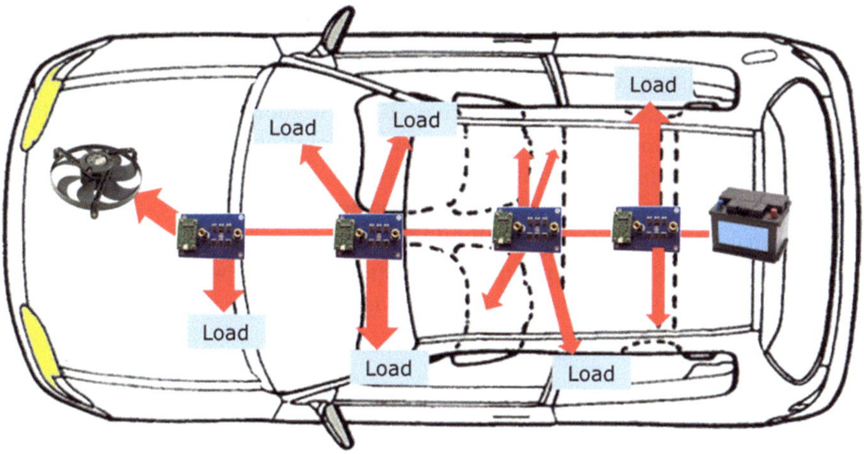

Fig. 8.9 Decentralized backbone system

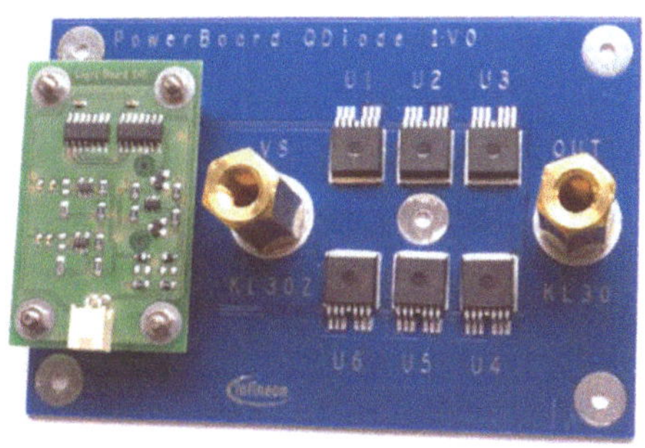

Fig. 8.10 Example of a subsystem—demonstrator by Infineon

have a very low current consumption. On vehicle level the total current consumption has to be in the mA range. For a single switch this means a current consumption of less than 30 μA. To have an optimal use of the system current sensing and wire protection has to be active during key-off as well. In off state such a current consumption is no challenge. Today's products have already a quiescent current significant lower than 1 μA typical in the off state.

At high currents the energy stored in magnetic fields of the wire inductance can be tremendous. In case of emergency switch off caused by a short circuit or overload, during turn off the semiconductor device must be able to dissipate the energy of the wire inductance. Therefore a high energy capability of the power semiconductor for single or repetitive pulses is a significant advantage. An alternative is the use of external freewheeling or suppressor diode which is additional cost to the system.

Infineon provides here the highly integrated PROFET™ switches which incorporates a broad range of smart features like diagnose and protection. The intelligent power high side switches consist of a DMOS power transistor and CMOS logic circuitry for complete built-in protection and diagnosis. They offer protection against e.g. overload, over temperature, short circuit for all kinds of automotive applications. The family provides switches with an ON resistance down to 1 mΩ which is applicable for load currents up to 40 A DC typical.

In case of higher load currents discrete MOSFETs are suitable as switches. To control MOSFETs an external driver IC is needed. One drawback of the discrete solution is the increased board space on printed circuited board. Further the implementation of diagnosis, protection and robustness is more difficult to achieve. Modern MOSFETs are optimized for low on resistance but have low repetitive energy capability in comparison to PROFET™ switches.

In case of reverse blocking requirement one additional switch at input may be required. This switch is needed only once per module.

8.2.5 Comparison of Different Components and Systems

In Tables 8.1 and 8.2 a summary of all properties at component and system level can be found.

Table 8.3 shows the description of rating:

8.3 Conclusions

For carmakers power distribution is "the" hot topic regarding power net architectures. They are forced to adapt their power network to the new requirements.

Semiconductors will be the enabler for this change. Advantages are optimal usage of wires, removal of accessibility requirements, small volume, enhanced diagnosis, low idle currents and high reliability.

Although semiconductors can be more expensive than relays and fuses at the component level, the change on system level will pay off due to savings in wiring harness, weight and power dissipation.

The main challenges for semiconductors are increased power density, precise diagnosis, high reliability, low idle/quiescent currents and acceptable costs.

Table 8.1 Comparison on component level

Component	Cost	Weight	Power losses in ON state	Peak current capability	Reliability	Diagnostic capability	Self-protected	No of switching cycles	Reverse blocking capability
Fuse	++	+	++	++	–	– –	++	++	– –
Mechanical relay	+	o	– –	++	–	– –	– –	– –	++
Mechanical bi-stable relay	o	o	++	++	–	– –	– –	– –	++
Solid state relay	o	++	+	+	++	++	+	++	– –
Solid state relay reverse blocking	–	++	+	+	++	++	+	++	++

Table 8.2 Comparison on system level

Type of power network	Total system cost	System weight	Flexibility in accessibility	Usage of wire	Wiring harness weight	Power losses in ON state	Reliability	Diagnosis capability	No of switching cycles	Reverse blocking capability
Classical centralized mechanical relays + fuses	+	–	––	––	––	––	––	––	––	++
Enhanced classical centr. mechanical and solid state relays + fuses	+	–	––	––	––	–	–	––	–	o
Semiconductor based centr. mechanical and SS relays	+	+	+	++	+	+	++	++	++	o
Semiconductor based decentralized SS relays only	++	++	++	++	++	+	++	++	++	o

Table 8.3 Description of rating

Symbol	++	+	o	–	––
Rating	Very good	Good	Neutral	Poor	Very poor

References

1. [European Environment Agency – Paper: CO2 emissions performance of car manufacturers in 2012] Regulation (EC) 443/2009. To be found on www.eea.europa.eu
2. Journal ATZ Elektronik issued 3/2013. Architekturstudie zur Leistungsverteilung im KFZ. Dr. Alfons Graf and Dipl.-Ing. Jürgen Kositza Infineon Technologies, Dipl.-Ing. Tobias Raithel Daimler AG
3. 8th International Conference Vehicle Electronics: Fuse Replacement with Smart Power Semiconductors. Dr. Alfons Graf, Hannes Estl

Chapter 9
High-Voltage Fast-Switching Gate Drivers

Bernhard Wicht, Jürgen Wittmann, Achim Seidel, and Alexis Schindler

9.1 Introduction

Gate drivers are key circuit blocks with growing importance for various automotive applications like fast switching DC-DC conversion and motor bridge drivers. New challenges arise from the introduction of the 48 V automotive board net and growing e-mobility with battery voltages in the range from 12 to 400 V. Increasing voltage requirements have influence on circuit design for power train control, power conversion between high- and low-voltage batteries, safety electronics and more. At the same time, the growing amount of electronics in cars demands for compact and cost efficient solutions with a high integration level.

More than hundred switched-mode converters with an output power of typically less than 10 W are installed in a car, for example. Converters have to be small in size and low in cost. This is achieved by increasing the switching frequency to the multi-MHz range, as the passive filter components of switching converters scale down. This results in very small on-times of the power switch [1].

Compact high-voltage converters demand for highly integrated gate drivers with galvanic isolation at even faster transition slopes. With wide bandgap semiconductors, such as GaN or SiC, converters with switching frequencies in the MHz-range and voltage slopes of >100V/ns are already achieved. This raises challenges also for the gate driver design, including compact layout with low parasitic inductances as well as robust transmission of gate driver control signals over a galvanic isolation barrier with small and matched propagation times [2].

With increasing switching frequency and steeper switching edges, electromagnetic compatibility (EMC) is a major concern in automotive. Gate drivers with slope control optimize EMC while maintaining good switching efficiency.

B. Wicht (✉) • J. Wittmann • A. Seidel • A. Schindler
Robert Bosch Center for Power Electronics, Reutlingen University, Reutlingen, Germany
e-mail: bernhard.wicht@reutlingen-university.de

© Springer International Publishing Switzerland 2017

A. Baschirotto et al. (eds.), *Wideband Continuous-time ΣΔ ADCs, Automotive Electronics, and Power Management*, DOI 10.1007/978-3-319-41670-0_9

After a brief overview of gate driver fundamentals in Sect. 9.2, Sect. 9.3 covers fast-switching drivers for 12 V and 48 V board net applications. Highly integrated gate drivers, suitable for high-voltage applications up to 400 V and above, are addressed in Sect. 9.4. The emphasis of Sect. 9.5 is on EMC and EMC optimized gate drivers.

9.2 Gate Driver Fundamentals

Driver Configurations and Building Blocks Figure 9.1 shows the fundamental power switch and driver configurations. Due to the lower on-resistance, the majority of power switches are n-type devices. This is in particular true in automotive applications. Depending on the location of the power switch, a low-side switch (Fig. 9.1a) and a high-side switch (Fig. 9.1b) can be distinguished. Both switches together form a half-bridge (Fig. 9.1c) and two half-bridges a full-bridge, also called H-bridge. Figure 9.1c indicates possible typical applications, an inductive DCDC converter and a DC motor driver, respectively.

A complete driver design includes several circuit blocks, some are highlighted in Fig. 9.1a, b. The control logic delivers the turn-on or turn-off signal for the power stage, usually in a pulse width modulated fashion. A level shifter converts the driver control signal from the low-voltage domain into the driver voltage domain. In a high-side driver, the control signal is shifted up with reference to the input supply voltage V_{IN} of the power stage. The high-side driver may include a galvanic isolation for safety reasons (in case of high-voltage) and as part of the signal transfer. The gate supply provides the gate overdrive voltage V_{drv}. Many gate drivers utilize a linear regulator or a shunt regulator from V_{IN} to generate V_{drv}. For an n-type high-side power transistor, a high-side gate supply of $(V_{IN} + V_{drv})$ needs to be available in order to keep the power transistor turned on. This can be accomplished with a bootstrap supply or a charge pump. If the power stage is configured as a half-bridge, a dead-time control is mandatory to avoid damage at the power stage due to cross conduction.

Gate Driver Operation and Key Characteristics The gate driver has to turn on a power transistor with sufficient gate overdrive voltage V_{drv}. Figure 9.2a shows

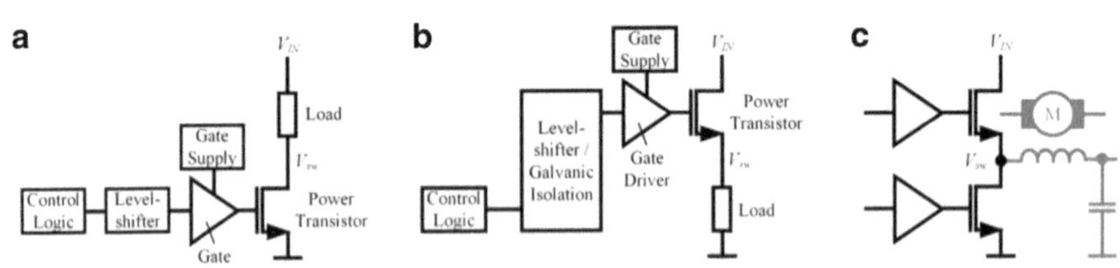

Fig. 9.1 Power stage and driver configurations: (**a**) low-side driver; (**b**) high-side driver; (**c**) half-bridge

the circuit principle of a low-side gate driver. To turn-on the power transistor, its gate gets connected to V_{drv}, while its gate is pulled to ground (or more precisely to the source potential of the power transistor) to turn it off, each with a finite on-resistance R_{drv} (which may be different in each path). Many applications use a CMOS inverter as a gate driver, especially for on-chip power stages with drive voltages $V_{drv} \leq 5$ V. However most discrete power transistors require $V_{drv} > 5$ V. PMOS devices in this voltage class, suitable for a driver design, are usually not area-efficient [3]. In some high-voltage technologies the PMOS transistor has up to 30 times larger resistance compared to the NMOS transistor. Therefore, an output stage with two NMOS transistors can be used [3–5]. A bootstrap circuit is required to provide the gate overdrive voltage for the NMOS transistor in the pull-up path. This will be covered in more detail in Sect. 9.4. While such configurations can be classified as hard switching drivers, few applications use current mode drivers, which essentially represent a current source output. Such stages are useful for EMC optimized drivers as discussed in Sect. 9.5.

The power transistor represents a capacitive load equal to the equivalent gate capacitance C_{gate} (see Fig. 9.2a). The larger the gate current I_{gate}, the faster the rate of change of the gate voltage V_{gate}. The dc characteristics in Fig. 9.2b relates the gate voltage V_{gate} to the current I_{gate}. While hard switching drivers are limited by

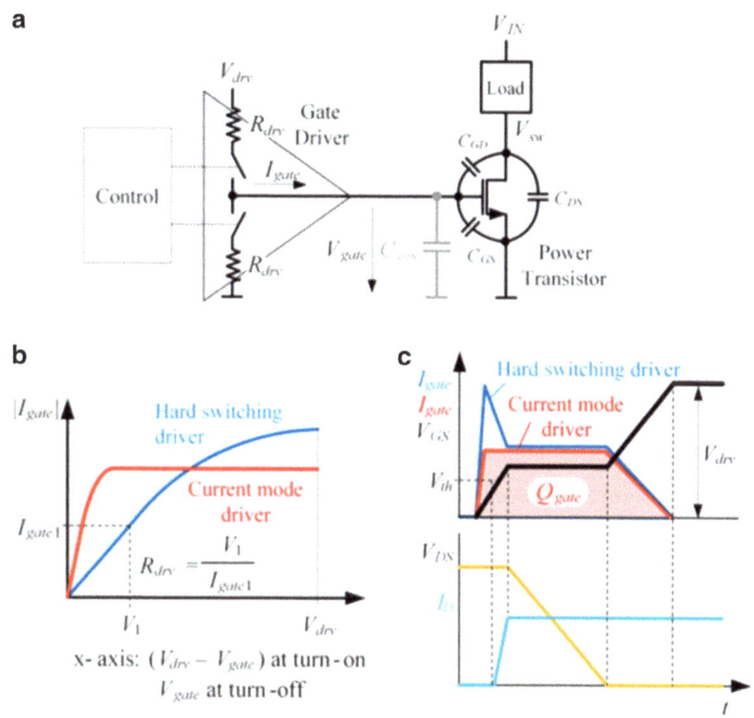

Fig. 9.2 Low-side gate driver: (**a**) equivalent circuit; (**b**) dc and (**c**) transient characteristics

their finite on-resistance R_{drv}, current source drivers provide a nearly constant gate current over a wide range of V_{gate}. If the x-axis is chosen as indicated in the graph, the curve is valid for both the turn-on and turn-off case.

The transition starts at $I_{gate} = 0$, reaches a maximum current after some delay and goes back to $I_{gate} = 0$ after the transition has finished. This is illustrated in Fig. 9.2c for the two driver types. The absence of the current over-shoot represents a major advantage of the current mode driver with respect to EMC. This will be further explored in Sect. 9.5.

Optimization for Speed, Power, Area Fast switching with low propagation delay can be achieved by cascaded drivers. This approach has been used originally for CMOS digital drivers to drive large off-chip (capacitive) loads, e.g. for I/O buffers. The driver is designed such that the W/L-ratio of the transistors increases from stage to stage. The most common approach is based on a fixed scaling factor α [6, 7]. A minimum delay can be achieved by optimizing the scaling factor α and the number of stages n with the general relationship expressed by

$$\alpha = \sqrt[n]{\frac{C_{gate}}{C_{inv}}}, \quad n = \frac{\ln \frac{C_{gate}}{C_{inv}}}{\ln \alpha} \quad . \tag{9.1}$$

While the model in [7] proposes an ideal scaling factor of $\alpha = \mathrm{e} \approx 2.71$, practical designs achieve a minimum delay for typical values of $\alpha = 3 \ldots 6$.

The optimization for minimum delay does not necessarily correspond to minimum power loss in the driver. The shoot-through current and dynamic gate charge losses in each inverter stage contribute to the overall power losses. Loss optimization typically results in less number of stages (larger scaling factor) compared to speed optimization. Usually power optimization comes along with reduced layout area, hence losses and area can be minimized concurrently.

dv/dt Robustness Once the driver is optimized for speed and/or power, there is one more design goal to be fulfilled. The driver needs to be strong enough to be robust against dv/dt transients. Referring to Fig. 9.2a, if the power transistor is turned off, the switching node transient causes a capacitive charging of C_{GD}. The pull-down path of the driver needs to be strong enough, so that the gate-source voltage of the power transistor does not reach the threshold voltage, i.e. $R_{drv} \leq V_{th} / \left(C_{GD} \frac{\partial V_{DS}}{\partial t} \right)$. If this condition requires to increase the driving strength of the final stage (i.e. w_{n-1}), the scaling factor α recalculates to $\alpha = \sqrt[n-1]{\frac{W_{n-1}}{W_1}}$ with the width W_1 of the first stage inverter.

9.3 Fast Switching Drivers

High-speed requirements for gate drivers are often derived from low-power switching converters in automotive systems. One battery supplies a large amount of applications like electrical motor control, safety, infotainment, lighting. Supply wires are often long and converters are implemented in the applications at the point of load to supply micro-controllers, control circuits, interfaces and sensors. Converters have to be small in size and low in cost. This is achieved by increasing the switching frequency to the multi-MHz range, as the passive filter components of switching converters scale down.

9.3.1 Fast Switching Gate Driver Stage

A conversion from 50 to 5 V at a switching frequency of 10 MHz requires an on-time of the switch of less than 10 ns with rise/fall times in the range of 1 ns. This also minimizes frequency dependent transition losses. The on-time pulse has to be transferred properly to the power switches by the level shifter and gate driver. The whole driver and in particular the level shifter have to be robust against coupling currents caused by fast switching transients. A cascaded gate driver as described in Sect. 9.2 is suitable to achieve these high-speed requirements. An implementation of a high-speed gate driver is shown in Fig. 9.3.

For $C_{inv} = 7\,\text{pF}$ and $C_{gate} = 145\,\text{pF}$ the driver in Fig. 9.3 provides a good trade-off between die size and propagation delay for $n = 6$ inverter stages with a scaling factor of $\alpha = 5.24$ per stage [8]. To improve dynamic losses, which become dominant towards higher switching frequencies, cross currents have to be avoided, in particular in the last driver stage. This is accomplished by splitting up the driver stage into two branches, as shown in Fig. 9.3. This allows to control the PMOS and NMOS transistor of the last driver stage separately. There is no area penalty, since each branch has to drive approximately half of the capacitive load. Cross currents in the last stage are eliminated by asymmetric sizing within each

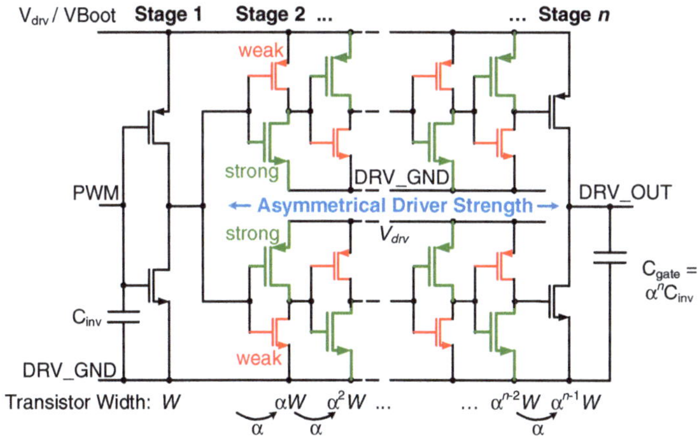

Fig. 9.3 Cascaded gate driver with asymmetrically sized inverter stages, optimized for propagation delay and current consumption

inverter stage [1, 9]. This assures that the PMOS in the last driver stage is always turned off before the NMOS turns on and vice versa. An asymmetry factor of 20 % is used in this design, i.e., the width of the strong transistors is increased and the width of the weak transistors is decreased by 20 % with respect to the nominal value (based on Eq. (9.1)). This way, the losses in gate drivers can be reduced by up to 25 %, confirmed by simulations.

9.3.2 Comparison of NMOS and PMOS Power Switch

Figure 9.1 has shown the use of an NMOS transistor as a high-side switch, which results in lower on-resistance and lower area consumption compared to a PMOS switch. On the other hand, its gate supply based on a charge pump or bootstrap circuit requires a higher effort and adds more complexity. The PMOS power switch generates also less switching noise, because its gate voltage receives a swing in between two constant voltages, *VIN* and $HSGND = VIN - 5\,V$, which is typically generated by an on-chip linear regulator. The main requirement for the PMOS level shifter is a fast switching speed and to propagate short on-time pulses of a few ns. This can be achieved with current mirror based [1, 10] or capacitive level shifters [11].

For an NMOS high-side transistor as shown in Fig. 9.1b, the source of the transistor is at the switching node *VSW*, which becomes also the ground reference of the gate driver and the high-side portion of the level shifter. The gate supply *VBoot* is above *VSW* by V_{drv} (typically 5 V) to safely turn on the NMOS transistor if $VSW \rightarrow VIN$. *VBoot* is generated by a bootstrap circuit [12] or by a charge pump. Figure 9.4 shows how *VBoot* follows the switching transition at *VSW*. Slopes larger than 20 V/ns charge/discharge the parasitic capacitances of the high-side isolation, resulting in large peak currents. Coupling currents in the mA range get superimposed to the level shifter signal currents, which are in the lower μA range. Existing level shifter concepts for PMOS transistors cannot directly be utilized for NMOS transistors, as they are sensitive to in-coupling and would cause faulty switching during fast high-side transitions. They would also not be fast enough for multi-MHz operation, as will be discussed in the next section.

9.3.3 Conventional Level Shifters

Several fast switching level shifters for NMOS power transistors have been published [13, 14], which achieve a very fast propagation delay, but a transfer of short PWM on-time pulses in the low nanoseconds range is not supported. In conventional level shifter concepts, the signal is typically transferred by a high-voltage switch on the low-side, which creates a voltage drop on the high-side across a resistor or a current source, as shown in Fig. 9.5. This voltage drop is detected on the high-side

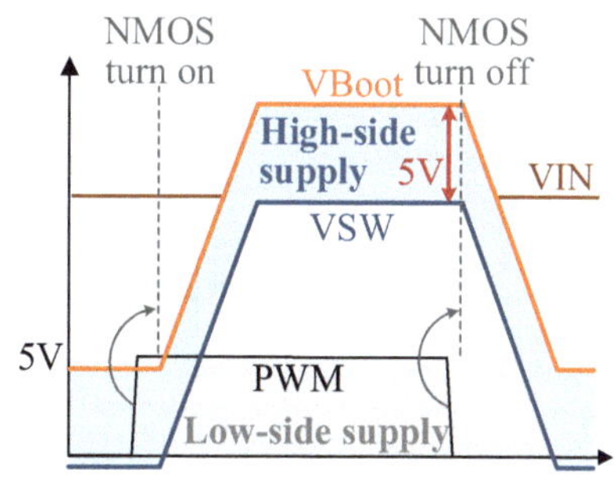

Fig. 9.4 High-side voltage domain for controlling NMOS power transistors with a floating high-side

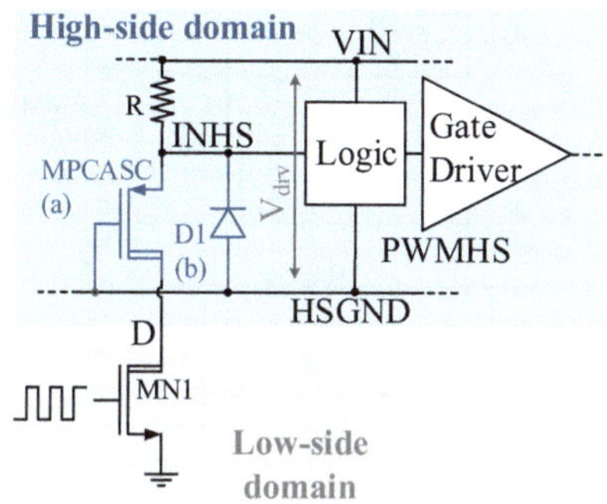

Fig. 9.5 Conventional level shifter with PMOS cascode (**a**) and diode clamping (**b**)

by a digital circuit, e.g. a latch. As the high-side circuits use low-voltage transistors, the voltage drop across the resistor or current source has to be clamped to not violate the maximum ratings. Figure 9.5 shows two ways of commonly used clamps, based on a cascode (a) and diode (b), respectively. Both clamping components need to be low resistive and, hence, add significant parasitic capacitance to the *INHS* node. The charging time is in the range of 20 ns to >100 ns in a 180 nm BiCMOS technology, for instance. Pulses within this time are completely filtered. Capacitive level shifters [15] provide an alternative option for fast signal transfer. However, they also need some kind of clamping and suffer from large parasitic coupling currents during high-side transitions. In conclusion, the speed of conventional level shifters is too low for multi-MHz switching.

9.3.4 High-Speed Level Shifters

For the control of a PMOS power transistor, the level shifter depicted in Fig. 9.6a is suitable and provides several advantages [1, 8, 16]. A symmetrical single-stage amplifier with its output stage referred to *HSGND* generates a rail-to-rail control signal *PWMHS* on the high-side. To achieve the required delay, a small 1 ţA pre-bias current keeps the bias voltages of the current mirrors above the threshold, reducing the voltage swing during switching significantly. This way, a total propagation delay in the level shifter in the range of 3 ns is achieved.

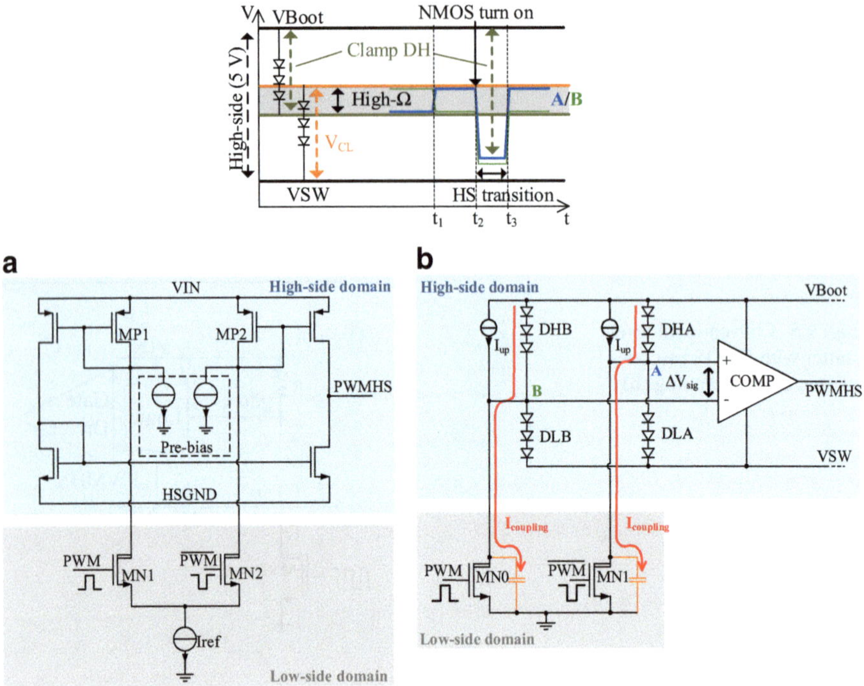

Fig. 9.6 High-speed level shifters (**a**) for control of a PMOS power transistor, and (**b**) control of an NMOS power transistor at floating high-side

Figure 9.6b shows the level shifter implementation [17], which provides a fast and robust control of a power stage with NMOS high-side transistor. High-voltage switches MN0 and MN1 create a differential voltage ΔV_{sig} between nodes A and B, which is detected by a high-speed comparator. The differential voltage range at the nodes A and B is limited by clamping diodes as shown in Fig. 9.6b. If the low-side switch MN0 or MN1 is turned on, the according node A or B is held at the lower boundary by the clamps DHA or DHB, respectively (see signals in Fig. 9.6b). Accordingly, when the low-side switch is turned off, the clamps DLA or DLB hold the voltage at A or B at the upper voltage boundary, while the clamp conducts the active load current I_{up}. The clamping voltages are designed such that none of the diode stacks is conducting in between the upper and lower voltage boundary. A high-impedance region is obtained, in which the differential signal between A and B can change its state at very high speed.

After the state change of *PWM*, the NMOS power transistor is turned on or off by the gate driver. A high-side transition of the switching node follows, in which the voltage rails *VBoot* and *VSW* experience a voltage step up to *VIN* with a slope of 20 V/ns and faster. The parasitic capacitances at the drain of MN0 and MN1 are charged. Even for small parasitic capacitances, coupling currents $I_{coupling}$ in the range of 2 mA are generated, which are superimposed to the signal currents in the range of 50 μA. At rising high-side transitions, the forward voltages across the clamps DHA and DHB cause the voltages at nodes *A* and *B* to fall significantly below the lower boundary of the high-impedance region. The overlapping diode clamps are designed such that the high-impedance region typically has a range of a few hundred millivolts and occurs in the center between *VBoot* and *VSW*. This allows the nodes *A* and *B* to drop by nearly two volts below the high-impedance region. With a large common mode input range of the comparator, a very large forward voltage across the clamps can be tolerated. Thus, the clamps can be designed small (low capacitance) to handle the full coupling currents. This is a major advantage, compared to clamps of Fig. 9.5, which occupy large area (hence, large capacitance). At falling high-side transitions, the coupling currents are clamped by DLA and DLB to *HSGND*, accordingly.

This concept allows to transfer PWM signals with on-time pulses smaller than 3 ns to the high-side with high robustness against the fast high-side transitions. Thus, the level shifter is suitable to operate at frequencies >10 MHz at large input voltage and high conversion ratios $VIN/VOUT$.

Besides the coupling into the signal path, large disturbances in sub-circuits on the entire IC can occur by coupling currents into the substrate, caused by parasitic capacitances of the high-side isolation well during fast switching transients. The high-side isolation (typically implemented as n-well on a p-substrate) is required to isolate the level shifter and gate driver from the low-voltage circuits on the substrate. To effectively dissipate the coupling currents, which can be in the range of hundreds of milliamperes even at switching transients below 10 V/ns, dedicated diverting structures have to be placed on the substrate. Back-side metalization, conducting trenches or p-guardrings are suitable [18].

9.4 High-Voltage Drivers

Gate drivers are essential for automotive applications in electric vehicles, running from voltages above 300 V. Figure 9.7 shows a typical setup of a high-voltage gate driver and power transistor. The area-efficient implementation of fully integrated gate drivers is discussed in Sect. 9.4.1. Section 9.4.2 covers galvanic isolation between the low-side driver control circuit and the high-side driver.

9.4.1 Driver Output Stage

Typical gate drive voltages of high-voltage power transistors are above 10V. In fully integrated gate drivers, a large portion of die area is occupied by the driver output stage, consisting of two or more transistors with voltage ratings for 15 V and above. NMOS transistors are preferred because of their lower RDSon per area, but they require a bootstrap supply for the gate overdrive. Placing a PMOS transistor in parallel to the NMOS pull-up transistor in the driver output avoids the need for a bootstrap supply. However, a parallel PMOS transistor still occupies significant area, as in most technologies only high-voltage NMOS devices (LDMOS) are optimized for high current density. Dependent on technology, an output stage with two NMOS transistors provides an area-efficient solution, utilizing the effect of high-voltage charge storing as part of an on-chip bootstrap circuit [19, 20]. The operation of a conventional bootstrap circuit is shown in Fig. 9.8a1. For simplified explanations, the diode forward voltages are assumed to be 0 V. If $IN = 0$ V and the level shifter output signal *LvlSh_OUT* is 'low' the driver is turned off and the node *OUT* is shorted to ground by *MN*2. *CBoot* is charged to 5 V by the supply voltage *V*1. By setting *IN* to high, *MN*2 turns off, and *MN*1 is switched on by *LvlSh_OUT* and the buffer *B*2. The *OUT* node rises to *Vdrv* while *DBoot* prevents *CBoot* from discharging to *V*1. *VCBoot* serves as floating voltage supply and has to provide the charge Q_{tot} for *B*2, the gate capacitance of *MN*1 and the level shifter. The gate charge for the actual power transistor is provided separately by *Vdrv*. A typical value for *Vdrv* is 15 V. The charge that is available from *CBoot* to achieve a voltage dip V_{dip} at *VCBoot* can be calculated as

$$Q_{tot} = CBoot \cdot V_{dip}. \tag{9.2}$$

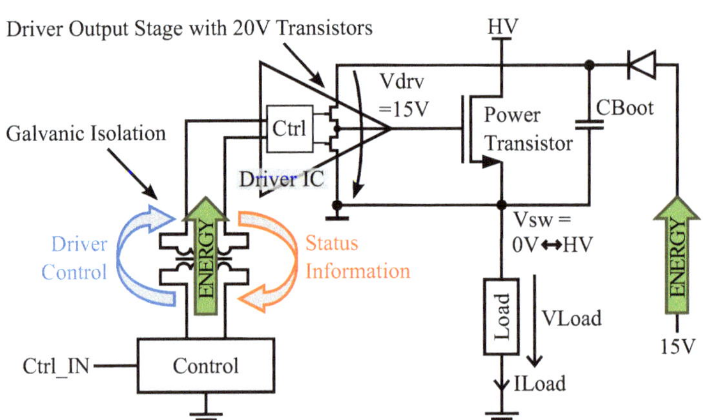

Fig. 9.7 Gate driver and power transistor

The MOSFET buffers $B1$ and $B2$ in Fig. 9.8 are tapered buffer stages similar to Fig. 9.3 as discussed in Sects. 9.1 and 9.3. They have a typical undervoltage level of 4–4.5 V [21], assuming a nominal gate-source voltage of 5 V for $MN1$, $MN2$. An upper voltage limit is given by its breakdown voltage which is 5.6 V in this case. Hence a bootstrap capacitor charged to 5 V can only be discharged by $Vdip = 0.5-1$ V. It has to be observed that in a real circuit $V1$ must be ~5.7 V to compensate the forward voltage of $DBoot$. The conventional bootstrap circuit suffers from small charge allocation in respect to the whole stored charge in the bootstrap capacitor. This represents a significant area limitation for an on-chip bootstrap capacitor.

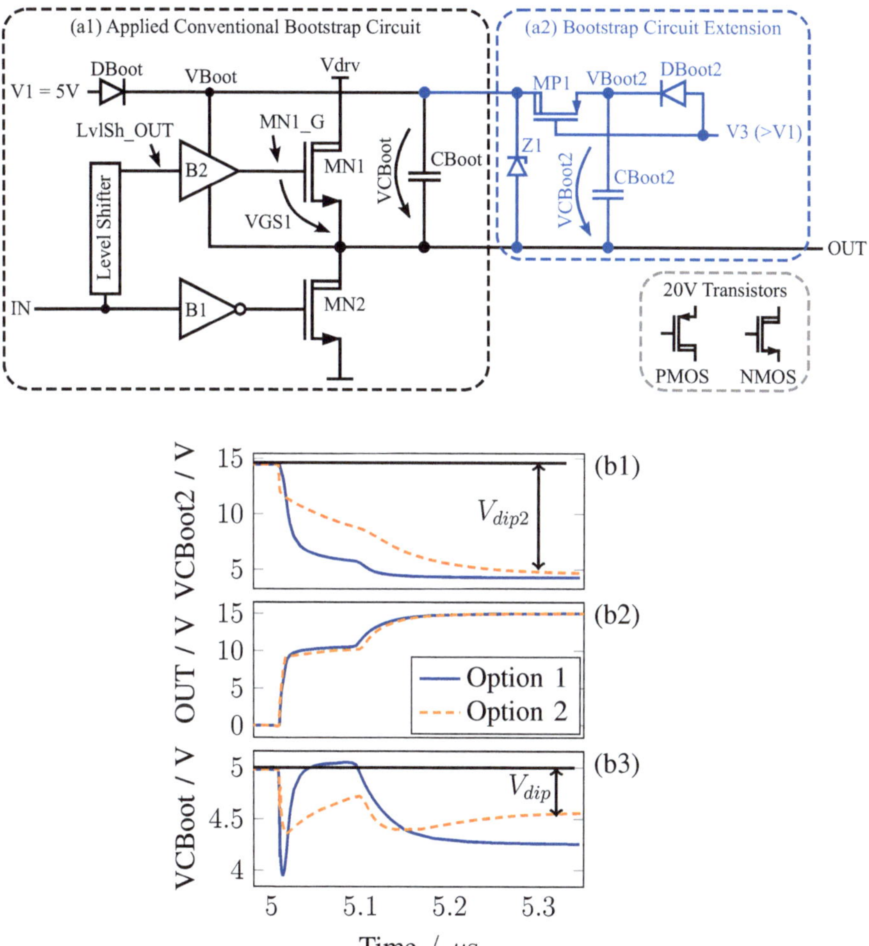

Fig. 9.8 Two NMOS transistor output stage buffer with (**a1**) conventional bootstrap circuit and extended by (**a2**) bootstrap circuit option 1. (**b**) Transient voltage signals according to the bootstrap circuit option 1 and option 2

The bootstrap concept in Fig. 9.8a2 [19, 20], is an extension of the conventional bootstrap circuit. Figure 9.8b (option 1) shows the corresponding simulated transient voltage signals. A second bootstrap capacitor *CBoot*2 is charged by *V*3, a higher voltage than *V*1, e.g., 15 V, Fig. 9.8b1. For *V*3 the same voltage as *Vdrv* could be used. If the output node *OUT* rises, MP1 conducts as *VBoot*2 exceeds *V*3 by more than a threshold voltage of MP1, Fig. 9.8b2. *DBoot*2 blocks the current from *CBoot*2 to *V*3 and *CBoot*2 automatically discharges to *CBoot*, to the gate node of *MN*1 and to the circuits supplied by *VCBoot* (Fig. 9.8b1 and b3) through *MP*1. A charge balance between *CBoot*, *CBoot*2 and any additional load capacitance (mainly the gate capacitance of *MN*1) occurs. In first order, *CBoot*2 discharges from a value of *V*3 to *V*1, e.g., from 15 to 5 V. *Z*1 protects the circuit at *CBoot* against overvoltage in case of failure, like an overcharged capacitor *CBoot*2. Since the large voltage swing V_{dip2} (Fig. 9.8b1) results in a high amount of charge (Eq. (9.2)) a significantly smaller bootstrap capacitor *CBoot* can be used. Area is saved, even with the addition of capacitor *CBoot*2.

A disadvantage of the circuit in Fig. 9.8 is that MP1 turns on after *OUT* rises. Before this, the required charge comes from *CBoot*, leading to a short voltage dip at *VCBoot* until *CBoot* is recharged from *CBoot*2, Fig. 9.8b3. A voltage dip larger than specified can influence circuit blocks supplied from *VCBoot*, such as faulty switching in the level shifter. In addition, the circuit of Fig. 9.8a2 requires a relatively large PMOS transistor *MP*1.

A second option of the driver output stage, shown in Fig. 9.9, solves these disadvantages (circuit option 3 in [19]). The corresponding transient voltage signals are shown in Fig. 9.8b (Option 2). *MN*3 directly connects *CBoot*2 to the gate node of *MN*1 after *MN*4 turns off, both controlled by the signal *Lvlsh_OUT* via the buffers *B*3 and *B2b*. Before the driver output voltage *OUT* rises, the charge for the gate node *MN*1_*G* is supplied directly from *V*3 without discharging *CBoot*2 and *CBoot*. This increases the voltage stability of *VCBoot* and reduces the total required charge from *CBoot*2 and *CBoot*, resulting in a smaller voltage dip *Vdip* in the end of the switching phase, Fig. 9.8b3. When *VGS*1 reaches the turn-off level of gate *NAND*1, *MP*4 turns on and *MN*1_*G* is fully charged by *CBoot*. While *VGS*1 rises, the gate-source voltage of *MN*3 decreases and is finally turned-off. *CBoot*2 is still connected to *CBoot* by *MP*1. The high-voltage PMOS transistor *MP*1 is optional, because the low-voltage transistor *MP*4 keeps *MN*1 in its on-state. Nevertheless, *MP*1 is advantageous for recharging *CBoot* after driver turn-on, Fig. 9.8b3.

Sizing guidelines for *CBoot* and *CBoot*2 are given in [19], considering worst case of operation and process corner. For circuit option 2, *CBoot* is calculated with $\sim$76 pF and *CBoot*2 with 19 pF to get a voltage dip < 1 V at *VCBoot*.

A comparison to a conventional bootstrap circuit can be based on Eq. (9.2). With $Q_{tot,max} = 246$ pC and $V_{dip,max} = 1$ V, *CBoot_conv* results in 246 pF. Considering the same worst case parameters as used for the calculations of *CBoot* and *CBoot*2, *CBoot_conv* results in 324 pF. A decrease of the overall capacitor area by about 70 % can be achieved in case that *CBoot*2 can be placed on top of *CBoot*, while *CBoot*2 can be implemented as high-voltage metal-metal capacitor and *CBoot* as a low-voltage poly-nwell capacitor. The metal-metal capacitor is assumed to have

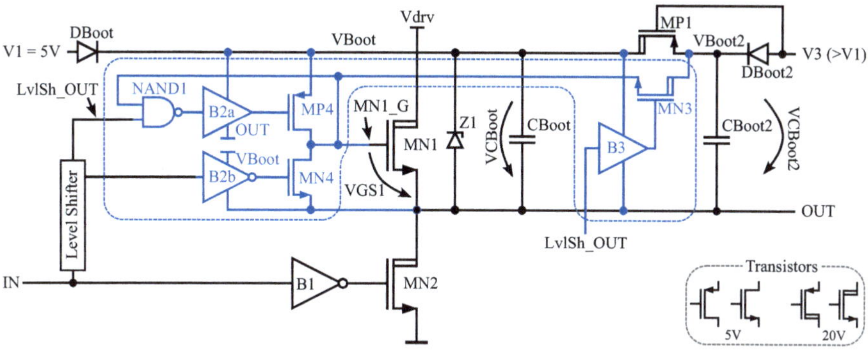

Fig. 9.9 Option 2 of the bootstrap circuit with an NMOS transistor connecting directly to the gate of MN1

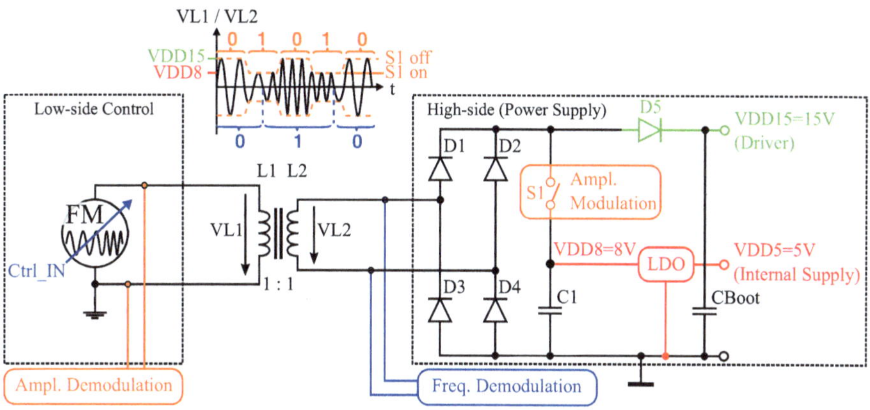

Fig. 9.10 Implementation of the transmission concept

a capacitance density of 25 % of a poly-nwell capacitor. For area comparison *CBoot_conv* was assumed to be a stacked capacitor, consisting of a metal-metal and poly-nwell capacitor, as well. Even if *CBoot*2 is placed next to *CBoot*, a decrease of >50 % in area is achieved, compared to a non-stacked conventional capacitor *CBoot_conv*.

9.4.2 Galvanic Isolation with Signal and Energy Transfer

For safety or robustness reasons, galvanic isolated gate drivers are required. There are mainly four physical ways to supply gate drivers with energy and control signals over the isolation barrier: (1) Inductively, with discrete or integrated transformers, (2) optically, with opto couplers or fibre optics, (3) electrically, with capacitors and

(4) mechanically, with piezo elements. For applications in a power range $< 100\,\text{kW}$, transformers and opto couplers are widely used [22]. Opto couplers provide a signal, but no energy transmission, and they have relatively large propagation times [23]. Transformers have the possibility for energy transfer and signal transmission. Separate transformers are typically used for each signal and energy channel [22, 24]. In [25], high performance energy and unidirectional signal transmission is realized by microwave circuits with high complexity and cost. An unidirectional capacitive signal transmission is proposed in [15] with parallel energy transfer via a transformer. With the goal to further reduce the size and component count, the approach in [26] uses the existing signal transformer for energy transfer in addition to a conventional bootstrap gate supply, as indicated in Fig. 9.7. The additional energy transfer eliminates the duty cycle limitation of conventional bootstrapping, as the driver does not need to be switched off periodically for bootstrap recharge. While bootstrapping is the main supply for high current peaks during the driver switching phase (gate charge of the driven power switch), the transfer via the transformer provides the energy to supply the high-side driver electronics and to compensate any leakage currents once the driver is turned on.

Figure 9.10 shows the implementation of energy and bidirectional signal transmission according to [26]. The corresponding voltage signals from a measurement are shown in Fig. 9.11. By alternating switch $S1$, the transferred energy of $VL1$ is distributed to two high-side supply rails, $VDD15 = 15\,\text{V}$ (for the gate overdrive of the external power transistor) and $VDD8 = 8\,\text{V}$ (for high-side circuit blocks), Fig. 9.10. The driver control signal *Ctrl_IN* from low to high-side is transmitted by a frequency modulated (FM) signal ($VL1$, 10/20 MHz), generated by a resonance circuit on the low-side. On the secondary side the signal is demodulated by a frequency demodulation circuit for driver control, switching a high-side power transistor with load (*VLoad*). As a key function, the power distribution switch $S1$ is utilized also for amplitude modulation (*AM_IN*) for a backward signal transmission, Fig. 9.10. If $S1$ is turned-off, $VL1$ oscillates with an amplitude of up to two times of $VDD15$. If $S1$ is turned-on, the amplitude is clamped to an amplitude of nearly two times of $VDD8$. The switching frequency of $S1$ must be chosen significantly lower than the lowest frequency of the FM signal (e.g. 1 MHz $<<$ 10 MHz). Alternating $S1$ according to the signal *AM_IN* enables a very energy efficient and well detectable modulation (*AM_recovered*) in combination with the energy supply. This is an advantage over conventional load modulation, which demands a trade-off between

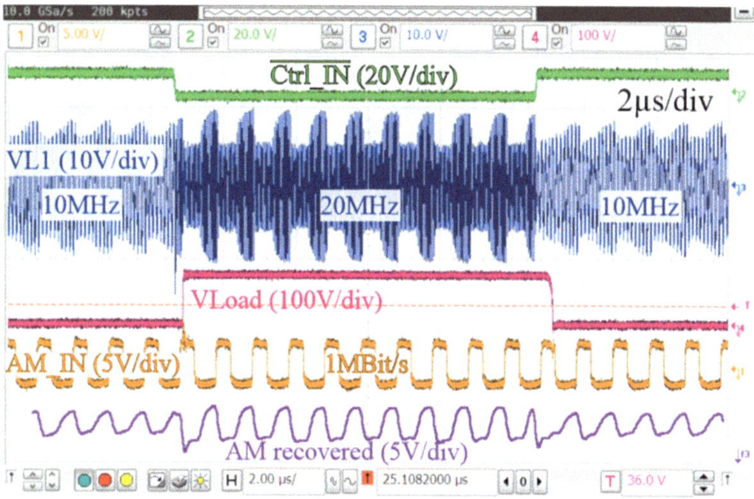

Fig. 9.11 Measured signal transmission over the transformer with a test setup shown in Fig. 9.7 ($Iload = 1.2\,\text{A}$, $HV = 100\,\text{V}$)

power efficiency due to power loss in the load resistor and good detectability of the modulated signals [27, 28].

For very fast switching applications, enabled by new fast switching devices like GaN devices, a short propagation delay and very good delay matching become increasingly important. Both parameters can be significantly improved by pushing the carrier frequency of the FM signal to higher frequencies or by applying pulsed signal transmission concepts.

9.5 EMC and Switching Losses

With the increase in switching speed, a reduction in switching losses and in PCB area can be achieved. At the same time, the electromagnetic emissions (EME) are vastly increased because of the fast transition speeds in the power switches. In the following subsections the EMC influencing factors in gate driver design and a trade-off between switching losses and EMC are discussed [29].

9.5.1 *EMC Influencing Factors*

The EME of a half-bridge circuit can be approximated with the fourier transform of a trapezoidal signal, Fig. 9.12a. This is shown in Eq. (9.3) with the simplification that the risetime τ_r is equal to the falltime τ_f [30].

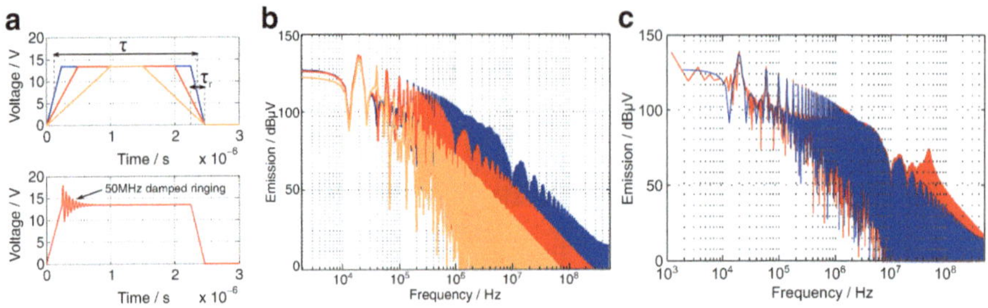

Fig. 9.12 (**a**) *Top*: Trapezoid signal with different rise times *Bottom*: Trapezoidal signal with 50 MHz ringing. (**b**) Spectral influence of different rise times of the trapezoidal signal. (**c**) Spectral influence of the sinusoidal overshoot

$$E_{dB} = 20\log\left(2A\frac{\tau}{T}\right) + 20\log\left|\frac{\sin(\pi\tau f)}{\pi\tau f}\right| + 20\log\left|\frac{\sin(\pi\tau_r f)}{\pi\tau_r f}\right| \tag{9.3}$$

The first term shows the DC magnitude, which is mainly influenced by the amplitude A of the signal and the on-time τ of the pulse width modulated (PWM) switching signal. The second term marks the first breakpoint, at which the energy contained in the signal will begin to fall with −20 dB/decade. According to (9.3) the second term is set by the on-time τ of the trapezoidal signal. As the amplitude A, the on-time τ and the PWM frequency f are given by the application, the first and the second term in (9.3) cannot be influenced by the gate driver. Only the third term remains to reduce the EME of the voltage transition. This term shows the second frequency breakpoint at which the signal begins to decrease with −40 dB/decade. To reduce the EME, the risetime τ_r of the trapezoidal signal has to be increased by reducing the gate current of the active MOSFET. This is shown by simulation in Fig. 9.12b. The result is a lower emission in the higher frequency range. An often not considered factor is the influence of the current transition on the EME spectrum. In a bridge setup, the commutation current and the parasitic inductances of the printed circuit board traces, interconnects, packaging and especially the bond wires of the semiconductors cause voltage overshoots and ringing, Fig. 9.12a. These effects are superimposed to the trapezoidal shaped voltage signal. This changes the EME at the ringing frequency of the signal as demonstrated in Fig. 9.12c by a Matlab®simulation (at 50 MHz ringing frequency in this example). It can be concluded that there are two main levers to optimize the EMC performance of a half bridge setup with the gate driver. With the voltage transition, the spectrum can be optimized for all frequencies above the second break point of the trapezoidal signal. By reducing the speed of the current transition, it is possible to reduce EME caused by the ringing effects of the commutating load current. Both measures take effect in the higher frequency range. The EME in the lower frequency ranges has to be reduced by effective filter circuits, because it is not possible to reduce them with the gate driver.

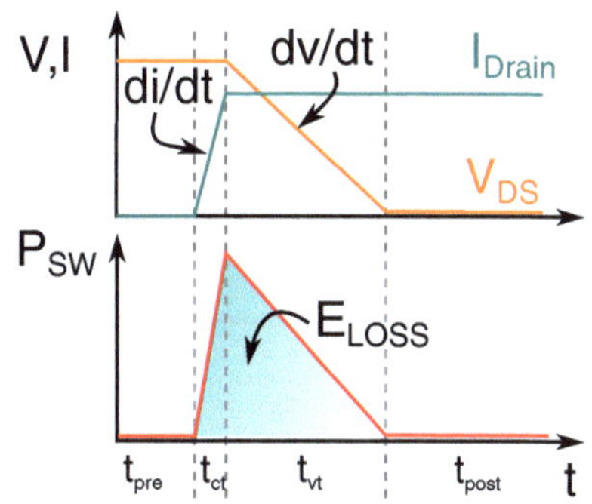

Fig. 9.13 Switching transitions for inductive switching and losses

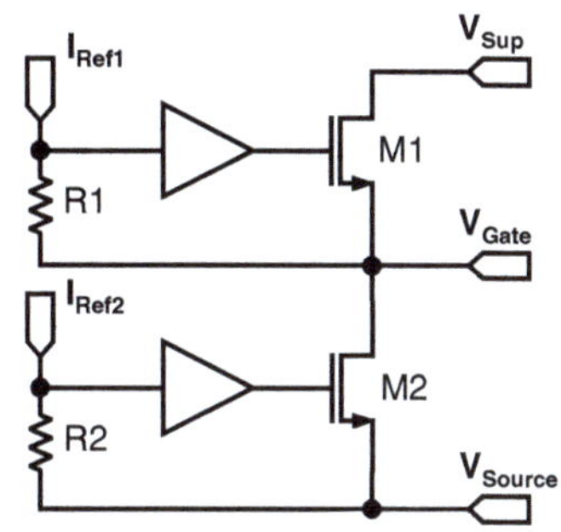

Fig. 9.14 Output stage of a current mode gate driver

9.5.2 *Switching Losses vs. EMC*

If di/dt and dv/dt are slowed down, the EME is minimized, but the switching losses in the switch(es) are maximized. The multiplication of V_{DS} and I_{Drain} forms a switching loss triangle as shown in Fig. 9.13. One slope of the triangle is given by the transition of the current and the other slope is defined by the transition of the voltage (see also Fig. 9.2c). Usually the current slope is much steeper than the voltage slope, i.e. the voltage transition is the main contributor to the switching losses. By controlling the di/dt and dv/dt, the gate driver can optimize the switching profile for better EMC performance or for lower switching losses. The profile can also be adjusted for optimized EME in an specific area of the spectrum by adjusting the transition speed with the highest influence in that area.

To be able to influence the transitions of MOSFET bridges separately, a gate driver with a highly variable output current has to be used. A topology with a voltage controlled current source (VCCS) output stage is shown in Fig. 9.14. The output stage can be used as a high-side or low-side driver and is able to switch between freely adjustable current levels with transition times of less than 10 ns.

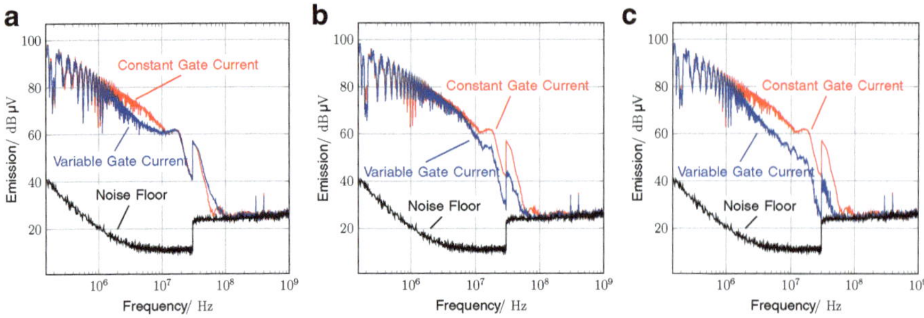

Fig. 9.15 EMC measurements with reduced gate current in (**a**) the voltage transition t_{vt}, (**b**) the current transition t_{ct}, (**c**) for both transition phases

9.5.3 Experimental Results

The driver in Fig. 9.14 was tested in a half bridge configuration driving a 300 µH air coil at a switching frequency of 20 kHz. The bridge voltage is $V_{Bat} = 13.5$ V and the load current is $I_{Drain} = 5$ A. EMC measurements were taken at the output node of the half bridge with the 150 Ohm method [31]. In the following measurements, the gate current is always set to 200 mA for the phases t_{pre} and t_{post}, Fig. 9.13. With a continuous gate current of 200 mA, the voltage transition takes 100 ns whereas the current transition is 20 ns. A large current spike of 3 A at the end of the current commutation is observed with a ringing frequency of 16 MHz, which corresponds to the peak in the EME spectrum in Fig. 9.15c. When the gate current is reduced to 20 mA for t_{vt} and t_{ct} the resulting voltage slope takes 300 ns and the current ringing is reduced from 3 to 1 A. This results in an improved EME as shown in Fig. 9.15c.

In the second setup, the gate current is switched from the initially constant 200 mA setting to 20 mA for the duration of the voltage transition t_{vt}, Fig. 9.16a. As expected, the EME is reduced for frequencies of 1 MHz and above, because of the shift in the second break point in the spectrum, Fig. 9.15a. The average EME reduction in this frequency range amounts to 10 dB. At higher frequencies, the spectrum is equal to the one without gate current reduction in t_{vt}, because of the emissions generated by the current transition.

To reduce the EME caused by the current transition, the gate current is now reduced to 20 mA only for t_{ct}, Fig. 9.16b. With this setup, the EME is reduced by 15–20 dB in the frequency range from 7 to 60 MHz, Fig. 9.15b. Slowing down the current transition time eliminates most of the ringing, caused by the parasitic inductances of the PCB. The switching losses of the four switching setups were also measured and analyzed, Fig. 9.17. As to be expected, the switching losses are highest with both transitions in low gate current mode and lowest with the constant high gate current setting. The transitions with only voltage or only current transition in low gate current mode both show a tradeoff between EME and switching losses. With a reduced gate current setting in the current transition the switching loss

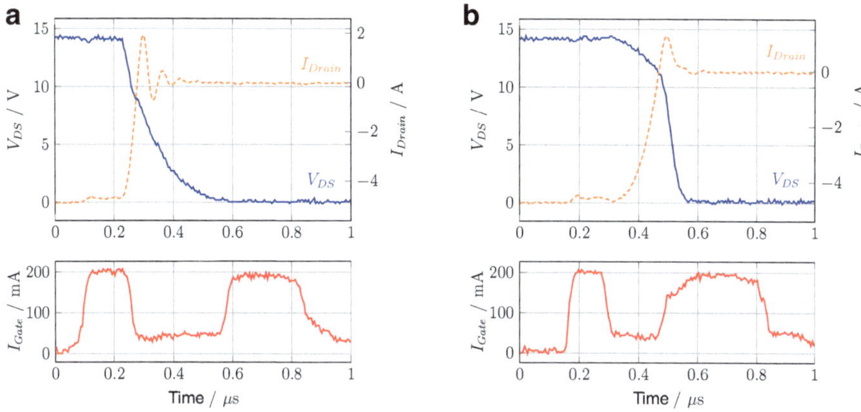

Fig. 9.16 Turn-on transition with reduced gate current in (**a**) the voltage transition phase t_{vt}, (**b**) the current transition phase t_{ct}

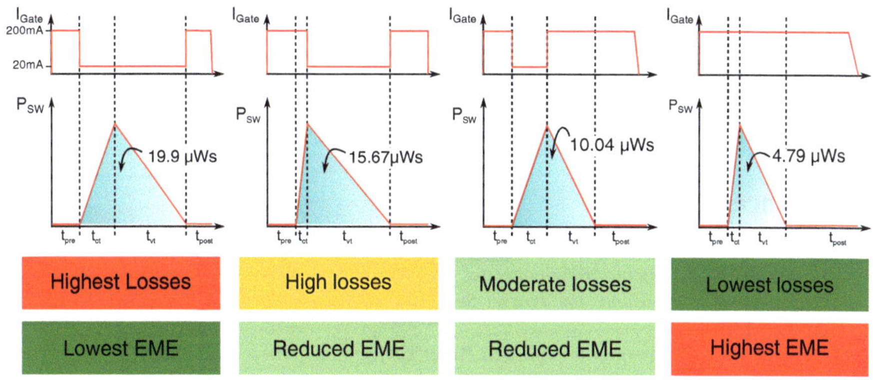

Fig. 9.17 Switching loss analysis of the four different switching cases

increase is moderate compared to the other options, while still reducing a large part of the EME. In the measurements the broadband EME between 7 and 60 MHz is reduced by up to 20 dB. At the same time the switching losses increased from 4.79 to 10.41 µWs. This results in an increase of 117 % compared to the constant current setting, but is 52 % lower than the switching losses with both transitions slowed down. The trade-off between EME and switching losses can be adjusted by the gate current settings for the individual transitions, if the gate driver can switch between current states fast enough. Reducing the transition speed of the di/dt is often the better trade-off between switching losses and EME, because the current transition is shorter and therefore has less impact on the losses. The decision, which transition to modify, is always highly influenced by the application and board layout. With a variable current gate driver the switching speed can be reduced during the transition, which enables significant EME reduction.

9.6 Conclusions

Gate drivers are key circuit blocks with growing importance for various automotive applications. They need to support increasing voltage levels, arising from the introduction of the 48 V board net and HV drive train with >300 V. Compact and cost efficient solutions require fast switching and highly integrated driver solutions. Cascaded drivers achieve high-speed operation and ensure low switching losses in the power stage. The driver itself can be further optimized for low power. In particular, asymmetric sizing eliminates cross-currents in the last driver stage. Fast switching is also enabled by appropriate level shifters. A 50 V level shifter is discussed, which achieves high-speed 3 ns minimum pulses and robust signal transfer for transition slopes of 20 V/ns.

The concept of on-chip high-voltage charge storing enables area-efficient, fully integrated high-voltage gate driver output stages. Combined signal and energy transfer via one single transformer is a way to further reduce size. The growing amount of electronics in cars and faster switching require more effort to meet EMC requirements. By means of a current mode driver, which can change the gate current within 10 ns, the EMC influence of the di/dt and dv/dt transitions have been studied. A trade-off between EMC and switching losses can be achieved.

References

1. J. Wittmann, B. Wicht, MHz-converter design for high conversion ratio, in *2013 25th International Symposium on Power Semiconductor Devices and ICs (ISPSD)* (2013), pp. 127–130
2. A. Lidow, *GaN Transistors for Efficient Power Conversion* (Wiley, Chichester, 2014)
3. Y. Xue, Z. Wang, L. Tolbert, B. Blalock, Analysis and optimization of buffer circuits in high current gate drives, in *2013 IEEE Transportation Electrification Conference and Expo (ITEC)* (2013), pp. 1–6
4. M. Huque, R. Vijayaraghavan, M. Zhang, B. Blalock, L. Tolbert, S. Islam, An SOI-based High-Voltage, High-Temperature Gate-Driver for SiC FET, in *IEEE Power Electronics Specialists Conference, 2007. PESC 2007* (2007), pp. 1491–1495
5. R. Greenwell, B. McCue, L. Tolbert, B. Blalock, S. Islam, High-temperature SOI-based gate driver IC for WBG power switches, in *2013 Twenty-Eighth Annual IEEE Applied Power Electronics Conference and Exposition (APEC)* (2013), pp. 1768–1775
6. H. Lin and L. Linholm, "An optimized output stage for MOS integrated circuits. IEEE J. Solid-State Circuits **10**(2), 106–109 (1975)
7. R. Jaeger, Comments on "An optimized output stage for MOS integrated circuits" [with reply]. IEEE J. Solid-State Circuits **10**(3), 185–186 (1975)
8. J. Wittmann, A. Barner, T. Rosahl, B. Wicht, A 12V 10MHz buck converter with dead time control based on a 125 ps differential delay chain, in *41st European Solid-State Circuits Conference (ESSCIRC)* (Graz, 2015), pp. 184–187
9. F. Hamzaoglu, M. Stan, Split-path skewed (SPS) CMOS buffer for high performance and low power applications. IEEE Trans. Circuits Syst. II: Analog Digit. Signal Process. **48**(10), 998–1002 (2001)

10. H. Ma, R. van der Zee, B. Nauta, An Integrated 80-V Class-D power output stage with 94% efficiency in a 0.14 μm SOI BCD process, in *2013 Proceedings of the ESSCIRC* (2013), pp. 89–92
11. Z. Liu, H. Lee, A 100V gate driver with sub-nanosecond-delay capacitive-coupled level shifting and dynamic timing control for ZVS-based synchronous power converters, in *2013 IEEE Custom Integrated Circuits Conference (CICC)* (2013), pp. 1–4
12. S. Finco, P. Tavares, A. Fiore De Mattos, M. Castro Simas, Power integrated circuit drives based on HV NMOS, in *2002 IEEE 33rd Annual Power Electronics Specialists Conference*, vol. 4 (2002), pp. 1737–1740
13. Y. Moghe, T. Lehmann, T. Piessens, Nanosecond delay floating high voltage level shifters in a 0.35 μm HV-CMOS technology. IEEE J. Solid-State Circuits **46**(2), 485–497 (2011)
14. G. Maderbacher, T. Jackum, W. Pribyl, S. Michaelis, D. Michaelis, C. Sandner, Fast and robust level shifters in 65 nm CMOS, in *2011 Proceedings of the ESSCIRC (ESSCIRC)* (2011), pp. 195–198
15. S. Zeltner, Insulating IGBT driver with PCB integrated capacitive coupling elements, in *2010 6th International Conference on Integrated Power Electronics Systems (CIPS)* (2010), pp. 1–6
16. J. Xu, L. Sheng, X. Dong, A novel high speed and high current FET driver with floating ground and integrated charge pump, in *2012 IEEE Energy Conversion Congress and Exposition (ECCE)* (2012), pp. 2604–2609
17. J. Wittmann, T. Rosahl, B. Wicht, A 50V high-speed level shifter with high dv/dt immunity for multi-MHz DCDC converters, in *ESSCIRC 2014 - 40th European Solid State Circuits Conference (ESSCIRC)* (2014), pp. 151–154
18. J. Wittmann, C. Rindfleisch, B. Wicht, Substrate coupling in fast-switching integrated power stages, in *2015 IEEE 27th International Symposium on Power Semiconductor Devices IC's (ISPSD)* (2015), pp. 341–344
19. A. Seidel, M. Costa, J. Joos, B. Wicht, Area efficient integrated gate drivers based on high-voltage charge storing. IEEE J. Solid-State Circuits **50**(7), 1550–1559 (2015)
20. A. Seidel, M. Costa, J. Joos, B. Wicht, Bootstrap circuit with high-voltage charge storing for area efficient gate drivers in power management systems, in *ESSCIRC 2014, Proceedings of the 40th European Solid State Circuits Conference* (2014), pp. 159–162
21. Fairchild Semiconductor Corporation, *Design Guide for Selection of Bootstrap Components*. Application Note AN-9052 (2008)
22. R. Herzer, Integrated gate driver circuit solutions, in *2010 6th International Conference on Integrated Power Electronics Systems (CIPS)* (2010), pp. 1–10
23. M. Munzer, W. Ademmer, B. Strzalkowski, K. Kaschani, Insulated signal transfer in a half bridge driver IC based on coreless transformer technology, in *The Fifth International Conference on Power Electronics and Drive Systems, PEDS 2003*, vol. 1 (2003), pp. 93–96
24. B. Chen, Isolated half-bridge gate driver with integrated high-side supply, in *IEEE Power Electronics Specialists Conference, PESC 2008* (2008), pp. 3615–3618
25. S. Nagai, Y. Yamada, N. Negoro, H. Handa, Y. Kudoh, H. Ueno, M. Ishida, N. Otuska, D. Ueda, A GaN 3x3 matrix converter chipset with drive-by-microwave technologies, in *2014 IEEE International Solid-State Circuits Conference Digest of Technical Papers (ISSCC)* (2014), pp. 494–495
26. A. Seidel, M. Costa, J. Joos, B. Wicht, Isolated 100% PWM gate driver with auxiliary energy and bidirectional FM/AM signal transmission via single transformer, in *2015 IEEE Applied Power Electronics Conference and Exposition (APEC)* (2015), pp. 2581–2584
27. G. Wang, W. Liu, M. Sivaprakasam, G. Kendir, Design and analysis of an adaptive transcutaneous power telemetry for biomedical implants. IEEE Trans. Circuits Syst. I: Reg. Pap. **52**(10), 2109–2117 (2005)
28. S. Atluri, M. Ghovanloo, Incorporating back telemetry in a full-wave CMOS rectifier for RFID and biomedical applications, in *IEEE International Symposium on Circuits and Systems, ISCAS 2007* (2007), pp. 801–804

29. A. Schindler, B. Koeppl, B. Wicht, EMC and switching loss improvement for fast switching power stages by di/dt, dv/dt optimization with 10ns variable current source gate driver, in *2015 10th International Workshop on the Electromagnetic Compatibility of Integrated Circuits (EMC Compo)* (2015), pp. 18–23
30. C.R. Paul, *Introduction to Electromagnetic Compatibility*, 2nd edn. (Wiley-VCH, Hoboken, 2006)
31. T. Steinecke, M. Bischoff, F. Brandl, C. Hermann, F. Klotz, F. Mueller, W. Pfaff, M. Unger, Generic IC EMC test specification, in 2012 Asia-Pacific Symposium on Electromagnetic Compatibility (APEMC) (2012), pp. 5–8

Chapter 10
A Self-Calibrating SAR ADC for Automotive Microcontrollers

Carmelo Burgio, Mauro Giacomini, Enzo Michele Donze, and Domenico Fabio Restivo

10.1 Introduction

Automotive microcontrollers make use of SAR ADC converters to manage a large variety of sensor input signals. In fact, this converter is able to achieve very good absolute accuracy and has the capability to manage, with proper multiplexing, several inputs with the same converter instance.

In the new generation of microcontrollers for automotive applications, we observe a significant increase of the required converter performances and the number of converters to be hosted, to achieve the required bandwidth and the capability of simultaneous sampling of several signals.

In order to avoid diverging cost of the converters, it is key to be able to reach the desired performance in a compact way.

Analog-to-Digital Converters (ADCs) require high linearity and so low voltage coefficient capacitors. A built in *self-calibration* and *digital-trim* algorithm correcting static mismatches in capacitive Digital-to-Analog Converter (CDAC) block is proposed.

C. Burgio (✉) • M. Giacomini (✉)
STMicroelectronics, Via C. Olivetti 2, 20864 Agrate Brianza, MB, Italy
e-mail: carmelo.burgio@st.com; mauro.giacomini@st.com

E.M. Donze • D.F. Restivo
STMicroelectronics, Via Tommaso Marcellini 8-L, 90129 Palermo, Italy
e-mail: enzo-michele.donze@st.com; fabio.restivo@st.com

© Springer International Publishing Switzerland 2017

A. Baschirotto et al. (eds.), *Wideband Continuous-time ΣΔ ADCs, Automotive Electronics, and Power Management*, DOI 10.1007/978-3-319-41670-0_10

10.1.1 Conventional SAR ADC Calibration/Trimming

Many techniques can be applied to compensate *random* mismatch errors in the matching-critical components for data converters. They can be divided into four categories: *trimming*, *calibration*, *switching-sequence adjustment*, and *dynamic element matching (DEM)*.

Trimming compensates the random mismatch errors by regulating the component parameters at the wafer stage. It requires accurate test equipment to continuously measure and compare trimmed parameters with their nominal values. There are mainly two types of trimming: one is to change the physical dimensions of the circuit. The second is to connect/disconnect an array of small elements using fuses or MOS switches [1, 2].

Calibration can be seen as an improved version of trimming since it characterizes errors on chip or board and subsequently corrects them. Yet, the obtained accuracy level is limited by the resolution and accuracy of error measurement circuits and correction signals [3, 4].

Switching-sequence adjustment is commonly used in layout designs to compensate the systematic gradient errors for some electrical parameter [5, 6].

DEM is another popular solution to the critical components random mismatch errors. It dynamically changes the positions of mismatched elements at different time so that the equivalent component at each position is nearly matched on a time average. Unlike the static random mismatch compensation techniques, DEM translates mismatch errors into noise. However, the translated noise is only partially shaped where the in-band residuals could possibly affect the data converters' signal-to-noise ratio (SNR). Furthermore, the output will be inaccurate at one time instant since DEM only guarantees matching on average [7, 8].

10.2 SAR ADC Overview

The SAR principle of operation is widely and perfectly known: essentially, it performs a binary search algorithm to converge to the input signal. At the heart of the SAR there are two components: a comparator and a DAC.

The DAC generates a "guess" of the analogue level and the comparator evaluates the difference between the DAC output and the actual, analogue input (Fig. 10.1).

Very often the Track&Hold structure is merged with the DAC for implementation efficiency and holding simplicity.

The sequencing logic and SAR register sets the next "guess" level, according to a binary search algorithm, in order to match the DAC output and the analogue input voltage. At the end, this process generates the digital code correspondent to the analogue input.

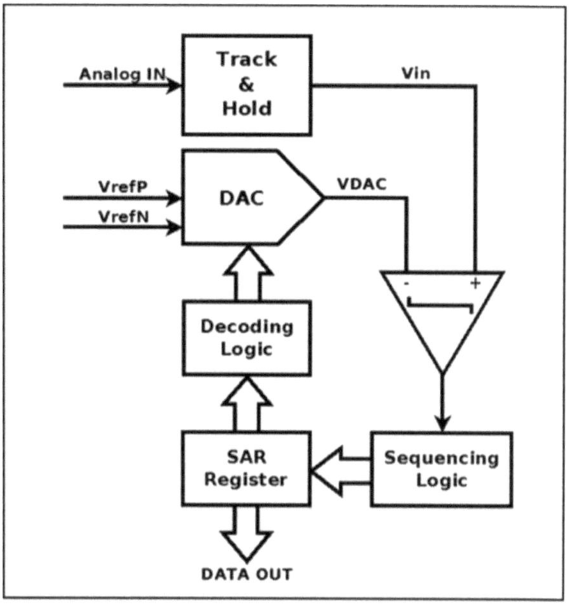

Fig. 10.1 SAR ADC standard architecture

There is a strict correlation between the conversion result and the accuracy of the DAC converter: any non-ideality in the DAC components has an impact in the behavior of the SAR ADC system.

The main requirements of the comparator section are an offset compatible with the accuracy required, a noise contribution in line with the converter resolution and an evaluation speed adequate to the conversion timings.

10.2.1 DAC Architecture

A set of binary weighted physical elements, normally capacitors or resistors, is the most straightforward implementation of the DAC component. This implementation is putting heavy constraints on the accuracy of the elements especially if good linearity and the absence of missing codes have to be granted.

A different DAC organization can reduce the matching needs: a structure containing a set of identical elements thermometer coded implements the *msb* bits by while binary coded elements implement *lsb* bits.

This approach, while granting by construction the monotonicity of the thermometric DAC and reducing the matching requirements to ensure a good DNL at the transition between thermometric and binary DAC, requires the usage of a much wider set of switches to control the various elements that can have a not negligible impact on the area of the converter. The partitioning of the DAC elements in a number of a thermometric bits (*msb*) and binary bits (*lsb*), depends on the accuracy to be reached by the converter, on the matching parameters of the unit element (technology parameter) and on the area taken by the switches portion.

The selection of this kind of arrangement (driven by accuracy requirements) automatically implies a series of consequences that we use, favorably, in our calibration process.

In our implementation, the DAC has been fully realized with capacitors given the stringent requirement for speed and reference consumption. In fact, avoiding the usage of resistors, remove any DC current component on the reference and the settling time of the DAC will depend on the time constant associated to the switches and the capacitors: switches resistance can be lowered without affecting the reference consumption.

The proposed architecture implements a six bits thermometric DAC and a six bits binary DAC.

10.2.2 Thermometric and Binary Elements Organization

As said before the thermometric organization provides advantages in terms of linearity at the expense of a certain redundancy in the circuitry to select the elements. This redundancy normally is not exploited because the thermometric sequence is hard coded in the decoding logic.

The principle of the calibration method presented in this paper consists in the smart usage of the intrinsic redundancy offered by the thermometric decoding. The sequence of the thermometric elements is not hard coded but can be changed arbitrarily.

Notice that in our implementation the thermometric DAC is built by a complete set of elements able to generate all the $2^{\mathrm{NbitTh}}+1$ (where $\mathrm{Nbit}_{\mathrm{Th}}$ is the number of bits thermometric coded) reference levels from V_{refN} to V_{refP}. The reason for this choice will become clear later on.

Therefore, the *binary* DAC will not be realized as a fully additive DAC, but will be able to provide both positive and negative voltage values as well.

Figure 10.2 represents a conventional approach for the generation of the references using a thermometric and a binary DAC arrangement while Fig. 10.3 describes the decoding strategy used for the proposed DAC.

As it can be noticed, another redundancy is generated in our implementation because, in principle, we have the possibility to generate voltage references even above V_{refP}. We will use favorably also this kind of redundancy in the measurement process.

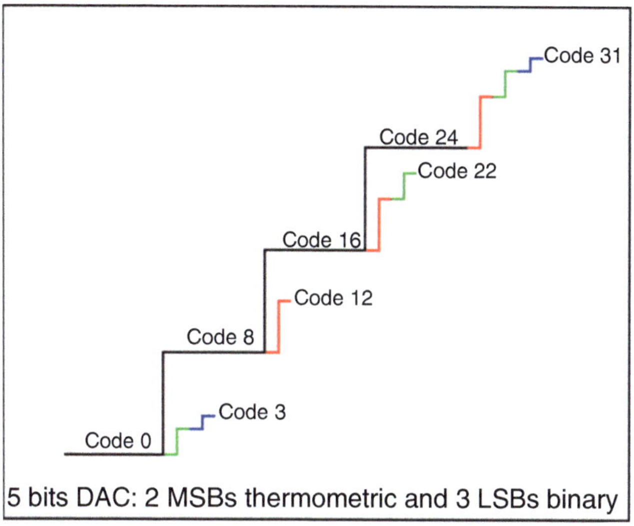

Fig. 10.2 Conventional approach for thermometric and binary DACs assembly

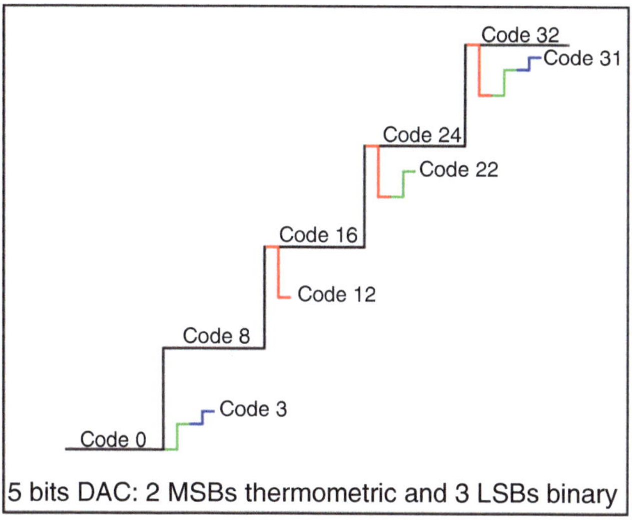

Fig. 10.3 Approach used in this paper for thermometric and binary DACs assembly

10.3 Thermometric Elements Calibration

Any calibration process is based on the capability to measure the system to be calibrated: in our case it is the set of unit capacitors constituting the thermometric DAC but the concept equally works with different physical elements as well (e.g.: current sources).

The accurate evaluation of the values of the thermometric elements is, without doubt, a critical step in the calibration procedure: wrong measurements will produce, invariably, bad calibration results.

To be effective from an industrial point of view, the final calibration process has to satisfy two basic requirements:

1. no area relevant, extra analogue HW has to be designed and implemented in silicon for the calibration feature
2. no high-accuracy references (internal and/or external) have to be requested by the calibration procedure

10.3.1 Thermometric Elements Measurement

We can easily recognize that with the hardware in our hands and without the help of any external equipment, we can measure, against a given arbitrary voltage reference, the difference between any of the two capacitors of the thermometric set.

In fact, what we really need to know is not the absolute value of each thermometric element (i.e. the capacitance values in fF), but the difference between every element and its 'ideal' value. In reality, as we will see later on, the difference between each thermometric element and *only one of them* chosen arbitrarily as a reference is sufficient to carry on the calibration procedure.

This is achieved by performing a simple SAR conversion of an error voltage generated *on purpose* by using the two capacitors under measurement.

Let's sample the voltage V_{refP} on an arbitrarily chosen thermometric element, $C_{Th\text{-}ref}$, while sampling V_{refN} on all the other elements.

After the sampling phase is ended, so the signal is frozen on the thermometric capacitors, we connect the $C_{Th\text{-}ref}$ capacitor to V_{refN} while connecting the capacitor to be measured, $C_{Th\text{-}k}$, to V_{refP}.

The voltage at the input of the comparator will change by an amount equal to

$$\Delta V_k = \frac{C_{Th-k} - C_{Th-ref}}{\sum_{j=1}^{N_{th}} C_j} * \left(V_{refP} - V_{refN}\right) \tag{10.1}$$

At this point, the error signal generated can be converted in a digital code by means of the binary DAC. Obviously ΔV_k can be either positive or negative but, as we said before, the binary DAC can provide both positive and negative output voltage levels as well.

To improve the resolution of the measurement, the reference used by the binary DAC can be scaled down (V_{ref}P_Cal): in our implementation, the scaling factor can be 8 or 16 providing an equivalent measurement resolution of 15 or 16 bits if compared to the resolution of the final converter. It has to be noted that the accuracy

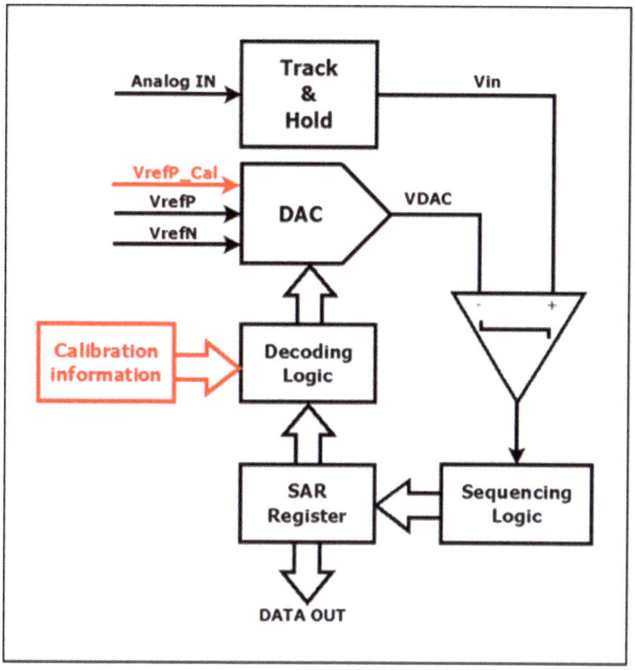

Fig. 10.4 Self-calibrating SAR ADC proposed architecture

of the calibration voltage reference has no impact on the accuracy of the calibration process so the area required to implement it is negligible compared to the area of the SAR ADC.

To enable the calibration process the structure of the SAR ADC needs to be modified according to Fig. 10.4. The overhead of the calibration process is very low, being confined to a small analog add-on and a digital block modification that in a 40 nm technology is for sure negligible.

In the mathematical computation terms, let us call T_k the generic thermometric element.

The 'ideal' value is simply the average value of the elements, because in case all the T_k would have the same value, they realise a perfect thermometric DAC.

$$T_{ave} = T_{ideal} = \frac{\sum_{k=1}^{N_{th}} T_k}{N_{th}} = \frac{T_{tot}}{N_{th}} \tag{10.2}$$

where N_{th} is the number of thermometric elements.

As described before, with the HW available it is easy measure the following quantity, $E_{j,k}$:

$$E_{j,k} = \frac{T_j - T_k}{T_{tot}} \qquad j,k \in \{1 \ldots N_{th}\} \tag{10.3}$$

This quantity is the basic element of the calibration process.

Selecting *arbitrarily* as reference for all measurements $T_k \equiv T_{ref}$, the value of Eq. (10.3) is evaluated for all the thermometric elements, T_j, using always the fixed chosen element T_{ref}.

Other, different, measurement policies can be adopted, but we will develop the next steps with the above assumption.

At the end of the measurement process, we have a numbered set of N_{th} values, E_j, representing the starting point of the calibration algorithm.

$$E_j = \frac{T_j - T_{ref}}{T_{tot}} \qquad \forall j,\ j \in \{1..N_{th}\}\,; \quad ref \in \{1..N_{th}\} \tag{10.4}$$

Again, T_{ref} is a chosen element of the array.

The choice of this 'reference' element, T_{ref}, is not univocal and it affects numerically the computations to be carried out during the calibration process. Notwithstanding, it is possible to demonstrate that it has no impact on the result.

The E_j itself does not represent directly the difference of the value of each thermometric element T_j from T_{ideal}, but it is possible to make a convenient use of it during the calculations.

10.3.2 Measurement Procedure

The measurements of the values of E_j are carried out using the binary section of the SAR ADC; the technology guarantees for this subsection a high degree of accuracy and we take advantage of it.

Moreover, the values to be measured are intrinsically very small. We are measuring the difference between two thermometric elements, assumed very similar by technology. This to say that a limited dynamic range for the measurement is needed and the SAR binary subsection dynamic range is adequate for that task.

Notwithstanding, a single measure of E_j is not enough accurate due to the intrinsic noise of the comparator and the quantization error. An average on a certain number of measurements is mandatory.

In case of an ideal measurement environment, the repetition of the measure does not give any real advantage: the code readout from the ADC is invariably the same (Fig. 10.5).

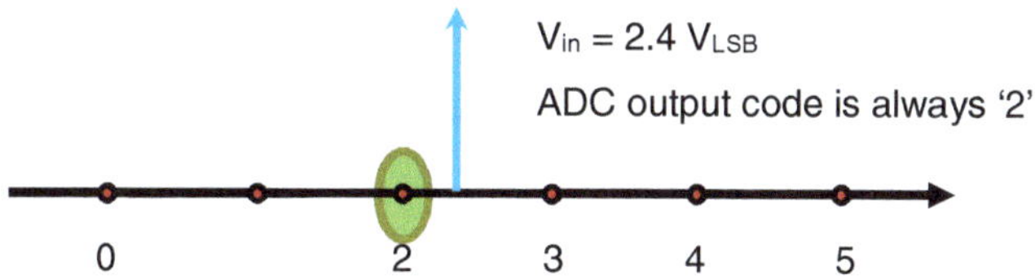

Fig. 10.5 Noiseless ADC conversion

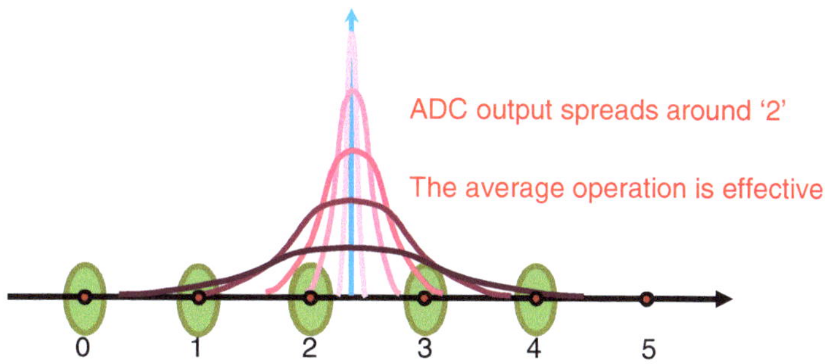

Fig. 10.6 Noisy ADC conversion

However, the situation is radically different if a limited, but not negligible, amount of noise is present. In such a case, the readout of the measurements will spread around a certain interval and the average action is effective (Fig. 10.6).

The more noise is present, the more codes are exercised, while an excess of noise can damage the measurement process.

In fact, there is an optimum level of noise given the number of measurements available for the average or vice-versa. Given the noise level it is possible to fix the optimum number of measurements (Fig. 10.7).

It is evident that the accuracy of the measurement depends on the noise level, given the number of samples used in the average. In this case, a noise around 0.5 *lsb* is enough to guarantee a precision of almost 0.02 *lsb* after an average on 4096 samples. In fact the 'signal', in this case a constant value, increases linearly with the number of measurements used in the average, N_{ave}, whilst the noise, adding in power, increases like sqrt (N_{ave}). At the end the 'signal' to noise ratio increases like

$$\frac{N_{ave}}{\sqrt{N_{ave}}} = \sqrt{N_{ave}} \tag{10.5}$$

Every doubling of the number of samples the accuracy increases by 3 dB, i.e. ½ *lsb*.

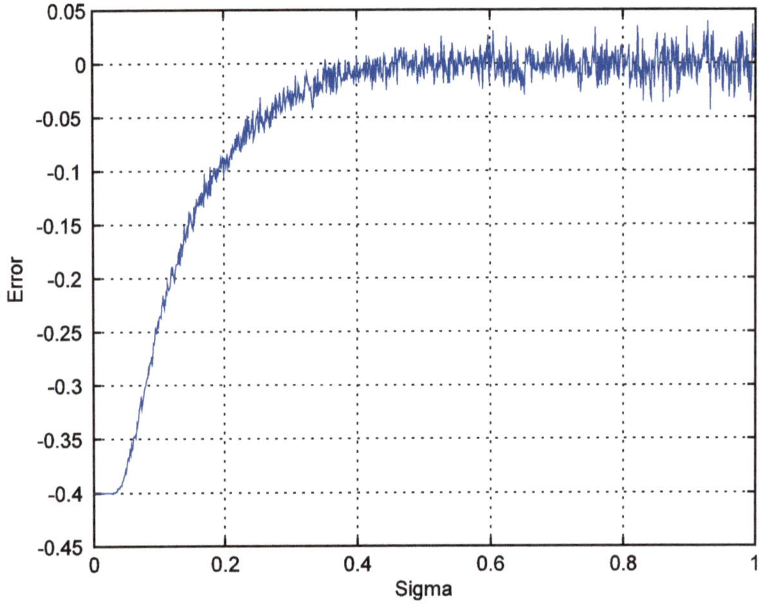

Fig. 10.7 Error after 4096 samples averaging vs RMS noise (in lsb)

10.3.3 Calculations and Data Processing

Let us assume there are N_{th} thermometric elements in a DAC converter.

The output voltages generated by this DAC for every input code are:

$$V_k = \frac{\sum_{j=1}^{k} T_k}{T_{tot}} V_{ref} \qquad k \in \{1..N_{th}\} \tag{10.6}$$

It is implicit that for the ground level (code 0) no elements are used and $V_0 = 0$; for the last level, where all elements are used, $V_{Nth} = V_{ref}$, independently on the values of the T_k. Ground level and full scale level are always, by definition, correctly positioned, i.e. their INL is 0.

It is a matter of fact, due to the technology limitations, that the T_k have different values, each of them being affected by an error respect to the ideal value, T_{ideal}

$$T_k = T_{ideal} + t_k \qquad k \in \{1..N_{th}\} \tag{10.7}$$

where t_k is the deviation of element from the ideal value or, otherwise stated, the DNL_k. By consequence, the DAC output voltage level becomes:

$$\frac{V_k}{V_{ref}} = \frac{\sum_{j=1}^{k}\left(T_{ideal} + t_j\right)}{T_{tot}} = \frac{\sum_{j=1}^{k} T_{ideal}}{T_{tot}} + \frac{\sum_{j=1}^{k} t_j}{T_{tot}} \qquad k \in \{1..N_{th}\} \tag{10.8}$$

Note that, by definition, for full scale level the following is valid:

$$\frac{\sum_{j=1}^{N_{th}} T_j}{T_{tot}} = \frac{\sum_{j=1}^{N_{th}}\left(T_{ideal} + t_j\right)}{T_{tot}} = \frac{N_{th} T_{ideal}}{T_{tot}} + \frac{\sum_{j=1}^{N_{th}} t_j}{T_{tot}} = 1 \quad \Rightarrow \quad \sum_{j=1}^{N_{th}} t_j = 0 \tag{10.9}$$

The sum of all the DNL errors is 0, as it has to be by INL definition.

The error between every level and the perfect level, i.e. the INL_k, is:

$$INL_k = \frac{\sum_{j=1}^{k}\left(T_{ideal} + t_j\right)}{T_{tot}} - \frac{\sum_{j=1}^{k} T_{ideal}}{T_{tot}} = \frac{\sum_{j=1}^{k} t_j}{T_{tot}} \qquad k \in \{1..N_{th}\} \tag{10.10}$$

By definition:

$$INL_0 = INL_{N_{th}} = 0 \tag{10.11}$$

Here the main point: the INL_k depends not only on the t_k values, but on their ordering as well and the proposed implementation takes advantage of this property: changing the 'sequence of firing' of the T_k elements, the INL characteristic changes as well. The problem is the identification the best permutation of the thermometric elements in order to minimize the INL error.

Using the measurement results, E_k, it is possible (see Appendix), to evaluate both DNL and INL errors:

$$DNL_k = E_k - \frac{\sum_{all} E_n}{N_{th}} \tag{10.12}$$

$$INL_k = \sum_{j=1}^{k} E_j - k\frac{\sum_{all} E_n}{N_{th}} \tag{10.13}$$

It is worth pointing out that both INL and DNL expressions do not imply the knowledge, as previously claimed, of the selected reference element T_{ref}.

Every permutation of the thermometric elements generates a different INL characteristic for the DAC. Actually this is a huge number of possibilities ($N_{th}!$) even for relatively low numbers of thermometric elements and an exhaustive search, trying all possible permutations, is practically impossible.

10.3.4 Calibration Algorithms

At this point, after the E_k measurements, it is possible to evaluate the INL characteristic of the DAC for all the possible permutations of the thermometric elements. The identification the best one, without perform an exhaustive search, is strongly desirable. Using metaheuristic optimization approaches (*Simulated-Annealing* and *Tabu-Search*), simulations show a remarkable probability of trapping in local minima.

A deterministic approach solves the issue.

10.3.4.1 Calibration Algo I

The discussion of the rationale behind the calibration algorithms (Algo) and the placement strategy for the thermometric elements follows in the next paragraphs.

Calibration Algo I: Discussion

In a 'perfect' DAC, every thermometric element has a 'perfect' value, T_{ideal}. In such a case, building conceptually the INL characteristic, when an element is placed in the position 'k', the characteristic increase by T_{ideal} and the resulting INL is 0. Actually, every element is affected by an error DNL_k and we have to place all of them; including the element having the worst DNL value, DNL_{worst}. Our ideal setup process, eventually, places this element as well, defining the best possible accuracy for INL. In fact, the best option we have is to place it exactly across the ideal staircase, fixing in this way the best INL performance of the current DAC at $DNL_{worst}/2$. Given the measurement results, we immediately know the best possible INL performance for the DAC under test. Now we need a strategy able to identify one (out of many) permutation satisfying the following relationship (Figs. 10.8 and 10.9):

$$\max\left(abs\left(INL_k\right)\right) \leq DNL_{worst}/2 \qquad \forall k, k \in \{1 \ldots N_{th-1}\} \tag{10.14}$$

Calibration Algo I: Elements Placement

Let us assume that the distribution law for the thermometric element is Gaussian around their average value, T_{ideal}; accordingly, a substantial number of them are placed near T_{ideal}, showing good accuracy.

By definition, the INL error value at the extremes (ground and full scale) is zero. So, starting from one of these points is possible 'to build', if not the optimal sequence, at least a very good approximation of it.

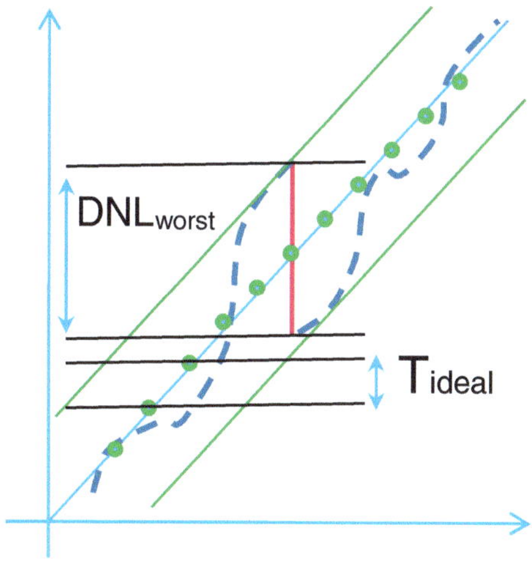

Fig. 10.8 General DAC characteristic

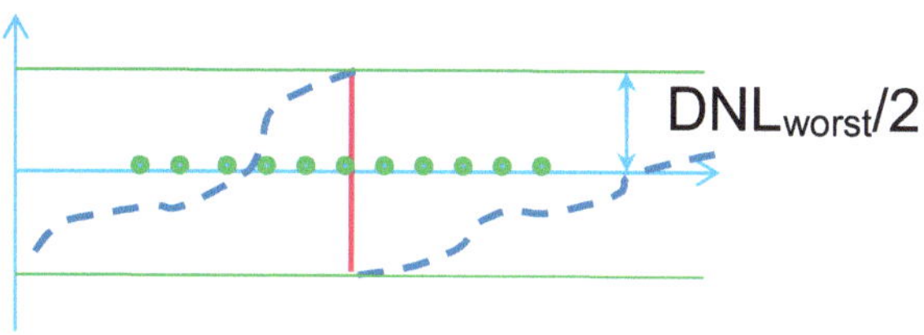

Fig. 10.9 Best possible INL profile

Here is the procedure:

0. Select ground (or full scale) as 'Current INL level'. Initialise the "Set of elements to be placed" with all elements.
1. Select the element with the worst DNL, DNL_{worst}, in the set of elements to be placed.
2. Taking advantage of the accurate elements having low DNL error, approximate as much as possible $DNL_{worst}/2$ value, paying attention never exceed +/- 0.5 DNL_{worst}
3. Place the low DNL error elements sequence, followed by the DNL_{worst} element, after the 'Current INL level'.
4. Remove the used elements from the "Set of elements to be placed".

   If the $DNL_{worst}/2$ value has been accurately approximated, the DNL_{worst} element will 'jump' from $DNL_{worst}/2$ to $-DNL_{worst}/2$.
5. Update 'Current INL level' to the level after DNL_{worst} element (the last one).
6. If the "Set of elements to be placed" is not empty, restart from step 1), otherwise INL is 0 (it has to be, by construction) and stop.

This procedure, consuming systematically the thermometric elements having high DNL error, takes advantage of the relative large number of elements with small DNL error. Most of the times, only elements with small DNL error will remain at the end of the process. The result is the 'stretching' at the end of the INL plot, as sketched in Fig. 10.11.

10.3.4.2 Calibration Algo II

Calibration Algo II: Discussion

The value of the DNL_{worst} element limits the "Algo I" calibration process performance. That element appears in the INL two times: the first one in the numerator of Eq. (10.6), when $k \geq$ '*worst*' index; the second one in the denominator, contributing to T_{tot} jointly with all the other elements. In the denominator it does not generate any trouble; it is a problem, conversely, in the numerator, where it introduces a big 'jump' in the INL characteristic of the DAC at $k =$ '*worst*'. Then a possible solution could be to use, in the numerator, only the elements having low DNL errors.

The following paragraph sketches the strategy adopted in order to avoid the presence of the elements with high DNL error in the numerator.

Calibration Algo II: Elements Selection and Placement

It is already established that

$$\sum_{all} DNL_k = 0 \tag{10.15}$$

Now we can split the set of thermometric elements into two subsets; let us call them for the time being 'set A' and 'set B':

$$\sum_{A} DNL_k + \sum_{B} DNL_k = 0 \tag{10.16}$$

or, equivalently:

$$\sum_{A} DNL_k = -\sum_{B} DNL_k \tag{10.17}$$

this means that, splitting the thermometric set into 2 subsets, the sum of subset 'A' will be always the opposite the sum of subset 'B', independently of the selection of the elements.

Now we can run the following procedure:

1. split the thermometric element set into two subset of $N_{th}/2$ elements
2. The first subset, let us call it the 'Good' set, contains the elements with lower DNL; the remaining elements belong to the second subset, called the 'Bad' set.
3. The 'Good' subset obeys to supplemental criteria: minimize the absolute value of the sum of the DNL of all the elements; by consequence, the 'Bad' set will have the same DNL sum value, with the opposite sign. Ideally, that value should be zero; in practice it is possible to get a value very near to that target, with a value lower than the majority, if not all, of the DNL errors.

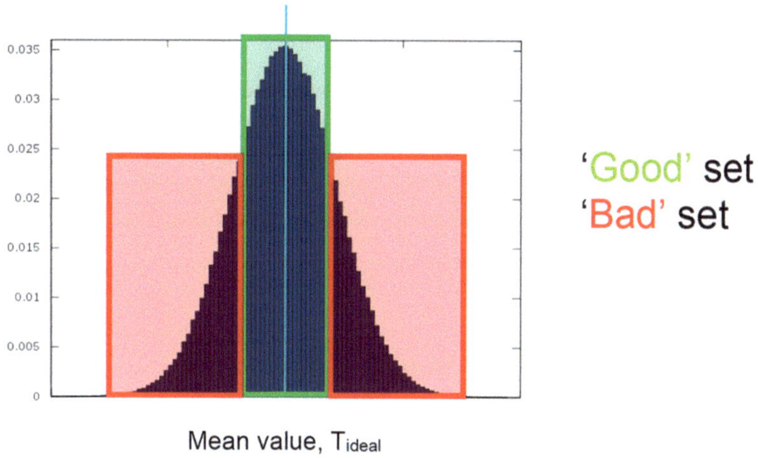

Fig. 10.10 Thermometric elements statistical distribution

4. The ordering of the elements of the 'Good' set can be established with an algorithm similar to 'Algo I' (Fig. 10.10)

   Then, for every DAC measurement, apply the following strategy:

1. Use, for the first half of the thermometric scale, only the 'Good' set
2. For the second half of the thermometric scale proceed to:
   (a) Substitute the 'Bad' set in place of the 'Good' set in the lower section of the array.
   (b) Use again the 'Good' set in the higher section of the array, but *with the elements used in the reversed order*.

The substitution of the 'Good' to 'Bad' is seamless if the sum of the 'Good' set is zero or very near to zero.

There is a price to pay for that procedure: the DAC characteristic is not monotonic 'by construction' any more. Notwithstanding, it is a matter of fact that, with a careful selection strategy, it is always possible to have an extremely good matching between the 'Good' and 'Bad' set, much better than 1 *lsb*, avoiding in this way the inference of non-monotonicity.

In this way:

1. Only the elements having low DNL error appear in the numerator as single step 'increment'. These DNL values are, now, the limiting factor for the DAC under test performance.
2. The discontinuity at half scale is minimized in the splitting procedure of the thermometric set in 'Good' and 'Bad' sets.

10.4 Statistical Simulations

A behavioral, statistical, MATLAB model evaluates the performances of the above calibration concept. That model takes into account all the main technology parameters, includes an accurate model of the analog part and the above described calibration algorithms. It predicts the behavior of the pre- and post-calibration INL characteristic for a single ADC and pictures the statistical behavior of a set of many ADC as well.

10.4.1 Calibration Algo I Statistical Simulation

Figure 10.11 represents, for a single ADC, the profile of the INL before (native) and after calibration with Algo I.

Figures 10.12 and 10.13 represent the simulation result on a population of 1024 elements. As evident in Fig. 10.13, the statistical distribution of the INL is compacted by roughly a factor of six.

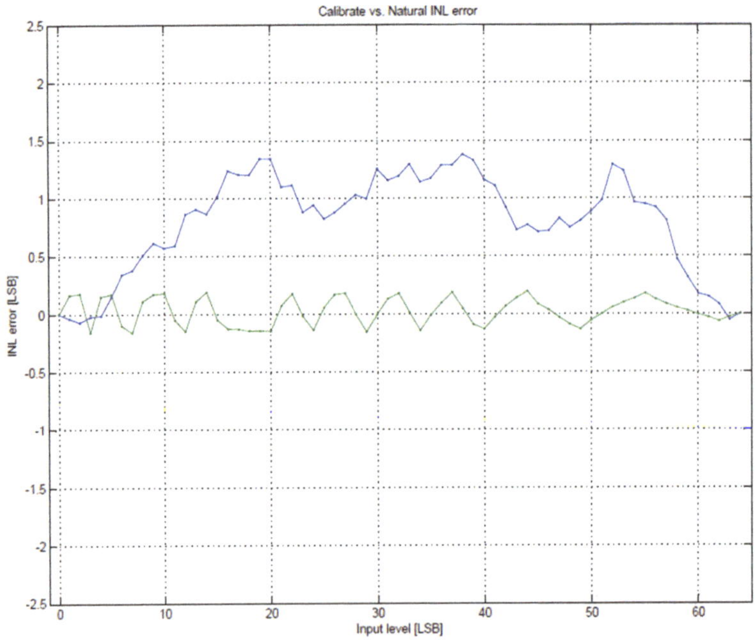

Fig. 10.11 Algo I—single ADC INL pre- and post-calibration behavior (simulation)

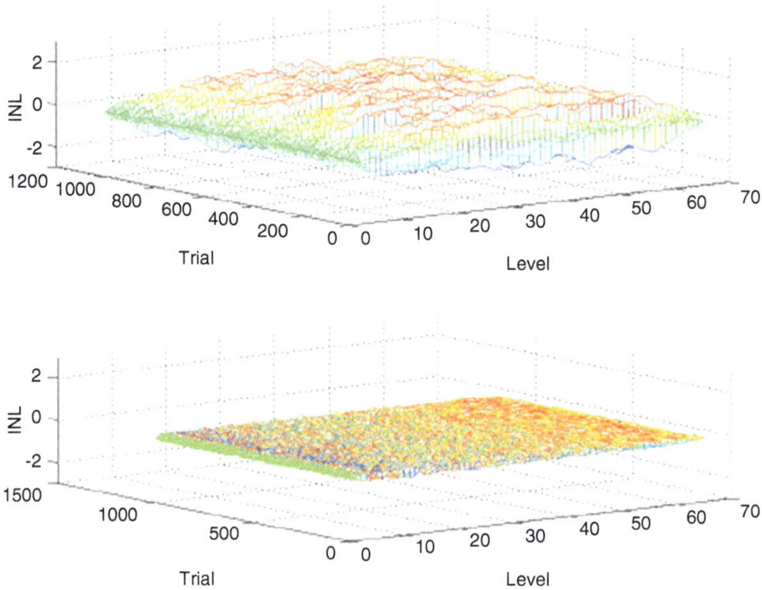

Fig. 10.12 Algo I—statistical pre- and post-calibration behavior (simulation)

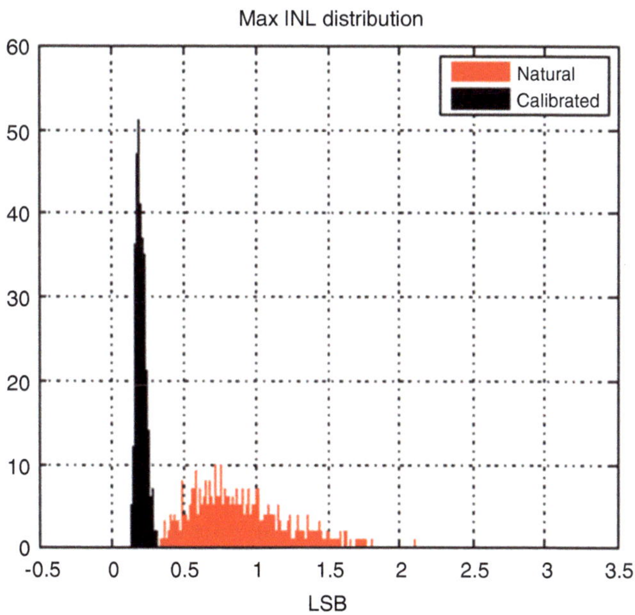

Fig. 10.13 Algo I—Max INL pdf pre- and post-calibration behavior (simulation)

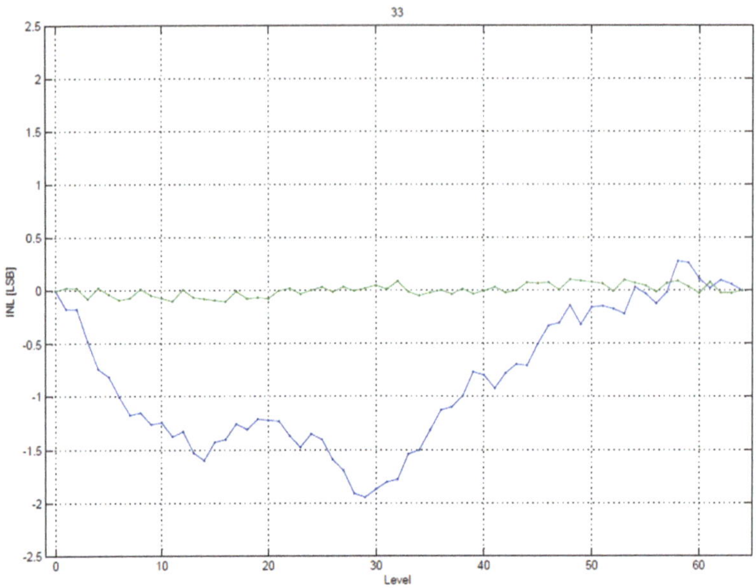

Fig. 10.14 Algo II—single ADC INL pre- and post-calibration behavior (simulation)

10.4.2 Calibration Algo II Statistical Simulation

Figure 10.14 represents, for one particular case of the statistical distribution, the profile of the INL before (native ADC) and after calibration with Algo II.

Figures 10.15 and 10.16 represent instead the result on a population of 1024 elements. The statistical distribution of the INL is even more compacted with respect to Algo I (improvement of a factor of eight), reducing significantly the minimum value. The worst case DNL of the thermometric elements is not the limit any more (Fig. 10.17).

10.5 Silicon Measurements

The following paragraphs carry out a comparison of the simulation results vs. the laboratory measurements. All the measurements refer to the same device.

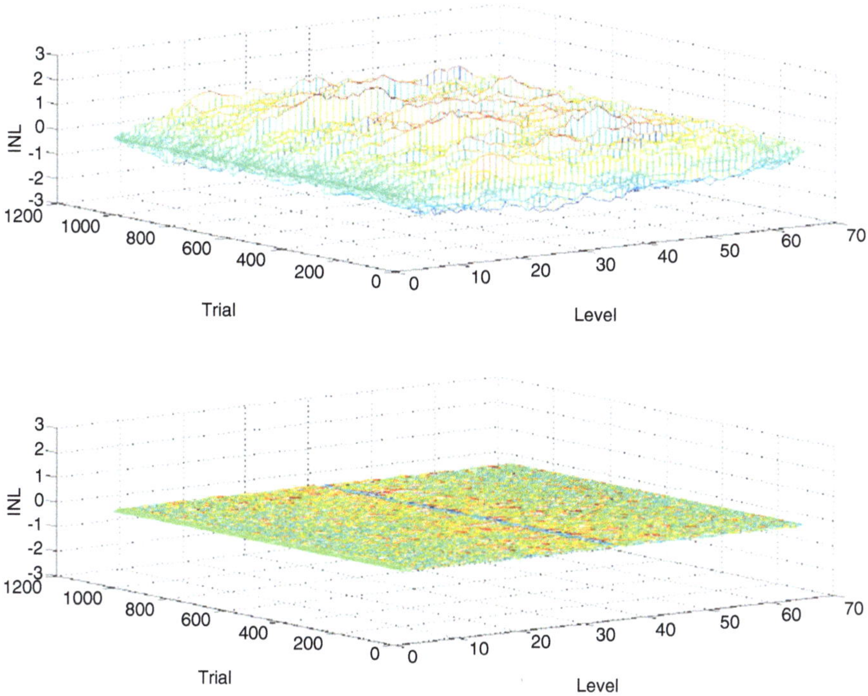

Fig. 10.15 Algo II—statistical pre- and post-calibration behavior (simulation)

10.5.1 Uncalibrated ADC Silicon Measurement

The first measure (Fig. 10.18) shows the native INL characteristic. Due to the initial, random, distribution of the thermometric elements values, the maximum INL error is larger than 1 LBS.

The root cause of the oscillatory behavior (∼0.15 *lsb*) is in the inaccuracy of the binary ADC implemented for the six *lsb* bits. Figure 10.19 shows the details of characteristic of the binary section of the ADC.

Using the self-measurement capability of the calibration procedure, it is possible to estimate the native INL characteristic of the device without the aid of any external instrument. Figure 10.20 compares such estimation with the INL measured in the laboratory.

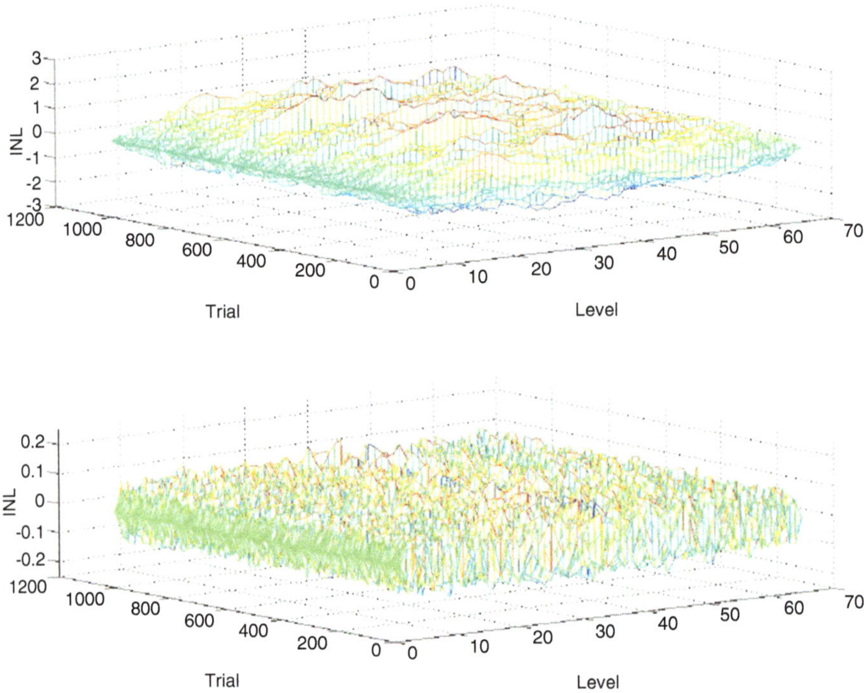

Fig. 10.16 Algo II—statistical pre- and post-calibration behavior detail (simulation)

10.5.2 Calibration Algo I Silicon Measurement

Figure 10.21 shows the complete INL characteristic of the device after self-calibration using Algo I. The typical shape of Algo I has a good matching with the behavioral simulation results (Fig. 10.8).

Figure 10.22 depicts the matching, after calibration, between the estimation and the actual measure of the INL of thermometric elements only.

10.5.3 Calibration Algo II Silicon Measurement

Figure 10.23 sketches the INL after self-calibration using Algo II. It is evident the peculiar symmetry around the central code of this calibration procedure.

Figure 10.24 depicts the matching, after calibration, between the estimation and the actual measure of the INL of thermometric elements only. Also in this case, the agreement between the calculated characteristic and the measured one is quite good. The deviation is in the range of 0.15 *lsb* @ 12Bit and we suspect this is mainly due to inaccuracy of the measurement of the ADC characteristic after calibration (Figs. 10.25 and 10.26).

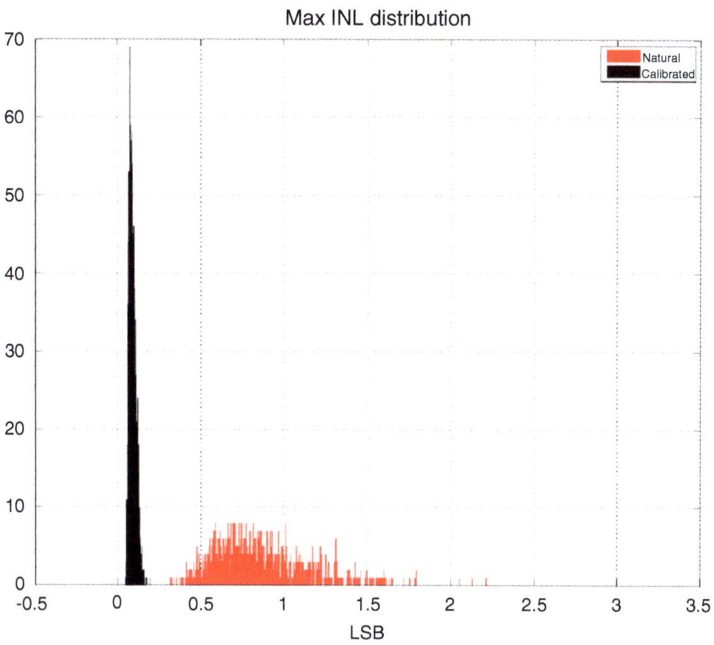

Fig. 10.17 Algo II—Max INL pdf pre- and post-calibration behavior (simulation)

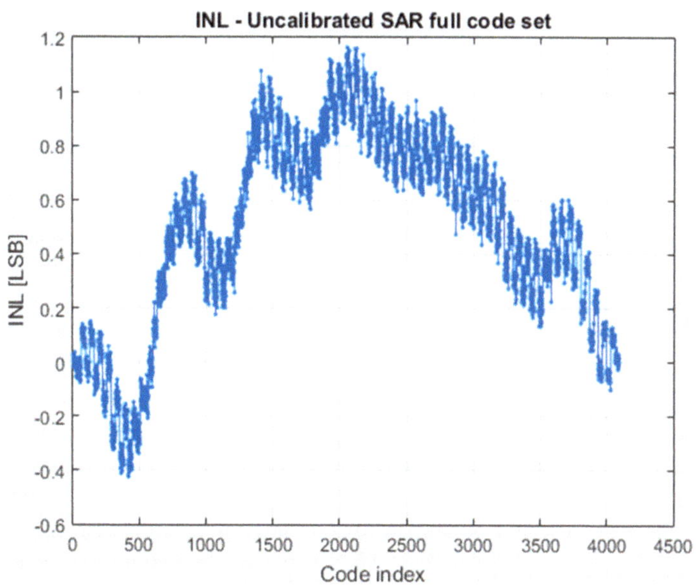

Fig. 10.18 Self-calibrated SAR INL measure before calibration

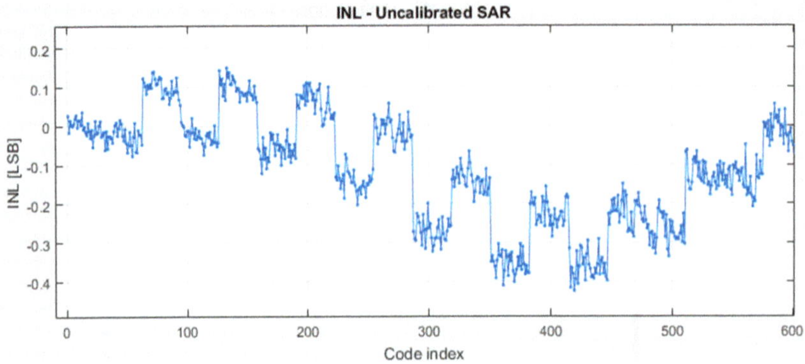

Fig. 10.19 Self-calibrated SAR INL measure before calibration (detail)

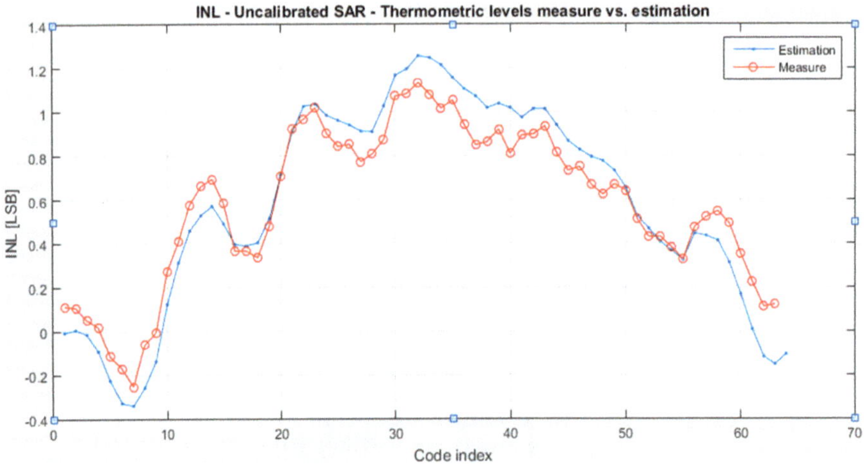

Fig. 10.20 Self-calibrated SAR INL before calibration. Thermometric elements only

10.6 Conclusions

A low cost, high performance calibration method for SAR ADC has been presented. The theory, the simulation and the experimental results show a perfect alignment, leading to an improvement of the performance up to a factor of eight reaching 0.15 *lsb* of INL for the calibrated thermometric elements.

The overhead in terms of silicon area is negligible and there is no need of external, high precision instruments. The quality of the calibration measurements is high enough to allow the calculation of the native and calibrated INL profile with an error in the range of 0.15 *lsb* @ 12bit.

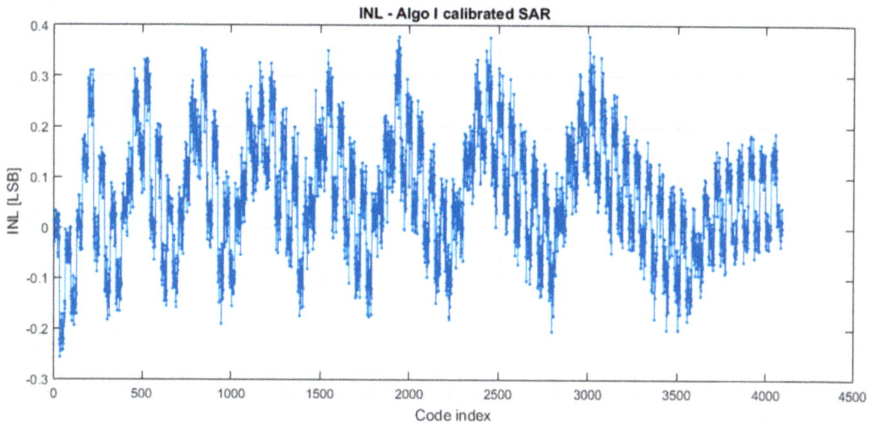

Fig. 10.21 Algo I—self-calibrated SAR INL measure after calibration

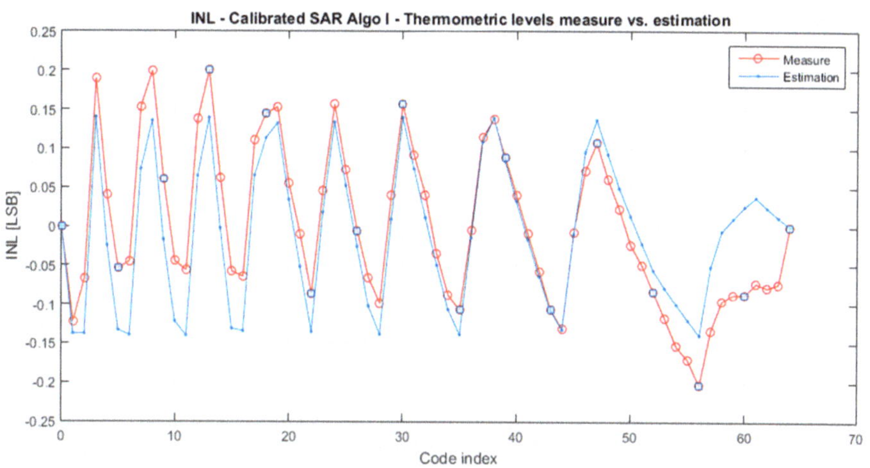

Fig. 10.22 Algo I-thermometric calibration: INL measure vs. estimation

A.1 Appendix

The detailed calculations for DNL_k and INL_k follow, using the measurements E_k.

Let us start from the quantities measured in the Analog Section:

$$E_j = \frac{T_j - T_{ref}}{T_{tot}} \qquad \forall j,\ j \in \{1..N_{th}\}; \quad ref \in \{1..N_{th}\}$$

Solving for T_j:

$$T_j = T_{ref} + E_j T_{tot} \qquad \forall j,\ j \in \{1..N_{th}\}; \quad ref \in \{1..N_{th}\}$$

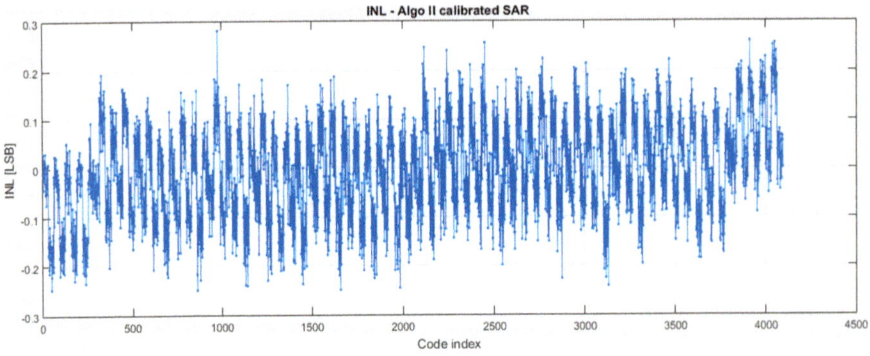

Fig. 10.23 Algo II—self-calibrated SAR INL measure after calibration

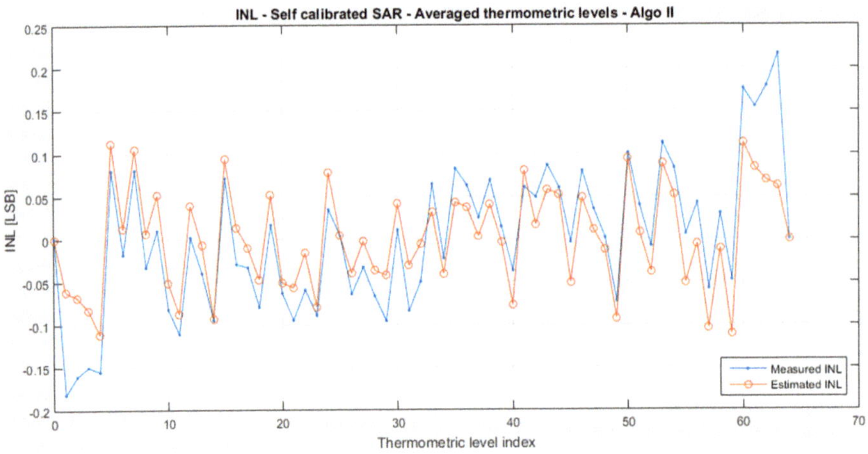

Fig. 10.24 Algo II-thermometric calibration: averaged INL measure vs. estimation

Let us evaluate:

$$T_{tot} = \sum_{j=1}^{N_{th}} T_j = \sum_{j=1}^{N_{th}} \left(T_{ref} + E_j T_{tot}\right) = N_{th} T_{ref} + T_{tot} \sum_{j=1}^{N_{th}} E_j$$

Using

$$T_{tot} = N_{th} T_{ref} + T_{tot} \sum_{j=1}^{N_{th}} E_j \quad \Rightarrow \quad \frac{T_{ref}}{T_{tot}} = \frac{1 - \sum_{all} E_j}{N_{th}}$$

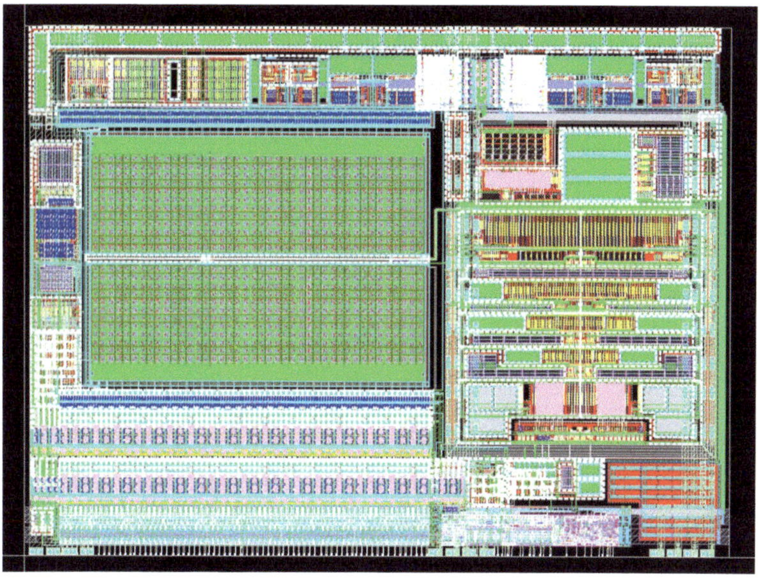

Fig. 10.25 Self-calibrated SAR layout

Fig. 10.26 ADC integration (40 nm technology)

It is possible now the evaluation of INL_k:

$$INL_k \equiv \frac{\sum_{j=1}^{k} T_j}{T_{tot}} - \frac{k}{N_{th}} = \frac{\sum_{j=1}^{k} \left(T_{ref} + E_j T_{tot}\right)}{T_{tot}} - \frac{k}{N_{th}} = \frac{kT_{ref} + T_{tot}\sum_{j=1}^{k} E_j}{T_{tot}} - \frac{k}{N_{th}}$$

$$= k\frac{T_{ref}}{T_{tot}} + \sum_{j=1}^{k} E_j - \frac{k}{N_{th}}$$

Substituting the previous result, we can eliminate T_{ref}:

$$INL_k = k\frac{T_{ref}}{T_{tot}} + \sum_{j=1}^{k} E_j - \frac{k}{N_{th}} = k\frac{1 - \sum_{all} E_j}{N_{th}} + \sum_{j=1}^{k} E_j - \frac{k}{N_{th}}$$

$$= \sum_{j=1}^{k} E_j - k\frac{\sum_{all} E_j}{N_{th}}$$

Moreover, by definition:

$$DNL_k \equiv INL_k - INL_{k-1} = \left(\sum_{j=1}^{k} E_j - k\frac{\sum_{all} E_j}{N_{th}}\right) - \left(\sum_{j=1}^{k-1} E_j - (k-1)\frac{\sum_{all} E_j}{N_{th}}\right)$$

$$= E_k - \frac{\sum_{all} E_j}{N_{th}}$$

References

1. A. Hastings, *The Art of Analog Layout* (Prentice-Hall, Englewood Cliffs, 2000)
2. D. Marche, Y. Savaria, Y. Gagnon, Laser fine-tuneable deep-submicrometer CMOS 14-bit DAC. IEEE Trans. Circuits Syst. I Reg. Pap. **55**(8), 2157–2165 (2008)
3. M. Taherzadeh-Sani, A.A. Hamoui, Digital background calibration of capacitor-mismatch errors in pipelined ADCs. IEEE Trans. Circuits Syst. II **53**(9), 966–970 (2006)
4. W. Liu, P. Huang, Y. Chiu, A 12-bit, 45-MS/s, 3-mW redundant successive-approximation-register analog-to-digital converter with digital calibration. IEEE J. Solid-State Circuits **46**(11), 2661–2672 (2011)
5. J. Bastos, A.M. Marques, M.S.J. Steyaert, W. Sansen, A 12-bit intrinsic accuracy high-speed CMOS DAC. IEEE J. Solid-State Circuits **33**(12), 1959–1969 (1998)

6. G.A.M. Van der Plas, J. Vandenbussche, W. Sansen, M.S.J. Steyaert, G.G.E. Gielen, A 14-bit intrinsic accuracy Q2 random walk CMOS DAC. IEEE J. Solid-State Circuits **34**(12), 1708–1718 (1999)
7. B.H. Leung, S. Sutarja, Multibit Σ-Δ A/D converter incorporating a novel class of dynamic element matching techniques. IEEE Trans. Circuits Syst. II **39**(1), 35–51 (1992)
8. E. Fogelman, I. Galton, W. Huff, H. Jensen, A 3.3-V single-poly CMOS audio ADC delta-sigma modulator with 98-dB peak SINAD and 105-dB peak SFDR. IEEE J. Solid-State Circuits **35**(3), 297–307 (2000)

Chapter 11
Advanced Sensor Solutions for Automotive Applications

Paolo D'Abramo, Alberto Maccioni, Giuseppe Pasetti, and Francesco Tinfena

11.1 Introduction

A sensor is a device whose input is a physical phenomenon and whose output is a quantitative measurement of that physical phenomenon. In the world of electronics, sensors provide the interface between physical or chemical quantities and electrical signals. Direct or indirect measurements are provided depending on whether the output of the sensor is the interesting quantity or it acts as an information carrier for another quantity. Position, velocity, linear and angular displacement, electrical currents are examples of physical signals that can be indirectly measured.

Nowadays sensor signal-conditioning electronics, which performs the necessary analog and digital signal processing, is often provided by integrated circuits and in many cases the sensor itself is fabricated using IC technologies (either on the signal processing die itself or on a separate die). Sensors with integrated electronics are a powerful support in many applications over a broad range of mankind activities, particularly in the automotive environment [1].

The importance of electronics in automotive is constantly increasing and the growth of the number of electronic devices in modern cars is in large part due to the high quantity of sensors that are required for reliable vehicle operation and for achieving higher levels of efficiency, safety and comfort [2]. In fact, sensors with integrated electronics are involved in the engine control and contribute to improve performance, increase efficiency, reduce fuel consumption and cut harmful emissions to fulfill the stringent emissions limits set by environmental pollution standards. They are also used in almost all the safety systems of vehicles, whether

P. D'Abramo (✉)
ams AG, Tobelbader Strasse 30, 8141 Premstaetten, Austria
e-mail: paolo.dabramo@ams.com

A. Maccioni • G. Pasetti • F. Tinfena
ams Italy Srl, Via Giuntini 63, 56023 Navacchio (Pisa), Italy

© Springer International Publishing Switzerland 2017

A. Baschirotto et al. (eds.), *Wideband Continuous-time ΣΔ ADCs, Automotive Electronics, and Power Management*, DOI 10.1007/978-3-319-41670-0_11

the active ones, which help to avoid accidents, or the passive ones, which help to reduce the effects of accidents. Particularly the sensors play an essential role in all the active safety processes for crash avoidance (e.g. brake assistance, traction, steering and stability control), in all the active advanced driver assistance systems and also in some passive crashworthy devices (like airbags). Furthermore there are many essential comfort components and many desirable comfort features that operate sensing the environment and interacting with the passengers (e.g. seat occupancy detection, gesture sensing).

Whatever is their function, in the automotive applications the sensors have to operate in one of the harshest environment for electronic components and, at the same time, they have to ensure excellent reliability and durability levels. In terms of electrical constraints, the sensors have to fulfill especially more and more stringent requirements for Electro-Magnetic Compatibility (EMC): they have to operate correctly when exposed to strong electromagnetic fields caused by other equipment and also emit very low levels of radiated and conducted disturbances that may interfere with other electronic equipment on car. In order to guarantee this, all sensor solutions for vehicles have to undergo many severe EMC tests and fulfill the limits defined by automotive OEMs and international standards (CISPR25, IEC61967, ISO7637 and ISO11452 just to mention the main ones).

Examples of physical and chemical quantities that need to be monitored in automotive are: proximity, position, linear and rotational speed, linear and angular displacement, acceleration, torque, temperature, pressure, electrical current, humidity, airflow, fluid level, fluid concentration, selective gas concentration.

Many advanced sensor technologies are used to sense these quantities. One of the most interesting is based on the measurement of the capacitance variation between two electrodes, which is caused by the variation of the quantity to be indirectly sensed. The transducers that use this measurement principle are in general referred to as capacitive sensors and they are able to detect many of the physical and chemical quantities indicated above [3, 4].

A key advantage offered by capacitive sensors is that indirect measurements are performed without any contact between sensing device and mechanical system. This allows achieving high performance with a relatively easy, robust and cheap construction and for this reason they are widely used in automotive.

On the other hand, depending on the architecture, they could be very sensitive to electro-magnetic fields or they could be an efficient emitting source of disturbances. In particular there are capacitive sensors (e.g. for seat occupancy detection) where at least one of the electrodes needs to have a significant physical extension and act as an antenna, picking up the external electromagnetic fields and/or emitting disturbances.

This poses a huge challenge on the design of the signal conditioning electronics for those sensors: the read-out circuitry has to withstand and suppress the unwanted electromagnetic fields collected by the sensor electrodes while the drive circuitry needs to minimize the emission of unnecessary electromagnetic components. The next chapters describe several architectures for capacitive sensor interfaces and the design aspects to be considered for achieving excellent electromagnetic immunity and emission performance as required in automotive applications.

11.2 Architectures for Capacitive Sensor Interfaces

Different architectures are possible depending on the sensor configuration and the characteristics of the sensor capacitance.

Sensor can be (Fig. 11.1) "grounded" i.e. one of the two terminals is permanently grounded or "floating" i.e. both the terminals are accessible and can be controlled. This constraint has to be taken into account when choosing the sensor interface architecture.

Moreover sensors can have different characteristics in terms of capacitive range, offset value, resistive component (series or parallel), frequency response and so on. Based on these and the specific requirements of each application, the architecture of the sensor interface with the optimal trade-off has to be chosen.

The simplest method to measure the capacitance is to charge it with a constant current and measure the time to charge it until a reference voltage [5, 6]. Figure 11.2 shows the principle implementation.

The capacitance value C_{sense} can be obtained measuring the time delay between the falling edge of the "CLK" signal and the rising edge of the "Comp_out" signal:

$$C_{sense} = Ibias * Delay/V_{Ref} \tag{11.1}$$

The robustness of this architecture can be improved using a reference capacitor and a symmetrical circuit. In this way, errors due to the variation of components, like bias current and reference voltage, can be reduced drastically since the common mode errors do not impact the measurement result. Figure 11.3 shows the improved architecture. In this architecture, the value of the capacitance can be derived from the duty cycle of the "OUT" signal:

$$DutyCycle = C_{sense}/\ (C_{sense} + C_{Ref}) \tag{11.2}$$

Main advantages of the above mentioned approaches are the low complexity, a very low power consumption and fast measurement.

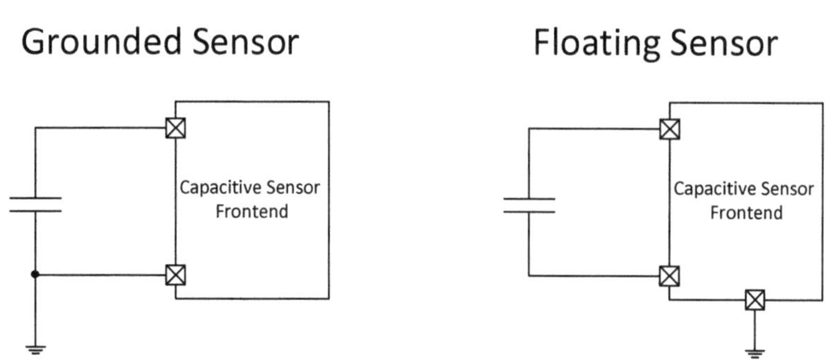

Fig. 11.1 Possible sensor configuration (grounded or floating)

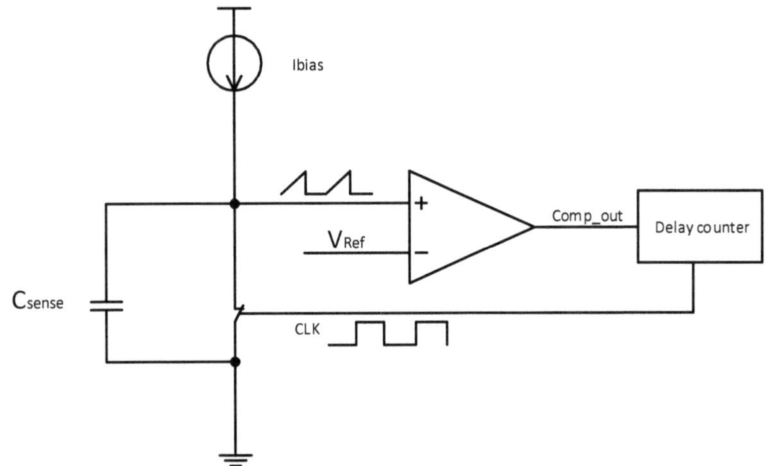

Fig. 11.2 Time measurement concept

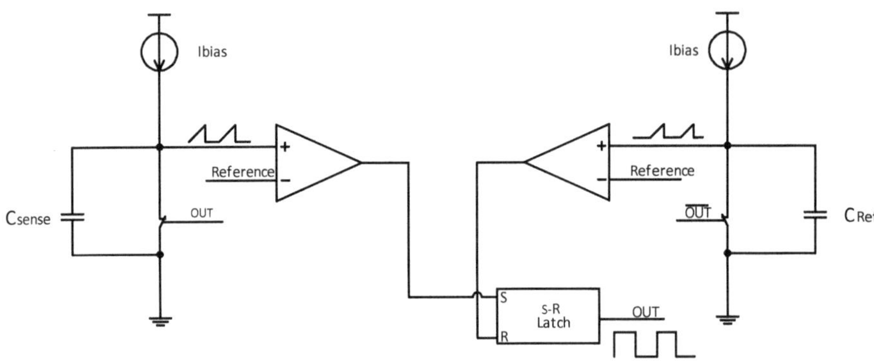

Fig. 11.3 Duty-cycle measurement concept

Unfortunately in terms of electromagnetic immunity, any disturbance on the electrodes affects the result of the measurement and it can only be minimized averaging multiple readings, with a negative impact on response time and power consumption.

In terms of electromagnetic emissions, the speed of the charge and discharge ramps should be as low as possible but this is constrained by the required response time; moreover, the lower the slope of the charging ramp, the more susceptible the front-end is to internal noise (high jitter) and external disturbers (low immunity). An additional constraint to the useful ranges of charge and discharge slopes are the parasitic resistive (series and parallel) and inductive components of the sensor.

A more sophisticated architecture makes use of a Sigma-Delta converter where the sensor capacitance acts as one of the modulator capacitor in the switched capacitor amplifier (Fig. 11.4) [7, 8].

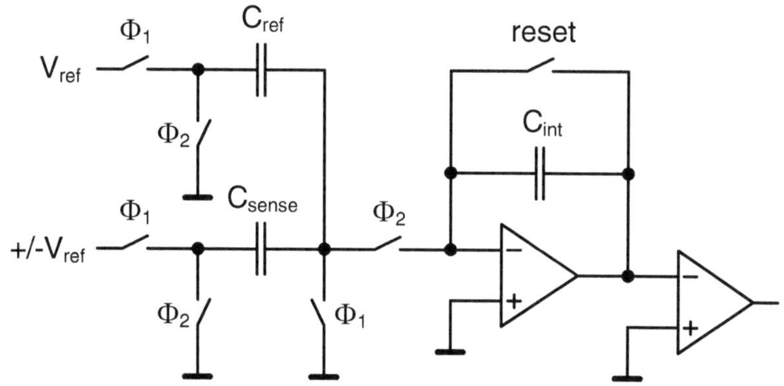

Fig. 11.4 Sigma-Delta switched-capacitor measurement concept

The sensor capacitance value can be calculated according to Eq. 11.3, where M is the mean value at output of the comparator and OSR is the oversampling ratio of the Sigma-Delta converter:

$$C_{sense} = \frac{M}{OSR} C_{ref} \tag{11.3}$$

Main advantages of this architecture are the low noise and the high dynamic range. The desired trade-off between accuracy, settling time and power consumption can be achieved choosing the appropriate oversampling ratio.

In terms of electromagnetic immunity the Sigma-Delta provides good filtering, while the fast commutations of C_{sense} terminals are still critical for the emissions.

The main disadvantage of this architecture is that it requires a floating sensor where both terminals of the capacitor can be driven. This is often not the case for many applications like proximity and gesture detection.

To achieve optimal EMC performance without the need of a floating sensor, the architecture shown in Fig. 11.5 can be used, where the capacitive component of the sensor impedance is measured forcing a known AC voltage and sensing with synchronous demodulation the quadrature current sourced by the driver.

The "drive path" provides the current required to drive the sensor electrode at the excitation frequency (chosen based on the application requirements and sensor characteristics) while keeping the spurious harmonic content at the minimum.

The quadrature component of the current sourced by the driver is sensed by the "pick-up path", which suppresses the potential interfering signals thanks to the synchronous demodulation and adequate filtering.

The next two chapters highlight the specific design aspects to be considered in this architecture in order to achieve optimal electromagnetic emissions and immunity performance respectively.

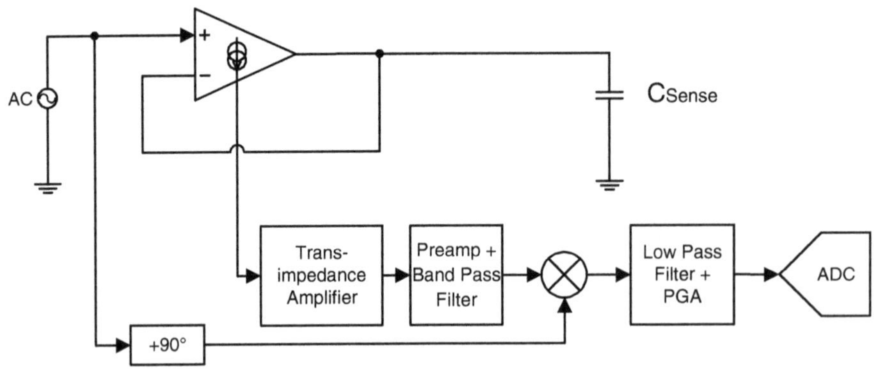

Fig. 11.5 Complex impedance measurement concept

11.3 Design Considerations for Electromagnetic Emissions

The critical part in terms of emissions is the drive path that generates the sinusoidal excitation of the electrode: on one hand the electromagnetic energy injected by the driver into the sensor is radiated by the electrode and on the other hand the AC current that the driver sinks from the supply may cause conducted emissions through the corresponding wire.

Ideally a pure sinusoidal excitation signal would limit the emission spectrum to a single tone at the selected operating frequency. This can be achieved quite well using an oscillator with high spectral purity as signal source. Unfortunately the application requires the programmability with very fine steps of the operating frequency over a wide range which would make such an implementation too large in area (multiple oscillators with large trimming structures are needed).

In order to fulfill those requirements within an acceptable silicon area, the excitation signal is generated by digital synthesis and a sinusoidal DAC.

The sinusoidal DAC gives a good approximation of the sinewave, but, because of non-idealities (e.g. quantization, mismatch of the levels, switching transients), the spectrum of the output signal contains harmonic components of the chosen operating frequency, which can be attenuated adding a low-pass filter between the output of the DAC and the input of the electrode driver.

For the particular application the best trade-off to achieve the required harmonic content while keeping current consumption and silicon area to an acceptable level needs a 16 phases DAC (Figs. 11.6 and 11.7). Special attention has to be taken care also in the DAC layout to minimize the sources of the above mentioned non-idealities.

The high frequency clock to generate the DAC phases is obtained using a multi-phase (16 in this case) PLL locked on the fundamental excitation frequency; using an oscillator at 16 times the sinewave frequency would require much higher current with a negative impact on the conducted emissions.

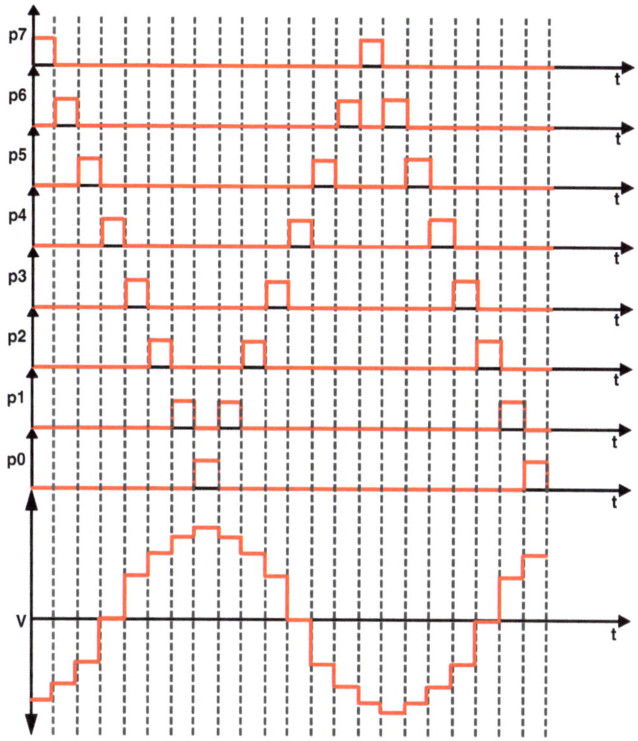

Fig. 11.6 Sinusoidal DAC operation in the time domain

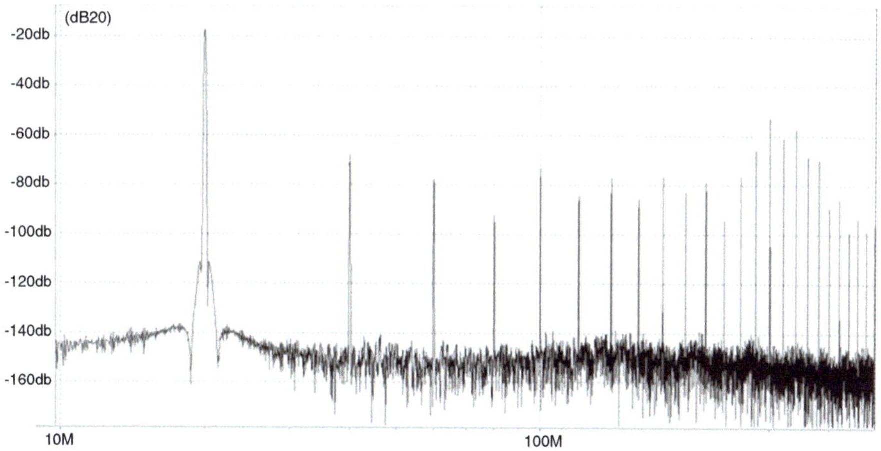

Fig. 11.7 Sinusoidal DAC output in the frequency domain

The sinusoidal DAC output is a low power signal which can't drive directly the electrode. An output driver is needed in order to provide the required current (5 mA for a 100 pF sensor driven with 400 mV at 20 MHz). In order to avoid introducing additional unwanted harmonic components, the design of the driver has to ensure that its linear operation region (avoiding saturation effects both at the input and the output stages) covers the dynamic range of the sinusoidal signal with a proper margin over process, supply voltage and temperature variations.

Last but not least, any parasitic couplings between blocks operating at different frequencies inside the IC have to be minimized. Those introduce crosstalk that can generate intermodulation products between two or more frequencies, leading to additional components in the emission spectrum: for example a parasitic coupling between the electrode excitation frequency and the fixed digital clock frequency will generate an unwanted component which is not a harmonic of the excitation signal itself.

It can be caused by many sources (e.g. parasitic resistance of the supply and ground connections, parasitic capacitive coupling via the substrate or the metal routing, parasitic coupling via common bias blocks) and appropriate measures have to be taken in both schematic and physical design: proper concept for the generation and distribution of reference voltages and bias currents, dedicated supply and ground routing from the pad to each sensitive block, shielding of sensitive signal routing.

The conducted emissions through the supply pins, generated by the drive path but also by other blocks like the digital part, are also a concern, which can be addressed using a LDO to generate the 5 V supply from the 12 V battery voltage with adequate load regulation to suppress all the high frequency components of the IC supply current.

Also the level of the excitation frequency itself may exceed the allowed emission limits. In this case a spread-spectrum scheme can be implemented to vary the operating frequency using a periodic function so that the signal energy is distributed over a wider band and the peak emission level is reduced. When using this, the pick-up path (especially the ADC sampling clock) needs to be synchronized to the periodic modulation signal, in order not to generate measurement ripples.

11.4 Design Considerations for Electromagnetic Immunity

The disturbances can reach the IC through either the sensing electrode or the power supply. The main function of the pick-up path is exactly to amplify and isolate the signal component (AC current in quadrature to the sensor excitation signal), which is then converted into the digital domain for further processing.

Regarding the disturbance collected directly by the sensing electrode, the inherent selectivity in the frequency domain of the architecture is used to reject unwanted input signals that are at different frequencies or not coherent with the excitation signal.

However, the previous sentence is true only if adequate measures are taken in the design to ensure that the all the blocks, for both the drive and the pick-up path, do not saturate even in presence of the worst external interferers.

In particular, the operating current of the electrode driver for a 100 pF sensor operated with 400 mV amplitude at 20 MHz is equal to:

$$Z_{sense} = 1/\left(2^{*}\pi^{*}f^{*}C\right) = 1/\left(2^{*}\pi^{*}20\ M^{*}100\ p\right) = 80\ \Omega \tag{11.4}$$

$$I_{out} = V_{out}/Z_{sense} = 400\ m/80 = 5\ mA \tag{11.5}$$

But in the given application experimental measurements show that during immunity tests up to 15 mA can be induced on the sensor electrode, three times higher than the operating current! This implies that, as first step to achieve optimal performance in terms of electromagnetic immunity, the driver has to be sized to handle up to 20 mA without saturating.

The same consideration applies to the whole pick-up path, which has to sense such a current still keeping a linear response. Only under these conditions, the high selectivity of the synchronous demodulator and related filters can guarantee an effective suppression of the interferers while extracting correctly the amplitude of the sensor excitation current.

The disturbances reaching the IC through the power supply are handled using a LDO with adequate PSRR over the relevant frequency range to generate the supply voltages of the different IC blocks, which need also to be optimized with respect to PSRR. Special care has also to be taken, as mentioned already, both in schematic and physical design to avoid crosstalk between blocks.

Of course the most appropriate trade-off between selectivity of the pick-up path and the other application requirements (e.g. response time, noise, temperature drift) has to be determined. In the specific application single digit fF measurement errors could be achieved with conversion times below 100 μs and current consumption below 100 μA (in low-power mode, required when the car is parked).

Last but not least, in case a strong but narrow-band disturber is within the pick-up receiving band, a frequency hopping scheme can be adopted switching dynamically the excitation frequency to a different one far enough from such an interfering signal.

11.5 Conclusions

Sensors are more and more widely used in many applications over a broad range of mankind activities, particularly in the automotive environment.

Among these, the capacitive sensors play a significant role because they can indirectly monitor many different physical quantities with the advantage of a contactless implementation.

In this work several architectures for capacitive sensor interfaces have been presented, with special focus on the design aspects to be considered for achieving excellent electromagnetic immunity and emission performance as required in automotive applications.

References

1. E. Marenzi, G.M. Bertolotti, F. Leporati, G. Danese, Capacitive sensors matrix for interface pressure measurement in clinical, ergonomic and automotive environments, in *Euromicro Conference on Digital System Design (DSD), 2013*, pp. 803–806, 4–6 Sept. 2013. doi:10.1109/DSD.2013.123
2. B. George, H. Zangl, T. Bretterklieber, G. Brasseur, Seat occupancy detection based on capacitive sensing. IEEE Trans. Instrum. Meas. **58**(5), 1487–1494 (2009). doi:10.1109/TIM.2009.2009411
3. R. Reichenbach, D. Schubert, G. Gerlach, Micromechanical triaxial acceleration sensor for automotive applications, in *12th International Conference on TRANSDUCERS, Solid-State Sensors, Actuators and Microsystems, 2003*, vol. 1, pp. 77–80, 8–12 June 2003. doi:10.1109/SENSOR.2003.1215257
4. Y. Hezarjaribi, M.N. Hamidon, A.R. Bahadorimehr, Pressure sensors based on MEMS, operating in harsh environments (touch-mode), in *Electronic Manufacturing Technology Symposium (IEMT), 2008 33rd IEEE/CPMT International*, pp. 1–5, 4–6 Nov 2008. doi:10.1109/IEMT.2008.5507791
5. J. Cho, S. Park, Capacitive sensor for automotive engine oil degradation using wireless network, in *International Symposium on Advanced Packaging Materials: Microtech, 2010. APM '10*, pp. 88–91, Feb 28, 2010–Mar 2, 2010. doi:10.1109/ISAPM.2010.5441375
6. F. Baronti, F. Lenzi, R. Roncella, R. Saletti, Distributed sensor for steering wheel rip force measurement in driver fatigue detection, in *Design, Automation & Test in Europe Conference & Exhibition, 2009. DATE '09*, pp. 894–897, 20–24 Apr 2009. doi:10.1109/DATE.2009.5090790
7. D.S. Lee, H.D. Tiwari, S.Y. Kim, J. Lee, H. Park, Y. Pu, M. Seo, K.Y. Lee, A highly linear, small-area analog front end with gain and offset compensation for automotive capacitive pressure sensors in 0.35-um CMOS. IEEE Sens. J. **15**(3), 1967–1976 (2015). doi:10.1109/JSEN.2014.2369471
8. L. Cardelli, L. Fanucci, V. Kempe, F. Mannozzi, D. Strle, Tunable bandpass sigma delta modulator using one input parameter. Electron. Lett. **39**(2), 187–189 (2003). doi:10.1049/el:20030146

Chapter 12
A Low-Power Continuous-Time Accelerometer Front-End

Piero Malcovati, Marcello De Matteis, Alessandro Pezzotta, Marco Grassi, Marco Croce, Marco Sabatini, and Andrea Baschirotto

12.1 Introduction

Smart sensor devices exploiting CMOS mixed-signal interface circuits are widely used in several applications, where continuous monitoring of physical, chemical, or biomedical variables is required [1–5]. These systems are often battery operated or even intrinsically self-powered and, therefore, maintaining a sufficiently large signal-to-noise ratio (*SNR*) while minimizing the power consumption is the main design issue.

Among smart sensor devices, MEMS capacitive accelerometers are extensively used in several fields, including, for example, automobiles, smartphones, and body networks [1, 3]. In all of these application the MEMS accelerometer capacitance variation has to be read out and digitized by a suitable analog front-end (AFE) circuit.

In this paper we present a complete AFE circuit for 3-axes MEMS capacitive accelerometers, which exploits a continuous-time (CT) constant-voltage approach for reading out the sensor capacitance. The CT topology used allows a significant power consumption reduction with respect to conventional switched-capacitor (SC) discrete-time (DT) solutions. In order to make the proposed CT approach viable, the proposed AFE circuit exploits a novel sensor biasing technique to achieve reliable

P. Malcovati (✉) • M. Grassi • M. Croce
Department of Electrical, Computer, and Biomedical Engineering,
University of Pavia, Pavia, Italy
e-mail: piero.malcovati@unipv.it

M. De Matteis • A. Pezzotta • A. Baschirotto
Department of Physics, University of Milano-Bicocca, Milano, Italy

M. Sabatini
Pirelli Tyre, Milano, Italy

© Springer International Publishing Switzerland 2017

A. Baschirotto et al. (eds.), *Wideband Continuous-time $\Sigma\Delta$ ADCs, Automotive Electronics, and Power Management*, DOI 10.1007/978-3-319-41670-0_12

operation even in the presence of strong sensor leakage currents as well as process, voltage, and temperature (PVT) variations.

The paper is organized as follows. Section 12.2 describes the basic operating principle of a MEMS capacitive sensor, as well as the possible AFE circuit topologies that can be used for reading out the sensor capacitance variations. The specifications of the proposed AFE circuit are derived in Sect. 12.3, while Sect. 12.4 describes in detail the AFE architecture and the various building blocks. Finally, the experimental results obtained from an integrated AFE circuit prototype are reported in Sect. 12.5, while in Sect. 12.6 we draw some conclusions.

12.2 MEMS Capacitive Sensors

In its simplest form, a MEMS capacitive sensor consists of a parallel-plate capacitor with a fixed plate and a movable plate, as shown in Fig. 12.1a. The measurand M_S (e.g. the acceleration) affects the distance x between the plates. Therefore, the capacitance of a generic MEMS capacitive sensor can be expressed as

$$C(M_S) = \frac{\epsilon_0 A}{x(M_S)} = \frac{\epsilon_0 A}{x_0 + \Delta x(M_S)}, \tag{12.1}$$

where A is the area of the smallest capacitor plate and ϵ_0 is the vacuum dielectric permittivity.

Considering that $C = Q/V$, as for any capacitive sensor, there are two possible ways to transform the capacitance variation into an electrical signal (a voltage V or a charge Q):

- the constant-charge approach, for which the output signal is a voltage ($V = Q/C$);
- the constant-voltage approach, for which the output signal is a charge ($Q = CV$).

Denoting with C_0 the MEMS capacitance in quiescent conditions, i.e. when $x = x_0$, and assuming linear the relationship between the measurand M_S and the deformation x ($\Delta x = -\kappa \Delta M_S$), which is actually true for $\Delta x \ll x_0$, we can calculate the output signals (ΔV and ΔQ) as a function of ΔM_S in the two cases.

With the constant-charge approach, the MEMS capacitor is initially charged to a fixed voltage V_B. If the capacitor is well insulated, the charge stored on it

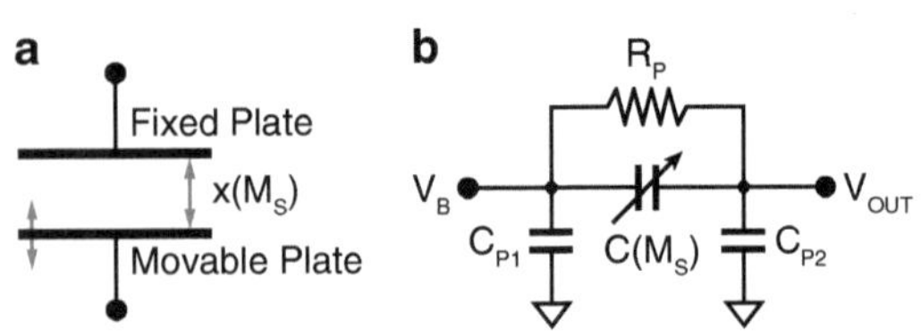

Fig. 12.1 Simplified structure (**a**) and equivalent circuit (**b**) of a MEMS capacitive sensor

($Q = C_0 V_B$) remains constant. As a consequence, the capacitance variation due to a variation of the measurand ΔM_S leads to a voltage signal (ΔV) given by

$$\Delta V = \frac{Q}{C(M_S)} - \frac{Q}{C_0} = \frac{Q\Delta x}{\epsilon_0 A} = -\frac{\kappa C_0 V_B \Delta M_S}{\epsilon_0 A} = -\kappa_V \Delta M_S, \tag{12.2}$$

where κ_V denotes the voltage sensitivity of the sensor.

With the constant-voltage approach, a fixed voltage V_B is applied and maintained across the MEMS capacitor. The capacitance variation due to a variation of the measurand ΔM_S, assuming $\Delta x \ll x_0$, leads to a charge signal (ΔQ) given by

$$\Delta Q = V_B \left[C(M_S) - C_0\right] \approx -\frac{V_B C_0^2 \Delta x}{\epsilon_0 A} = \frac{\kappa C_0^2 V_B \Delta M_S}{\epsilon_0 A} = \kappa_Q \Delta M_S, \tag{12.3}$$

where $\kappa_Q = \kappa_V C_0$ denotes the charge sensitivity of the sensor.

According to (12.2) and (12.3), both κ_V and κ_Q depend on the bias voltage V_B. Therefore, in order to increase the sensitivity and, hence, the *SNR* of the sensor, the value of V_B has to be at least of the order of few V.

In practical implementations, a MEMS sensor is not just a capacitor, but some additional parasitic components have to be taken into account. The equivalent circuit of an actual device is shown in Fig. 12.1b. Besides the variable capacitance $C(M_S)$, the equivalent circuit includes two parasitic capacitances C_{P1} and C_{P2}, connected between each plate of the MEMS sensor and the substrate, as well as a parasitic resistance R_P, connected in parallel to $C(M_S)$. The value of these parasitic components depends on the specific implementation of the sensor, but typically C_{P1} and C_{P2} are of the order of few pF, while R_P is in the GΩ range.

The AFE circuit for a MEMS capacitive sensor has to read out the electrical signal ΔV or ΔQ and convert it in the digital domain. Digital output is, indeed, a must in most applications, in order to gain a competitive advantage over conventional devices, in terms of area and cost at system level. Therefore, the AFE circuit for a MEMS sensor, whose block diagram is shown in Fig. 12.2, typically consists of some sort of preamplifier followed by an analog-to-digital converter (ADC).

The topology and the functionality of the preamplifier in a MEMS capacitive sensor AFE circuit is different depending on the approach used to read out the capacitance variation.

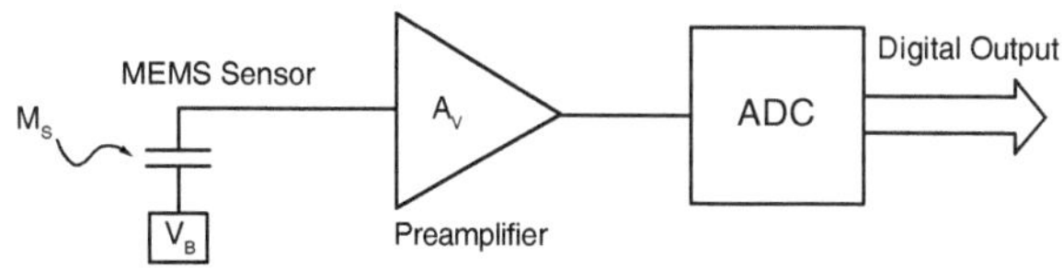

Fig. 12.2 Typical block diagram of the AFE circuit for a MEMS capacitive sensor

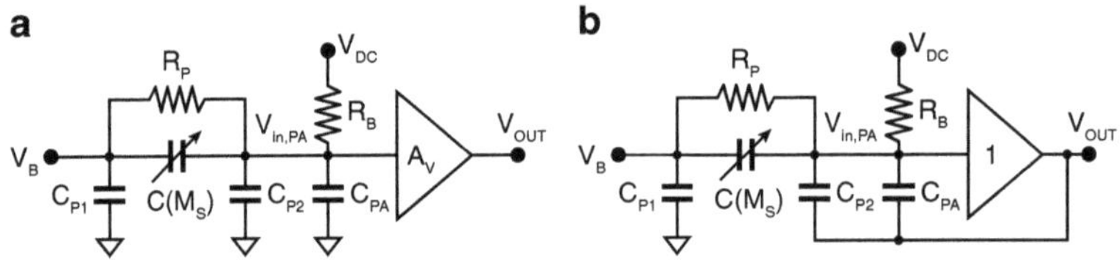

Fig. 12.3 Block diagram of the preamplifier for the constant-charge approach without (**a**) and with (**b**) parasitic capacitance bootstrapping

12.2.1 Constant-Charge Approach

With the constant-charge approach, the preamplifier has to buffer the sensor output voltage, eventually introducing some gain, providing a suitable signal, with low output impedance, to the subsequent ADC, as shown in Fig. 12.3a. In this case, the input impedance of the preamplifier has to be extremely high (larger than 10 GΩ), in order to guarantee that the charge stored on the sensor capacitance is maintained, while providing, at the same time, a suitable dc bias voltage at the preamplifier input node. The biasing network at the preamplifier input is, therefore, very critical and represents typically the most challenging part of the preamplifier design. The solutions usually adopted to implement R_B are based on inversely biased diodes or switched networks. Resistor R_B introduces a high-pass filter with cut-off frequency $f_{HP} \approx 1/(2\pi R_B C_0)$. The bandwidth requirement of the preamplifier with the constant-charge approach depends only on the bandwidth of the measurand (typically few kHz).

The parasitic capacitance at the preamplifier input (C_{PA}) is also particularly important, considering that the output voltage of the sensor ΔV, given by (12.2), in the presence of parasitic capacitances (both C_{P2} and C_{PA}), is actually attenuated, leading to

$$V_{in,PA} = \Delta V \frac{C_0}{C_0 + C_{P2} + C_{PA}} = -\Delta M_S \frac{\kappa_V C_0}{C_0 + C_{P2} + C_{PA}}. \qquad (12.4)$$

This attenuation can often be quite substantial, thus leading to a degradation of the actual sensor sensitivity and, hence, of the *SNR*. This problem can be mitigated by bootstrapping C_{P2} and, eventually, also C_{PA}, as shown in Fig. 12.3b. In this case, the voltage across the parasitic capacitances is maintained constant, independently of the signal, and, therefore, $V_{in,PA} \approx \Delta V$. In order to achieve proper bootstrapping, the gain of the preamplifier (or, at least, of the preamplifier first stage) has to be unitary and, hence,

$$V_{OUT} = -\kappa_V \Delta M_S. \qquad (12.5)$$

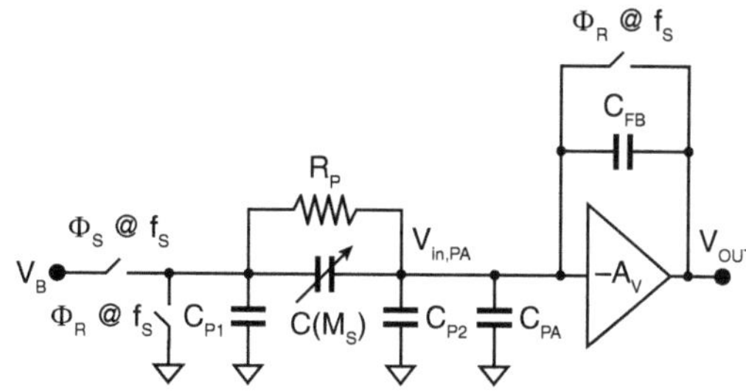

Fig. 12.4 Block diagram of the preamplifier for the constant-voltage approach

12.2.2 *Constant-Voltage Approach*

With the constant-voltage approach, the preamplifier has transform the charge ΔQ provided by the sensor, given by (12.3), into a voltage signal, with low output impedance, suitable to drive the subsequent ADC, as shown in Fig. 12.4. In this case, therefore, the preamplifier is actually a charge amplifier, whose output voltage is given by

$$V_{OUT} = \frac{\Delta Q}{C_{FB}} = \frac{\kappa C_0^2 V_B \Delta M_S}{\epsilon_0 A_V C_{FB}} = \frac{\kappa_Q \Delta M_S}{C_{FB}} = \kappa_V \frac{C_0}{C_{FB}} \Delta M_S. \tag{12.6}$$

The input charge ΔQ is obtained by connecting the sensor capacitance alternatively to V_B or to ground with switching frequency f_S. The feedback capacitance C_{FB} is also reset periodically at the same frequency. As a consequence, the bandwidth requirement of the preamplifier with the constant-voltage approach is determined by f_S, which has to be much larger (at least one order of magnitude) than the bandwidth of the measurand.

The voltage swing at node $V_{in,PA}$ with the constant-voltage approach is quite small, thus making the effect of the parasitic capacitances C_{P2} and C_{PA} negligible. Moreover, since $V_{in,PA}$, in this case, is a low-impedance node, the DC biasing of the preamplifier input is not critical.

12.2.3 *Constant-Charge vs. Constant-Voltage*

Apparently, the constant-voltage approach looks more attractive than the constant-charge approach, since both the parasitic capacitance effect and the biasing of the input node are less critical. However, the constant-charge approach has three significant advantages, which make it appealing, especially for ultra-low-power applications.

Firstly, according to (12.5) and (12.6), the output voltage V_{OUT} in the constant-voltage approach does not depend only on κ_V, as in the constant-charge approach, but also on the ratio C_0/C_{FB}, which is not well controlled (C_0 and C_{FB} are realized with different technologies and depend on different process parameters). Therefore, the overall sensitivity of sensor and preamplifier ($V_{OUT}/\Delta M_S$) features a much larger variation with the process in the constant-voltage approach, where the spread is due to κ_V, C_0 and C_{FB}, than in the constant-charge approach, where the spread is only due to κ_V.

Secondly, in the constant-charge approach, the voltage reference that generates V_B does not have to provide any current, besides at system startup, whereas in the constant-voltage approach it has to deliver the signal charge ΔQ also during normal operation. As a consequence, in the constant-voltage approach the voltage reference specifications are more stringent, especially in terms of output impedance and, hence, of power consumption, in order to avoid distortion of the signal, due to dynamic variations of voltage V_B correlated with the measurand ΔM_S.

Finally, and most importantly, with the constant-charge approach the preamplifier bandwidth requirement is at least one order of magnitude lower than with the constant-voltage approach, leading to a potentially lower power consumption.

12.3 System Specifications

In the considered application, the MEMS capacitive accelerometer can be modeled as a bridge of four variable capacitors [1]. According to (12.1), each capacitance value (C_V) as a function of the acceleration (a) is given by

$$C_V = \frac{\epsilon_0 A}{x_0 \pm \Delta x(a)} = \frac{C_0}{1 \pm \kappa_a a}, \tag{12.7}$$

where $C_0 = \epsilon_0 A/x_0 \approx 450$ fF is the capacitance value in the absence of acceleration (offset capacitance), while $\kappa_a = \frac{\Delta x}{x_0 a}$ is the raw sensitivity of the device (about 47×10^{-6} g^{-1}, considering a expressed in g).

Since $\kappa_a a \ll 1$, (12.7) can be linearized, leading to

$$C_V \approx C_0 \mp \kappa_C a, \tag{12.8}$$

where $\kappa_C = 0.016$ fF/g is the actual sensitivity of the capacitive MEMS accelerometer. Considering a full-scale acceleration value of ± 3000 g, the maximum sensor capacitance variation ($\Delta C_V|_{max}$) turns out to be as large as ± 48 fF. To achieve these performance parameters, the sensor area and volume have to be large, leading to fairly low parallel resistances (R_P in Fig. 12.1b) and, hence, to large dc leakage currents flowing through the sensor capacitors ($I_L \approx \pm 10$ pA for $V_B \approx 600$ mV, V_B being the sensor bias voltage). Moreover, these leakage current can be significantly different in the four sensor capacitors, leading to both common-mode and differential components at the input of the AFE circuit. The accelerometer stand-alone achieves a $SNR = 60$ dB in the 1-Hz-to-4-kHz bandwidth and a

Table 12.1 Accelerometer features

Parameter	Value
Bandwidth (B)	1–4 kHz
Full-scale acceleration ($\Delta C_V\|_{max}$)	±3000 g
Sensitivity (κ_C)	0.016 fF/g
Offset capacitance (C_0)	450 fF
Leakage currents (I_L)	±10 pA

Table 12.2 AFE circuit specifications

Parameter	Value
Technology	130-nm CMOS
Power supply voltage (V_{DD})	1.2 V
Differential full-scale output signal (V_{FS})	1.2 V_{pp}
Signal-to-noise ratio (*SNR*)	> 63 dB
Total harmonic distortion at full-scale (*THD*)	< −30 dB
ADC resolution	10 bits
ADC sampling frequency (f_S)	1 MHz
Power consumption	< 100 μW

$THD \approx -28\,\text{dB}$. Such distortion is mainly due the intrinsically non-linearity of the acceleration-to-capacitance conversion law given by (12.7).

The AFE circuit has to read out the sensor signal without degrading the *SNR* and the *THD*, while maintaining the overall power consumption below 100 μW, including both static and dynamic contributions. Moreover, the AFE circuit has to feature a low-frequency (<1 Hz) high-pass transfer function to reject low-frequency sensor artifacts (such as the leakage currents). The main sensor features and the resulting AFE circuit specifications are summarized in Tables 12.1 and 12.2, respectively [6].

12.4 Analog Front-End Circuit Design

Most CMOS AFE circuits for MEMS capacitive accelerometers available in literature [7–10] are based on the constant-voltage approach and exploit SC DT techniques, which are intrinsically robust against many sensors non-idealities (such as leakage currents), but, as discussed in Sect. 12.2, require larger power consumption than CT solutions based on the constant-charge approach.

Therefore, in some recent battery-operated applications [1, 11], where power consumption is the most critical specification, SC AFE circuits turn out to be too power hungry and, hence, CT AFE circuits potentially represent the most suitable solution, provided that some critical issues, such as sensor biasing and robustness against sensor leakage currents, are properly handled.

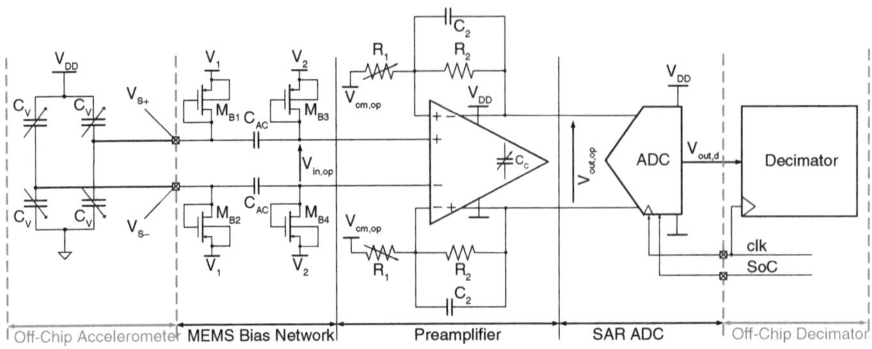

Fig. 12.5 Complete schematic diagram of the proposed AFE circuit

In this scenario, for the proposed AFE circuit we adopted a CT solution based on the constant-charge approach. The AFE circuit, whose complete schematic diagram is shown in Fig. 12.5, consists of the cascade of the following stages:

- the *MEMS bias network*, which guarantees constant charge in the sensor capacitors and prevents the sensor leakage currents from reaching the preamplifier input nodes;
- the *preamplifier*, which performs signal amplification, gain calibration, and anti-alias filtering;
- the *ADC*, which digitizes the sensor signal, exploiting oversampling;
- the *decimator*, which reduces the ADC sampling rate and concurrently increases the *SNR*.

12.4.1 MEMS Bias Network

The top plates of the accelerometer capacitors are biased using the AFE circuit power supply voltage ($V_{DD} = 1.2\,\text{V}$) and ground, thus avoiding the need for an additional voltage reference (eventually higher than V_{DD}), whereas the bottom plates are connected to voltage $V_1 = 600\,\text{mV}$ by means of transistors M_{B1} and M_{B2}, which feature very large resistance (>50 GΩ in nominal conditions). This approach allows a significant reduction of the system power consumption at the expense of a reduction of $V_B = V_{DD} - V_1 = 600\,\text{mV}$ and, hence, of the sensitivity κ_V given by (12.2), thus stressing the gain requirements of the preamplifier.

Transistors M_{B1} and M_{B2} guarantee charge conservation in the sensor capacitors, but, because of their very high impedance, in the presence of significant sensor leakage currents, they produce a large voltage shift at the sensor output nodes V_{S+} and V_{S-} (with both common-mode and differential components if the leakage currents are mismatched).

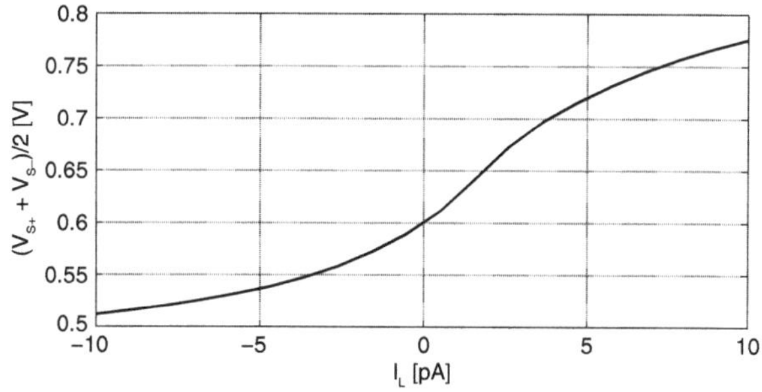

Fig. 12.6 Simulated sensor common-mode output voltage as a function of the sensor leakage current

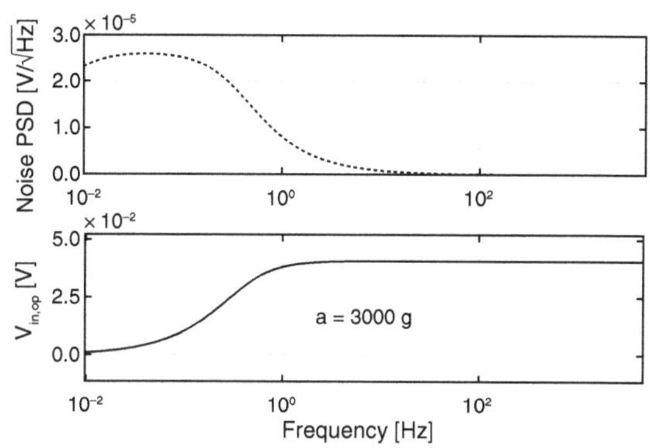

Fig. 12.7 Simulated noise *PSD* due to M_{B1}, M_{B2}, M_{B3}, and M_{B4}, referred to the preamplifier input nodes, and sensor signal transfer function

Figure 12.6 shows the simulated sensor common-mode output voltage $\frac{V_{S+}+V_{S-}}{2}$ with a sensor leakage current I_L of ± 10 pA. The leakage current, flowing through M_{B1} and M_{B2}, slightly reduces their impedance and changes their drain-source voltage, shifting the sensor common-mode output voltage (equal to $V_1 = 0.6$ V without leakage current) in the range from 0.52 to 0.78 V, which is unacceptable for the preamplifier, which has then to be protected against it.

For this reason, we introduced capacitors C_{AC}, which decouple at dc the preamplifier input nodes from the sensor output nodes. The preamplifier input nodes are then biased at voltage $V_2 = V_1 = 600$ mV, independently of the sensor leakage currents, by means of two diode-connected transistors (M_{B3} and M_{B4}), using the same technique adopted for biasing V_{S+} and V_{S-}. The resistance of M_{B3} and M_{B4} (R_B) together with C_{AC} determines the high-pass pole of the circuit $f_{AC} = \frac{1}{2\pi C_{AC} R_B} \approx 0.1$ Hz ($C_{AC} = 45$ pF).

This bias technique for both the sensor and preamplifier input nodes preserve also signal integrity in terms of *SNR*, since all the noise contributions from M_{B1}, M_{B2}, M_{B3}, and M_{B4} are low-pass filtered with a cut-off frequency equal to f_{AC}. Figure 12.7 shows the simulated noise power spectral density (*PSD*) due to M_{B1}, M_{B2}, M_{B3}, and M_{B4}, referred to the preamplifier input nodes, as well as the sensor signal transfer function. The noise *PSD* peak value is 25 $\mu V/\sqrt{Hz}$ at 0.1 Hz, then the *PSD* decreases, resulting in an overall in-band integrated noise at the preamplifier input nodes of $5\,\mu V_{rms}$. The signal at the preamplifier input nodes for full-scale acceleration is about 40 mV over the frequency range from 1 Hz to 4 kHz.

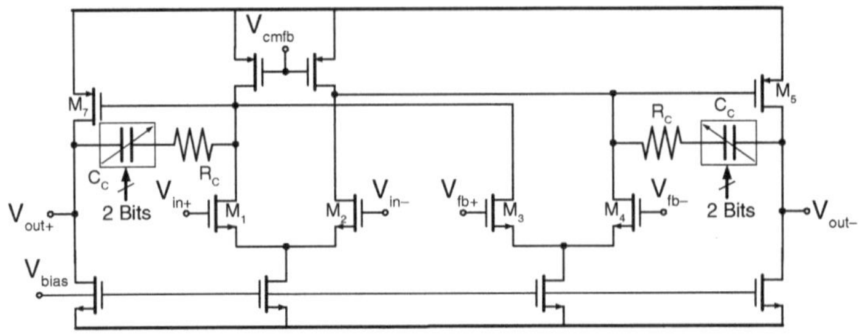

Fig. 12.8 Schematic diagram of the preamplifier

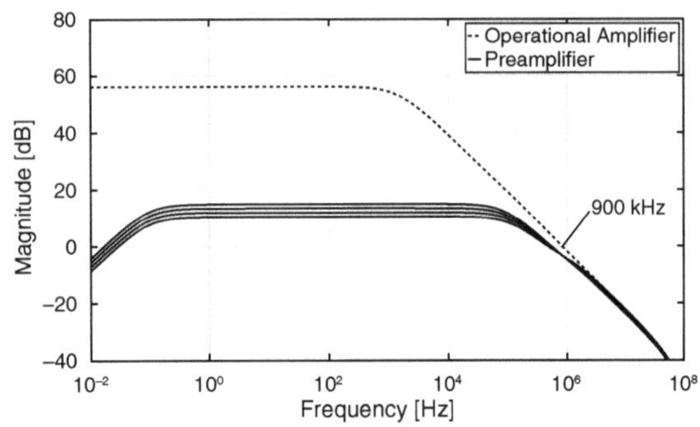

Fig. 12.9 Simulated frequency responses of the closed-loop preamplifier and of the open-loop operational amplifier

12.4.2 Preamplifier

The preamplifier infinite input impedance, needed to avoid parasitic current flow through M_{B3} and M_{B4}, requires the use an operation amplifier in non-inverting configuration. However, in order to effectively process the differential sensor signal a differential-difference topology has to be adopted, as shown in Fig. 12.8. In this topology, a differential stage (M_3 and M_4) is used to close the operational amplifier feedback loop through R_1, R_2, and C_2, thus determining the preamplifier gain, while a second differential stage (M_1 and M_2) receives the sensor differential signal.

The preamplifier pass-band gain is set to $1 + \frac{R_2}{R_1}$. In order to accommodate different accelerometer sensitivity values, while maintaining the same ADC differential input range (1.2 V_{pp}), programmable gain is implemented by selecting the value of R_1 (from 10.5 to 15 dB with 1.5-dB steps, i.e. with a 2-bit control word). To guarantee stability for the different gain levels, also the Miller compensation capacitance C_C is adjusted by the same digital control word.

The simulated frequency responses of the closed-loop preamplifier for different gain settings and of the open-loop operational amplifier are plotted in Fig. 12.9. The effect of the operational amplifier finite bandwidth at high frequency improves the aliasing rejection for the subsequent ADC, without degrading stability. The loop-gain magnitude and phase as a function of frequency are shown in Fig. 12.10 for minimum gain (10.5 dB, the worst case scenario for stability). A phase margin of

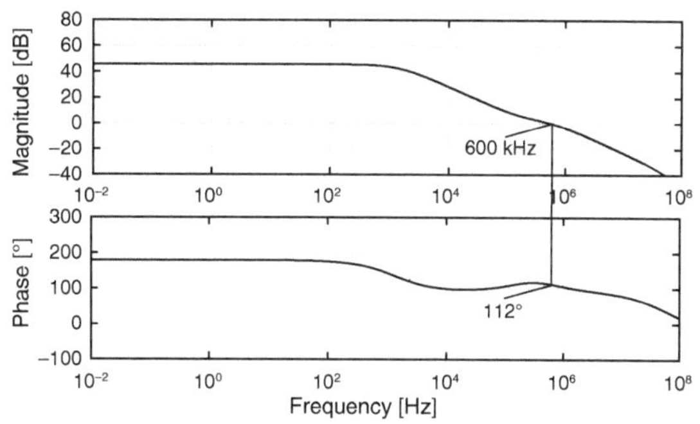

Fig. 12.10 Simulated *preamplifier* loop-gain magnitude and phase as a function of frequency

Table 12.3 Preamplifier features

Parameter	Value
Preamplifier gain (G_{PA})	15 dB, 13.5 dB, 12 dB, 10.5 dB
R_2	50 kΩ
R_1 @ $G_{PA} = 10.5$ dB	15 kΩ
C_2	16.5 pF
In-band input referred noise *PSD*	$95\ \text{nV}/\sqrt{\text{Hz}}$
Integrated in-band output noise	$50\ \mu V_{rms}$
Operational amplifier dc gain (A_0)	61 dB
Operational amplifier *GBW*	900 kHz
Maximum input offset voltage (V_{os})	0.75 mV
Total current consumption (I_T)	26.7 µA

112° has been achieved with a gain-bandwidth product (*GBW*) of 600 kHz. The simulations have been performed using an equivalent capacitive output load equal to 0.39 pF, which corresponds to the maximum input grounded capacitance of the following ADC. A summary of the most important preamplifier features is reported in Table 12.3.

The preamplifier current consumption is shared between input stage (6 µA, i.e. the minimum current required to properly bias the large input devices to guarantee the target *GBW*) and output stage (7 µA, i.e. the minimum current required to drive the subsequent ADC). This amount of current guarantees that the operational amplifier thermal noise is lower than the ADC quantization noise. The operational amplifier achieves a dc gain (A_0) of 61 dB, with a *GBW* of 900 kHz and a power consumption of 32 µW.

Preamplifier Offset

The CT AFE circuit topology adopted prevents the use of DT offset-cancelation techniques (such as auto-zero). As a consequence, we used very large operational amplifier input devices ($W = 100\,\mu\text{m}$, $L = 0.5\,\mu\text{m}$), which operate in the sub-threshold region. The output offset voltage of the operational amplifier is given by

$$V_{os,out} = V_{os}\left(1 + \frac{R_2}{R_1}\right), \tag{12.9}$$

where the input offset voltage V_{os} can be, in first approximation, calculated by using the mismatch parameter $\sigma_{V_{th}}$, available for the used 0.13-μm CMOS technology, according to

$$V_{os} \approx 3\frac{\sigma_{V_{th}}}{\sqrt{WL}} = 0.75\,\text{mV}. \tag{12.10}$$

Considering that the maximum preamplifier gain is 15 dB, the overall output differential offset is about 4.5 mV. Since the ADC differential input range is 1.2 V_{pp}, such offset can be effectively managed by the AFE circuit.

Preamplifier Noise

The large operational amplifier input devices, required to minimize the offset, also minimize the flicker noise contribution, which, in the considered application, is very important due to the limited bandwidth.

Therefore, the preamplifier output referred noise *PSD* ($V^2_{n,,out}$) is mainly due to the thermal noise of the resistors used for setting the closed-loop gain and of the transistors of both input stages of the differential-difference amplifier, whose transconductance is $g_m = 50\ \mu\text{A/V}$, resulting in

$$V^2_{n,out} \approx 8kTR_1G^2_{PA} + 8kTR_2 + \frac{64}{3}\frac{kT}{g_m}G^2_{PA}. \tag{12.11}$$

This noise *PSD* is limited by the low-pass frequency response pole defined by capacitor $C_2 = 16.5\,\text{pF}$, leading to an equivalent in-band integrated noise at the output of the preamplifier of approximately 50 μV_{rms}. Since the maximum differential output signal swing is 1.2 V_{pp}, the *SNR* turns out to be about 78.5 dB.

12.4.3 ADC

In the proposed AFE circuit, the ADC has to achieve a *SNR* larger than 63 dB, a total harmonic distortion (*THD*) lower than −36 dB, and minimum power consumption. The large difference between the *SNR* and *THD* specifications can be exploited to minimize the power consumption, by choosing an ADC topology in which only the *THD* depends on component matching, such as a successive-approximation register (SAR) ADC with capacitive DAC. In this ADC topology the mismatch among the DAC capacitive elements affects the *THD*, while the *SNR* depends only on the total capacitance of the DAC. This is not true for other ADC topologies, such as pipeline or algorithmic ADCs, while inherently high-performance ADCs, such as $\Sigma\Delta$ modulators, would consume too much power in the considered application. In view of this considerations, we chose the SAR topology to implement the ADC in the proposed AFE circuit.

In order to exploit the preamplifier frequency response as anti-aliasing filter, the sampling frequency of the ADC is fixed to $f_S = 1$ MHz. This leads to a large oversampling ratio (*OSR*), that can be exploited in the decimator (implemented in the companion off-chip microprocessor), in order to increase the *SNR*. Considering the signal bandwidth $B = 4$ kHz, the *SNR* is increased proportionally to the $OSR = \frac{f_S}{2B} = 125$, i.e. about 20 dB. For the target 63-dB *SNR*, therefore, only 43 dB are required at Nyquist-rate in the ADC. In order to keep a reasonable safety margin, we adopted a 10-bit SAR topology. In this case, the ADC power consumption is dominated by the reference voltage generator required for driving the capacitive DAC. Therefore, the total capacitance of the DAC has to be minimized. This has been achieved by exploiting a bridge configuration, whose only drawback is a *THD* degradation due to capacitance mismatch (which is not critical). The DAC capacitance minimization is also important to reduce the capacitive load of the preamplifier during sampling.

The schematic diagram of the implemented ADC, including the SAR structure and the buffered reference voltage generators (V_{ref+}, V_{ref-}, and V_{cm}), is shown in Fig. 12.11.

The 10-bit capacitive DAC consists of two 5-bit stages, coupled by a bridge capacitor (C_X) [12, 13].

The ADC operates with an internal clock (*clk*) activated by the start-of-conversion (*SoC*) signal. During the first clock cycle the input signal is sampled on the DAC capacitive array, while the next 10 cycles are used for the actual A/D conversion, based on the successive-approximation algorithm, resulting in a 1 MS/s maximum input-to-output data rate. The *S&H* digital signal is automatically generated every 11 clock periods by the SAR logic (whose clock frequency is, therefore, 11 MHz), provided that the *SoC* signal is active. This allows the ADC to be operated also with modulated conversion rate, as required in many sensor systems.

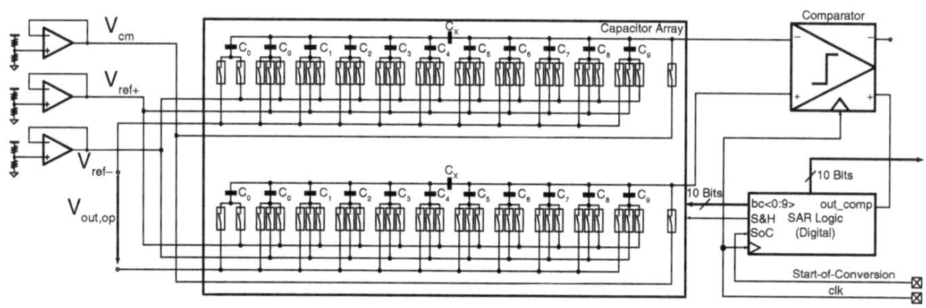

Fig. 12.11 Schematic diagram of the ADC

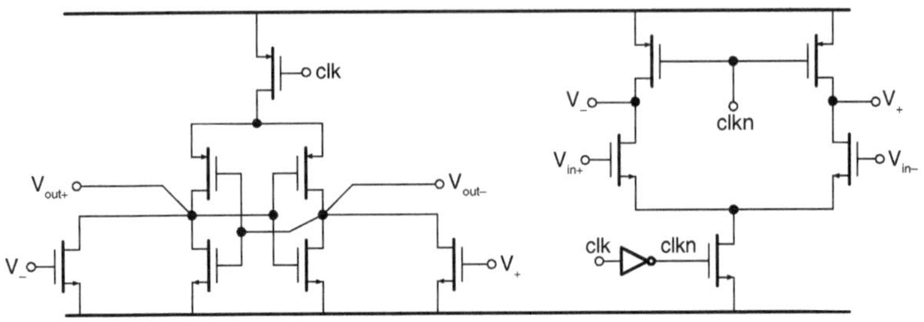

Fig. 12.12 Schematic diagram of the comparator

The comparator used for the successive-approximation algorithm, whose schematic diagram is shown in Fig. 12.12, is based on latched sense amplifier, featuring rail-to-rail output swing and negligible static power consumption [14]. In order to reduce the kickback noise, which typically affects latched comparators, the input stage has been implemented with a CT topology. In this way, the input signals are shielded from the switching comparator output stage.

The values of the capacitances used in the DAC array are given by

$$\begin{cases} C_i = 2^i C_0 \text{ for } i = 0, 1, 2, 3, 4 \\ C_i = 2^{i-5} C_0 \text{ for } i = 5, 6, 7, 8, 9 \\ C_X = \frac{32}{31} C_0 \end{cases}, \tag{12.12}$$

where C_0 is the unit capacitance. The total capacitance of the array is then given by

$$C_{TOT} \approx 2^6 C_0. \tag{12.13}$$

The equivalent load capacitance for the preamplifier and for the reference voltage generators turns out to be

$$C_L \approx 2^5 C_0. \tag{12.14}$$

This value is significantly lower with respect to a conventional binary-weighted capacitance array ($C_L \approx 2^{10} C_0$).

In order to avoid ADC performance degradation due to thermal noise, the value of C_L used for the sampling operation must comply with the minimum limit given by

$$\frac{2kT}{C_L} < \frac{LSB^2}{12} \Rightarrow C_L = 2^5 C_0 > \frac{24kT}{LSB^2} \approx 70 \text{ fF}, \tag{12.15}$$

where $LSB = \frac{1.2\text{ V}}{2^{10}} \approx 1.2\,\text{mV}$. Starting from the thermal noise limit given by (12.15), a numerical optimization has been carried out, considering the $THD < -36\,\text{dB}$ specification and the capacitor mismatch model in the used technology, given by

$$\sigma_{\frac{\Delta C}{C}} = \frac{k_C}{\sqrt{WL}}, \tag{12.16}$$

where $k_C = 0.6 \times 10^{-1}$ µm. As a result, keeping a safety margin for PVT variations, the selected unit capacitance is 12 fF. The resulting value of C_L is then about 390 fF (higher than the thermal noise limit). The thermal noise contribution, therefore, becomes about 146 μV_{rms}, which is well below the quantization noise (338 μV_{rms}).

The complete ADC, including on chip reference generators and buffers, consumes 60 µW (including static and dynamic power).

12.5 Experimental Results

The entire AFE circuit has been implemented in a 0.13-µm CMOS technology and operates from a single 1.2-V power supply, consuming 90 µW (32 µW in the preamplifier and 58 µW in the ADC, including dynamic power). The photograph of the chip, including three equal AFE channels for the three accelerometer axes, is shown in Fig. 12.13. The total chip area is 1.24 mm^2.

The functional scheme of the measurement setup is shown in Fig. 12.14.

The proposed AFE circuit has been designed taking into account the possibility to separately test the two main building blocks (preamplifier and ADC), by exploiting two test-points (i.e. *TP1* and *TP2* for preamplifier and complete AFE circuit, respectively).

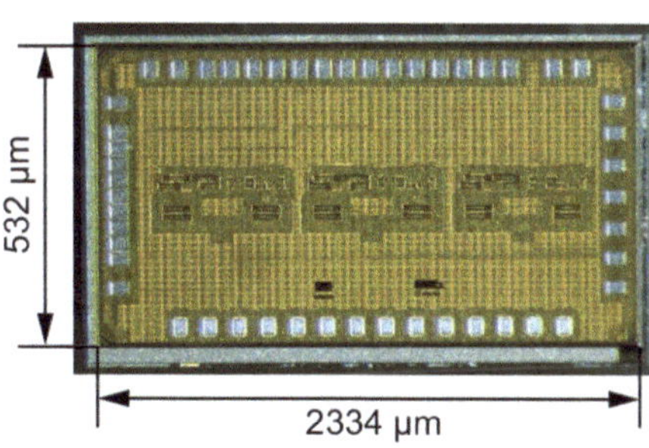

Fig. 12.13 Chip photograph

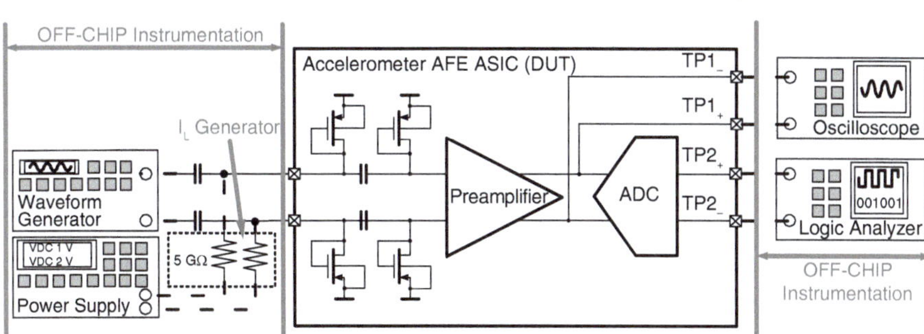

Fig. 12.14 Functional scheme of the measurement setup

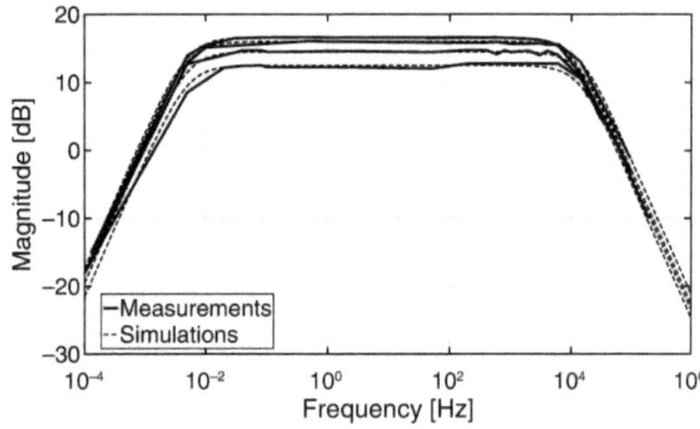

Fig. 12.15 Frequency response of the preamplifier for different gain settings

12.5.1 Preamplifier Characterization

For characterizing the preamplifier, the ADC clock signal (*clk*) and the *SoC* binary signal have been disabled, thus completely switching off the ADC. Under these conditions, the whole DAC capacitive array is connected to the preamplifier output nodes (i.e. $C_L = 0.39$ pF), which is the worst case scenario for the preamplifier capacitive load.

The measured preamplifier frequency response, obtained by applying a voltage signal directly at the preamplifier input nodes, is shown in Fig. 12.15. The circuit features a high-pass pole at a frequency lower that 0.1 Hz, thus demonstrating that indeed M_{B3} and M_{B4} implement a very high impedance. In this measurement, M_{B1} and M_{B2} are in parallel to the signal source and, therefore, they are short-circuited. The achieved in-band gain ranges form 10.5 to 15 dB, as expected.

In order to verify also the correct operation of M_{B1} and M_{B2}, the preamplifier frequency response at gain equal to 12 dB has been measured again by applying the input signal through two external decoupling capacitors and emulating the sensor leakage current through two 5-GΩ resistors connected to a voltage source. Both common-mode (CM) and differential mode (DIFF) leakage currents in the ±10 pA range are considered. The resulting frequency responses are shown in Fig. 12.16. Thanks to the ac coupling provided by C_{AC}, no significant variation of the high-pass pole frequency has been observed, thus demonstrating the capability of the AFE circuit to operate with large (even unbalanced) sensor leakage currents.

Figure 12.17 shows the measured transfer characteristic of the preamplifier. The achieved 1-dB compression point is 1.4 V_{pp}, which is 23 % higher than the output swing in the presence of full-scale acceleration (i.e. 1.2 V_{pp}). The preamplifier output spectrum at full scale is shown in Fig. 12.18. The achieved *THD* is lower than −45 dB.

The integrated output noise spectrum of the preamplifier is shown in Fig. 12.19. Over the considered 4-kHz bandwidth the achieved integrated noise is −76 dB_V, corresponding to 151 μV_{rms} (the noise level in Fig. 12.19 is 20 dB higher, since a low-noise amplifier has been used at the output of the preamplifier for interfacing the spectrum analyzer). The full-scale *SNR* at the preamplifier output is 69 dB.

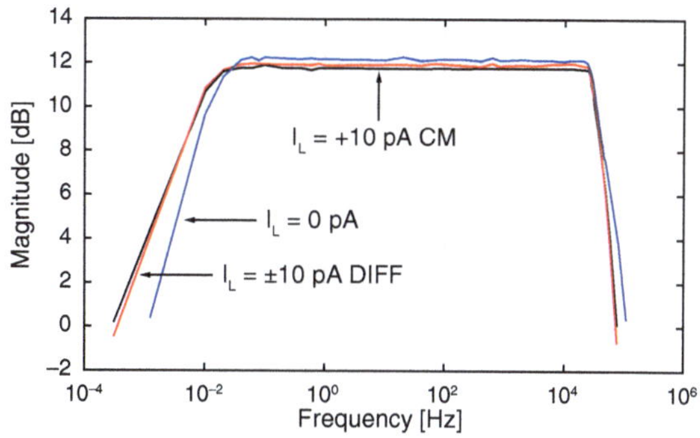

Fig. 12.16 Frequency response of the preamplifier for gain equal to 12 dB in the presence of common-mode (CM) and differential (DIFF) leakage currents

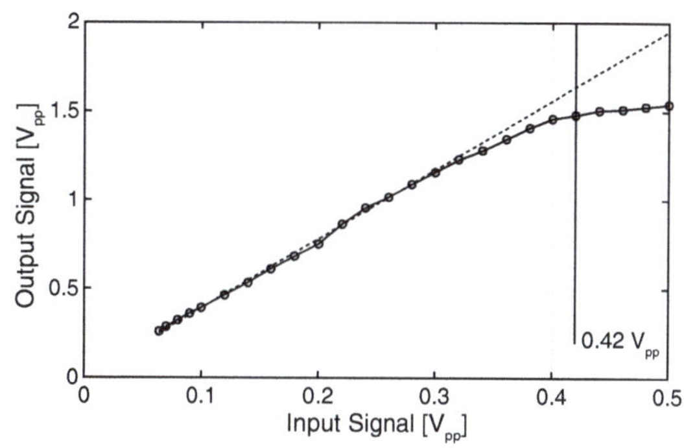

Fig. 12.17 Transfer characteristic of the preamplifier showing the 1-dB compression point

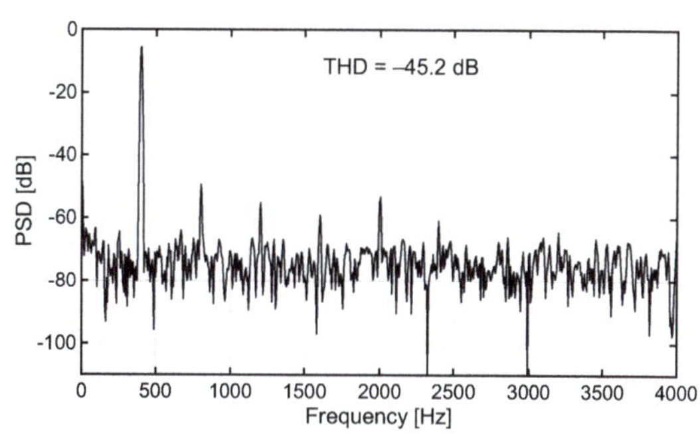

Fig. 12.18 Output spectrum of the preamplifier at full scale

12.5.2 ADC Characterization

For characterizing the ADC, the preamplifier has been switched off and a signal source has been connected to its output nodes (*TP1*), while a logic analyzer has been connected to the AFE circuit output nodes (*TP2*). The obtained ADC integral (*INL*) and differential (*DNL*) non-linearity plots are illustrated in Figs. 12.20 and 12.21, respectively. The peak *INL* is 8.1 LSB, while the peak *DNL* is 7.8 LSB. For a full-scale input signal the *THD* is approximately −45 dB, as expected.

The proposed AFE circuit has been designed to operate with an on-chip decimator and, therefore, the ADC output digital port driving capability has been

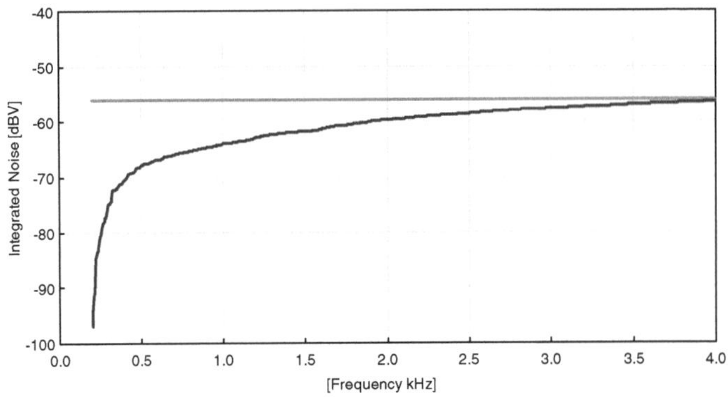

Fig. 12.19 Integrated output noise spectrum of the preamplifier

Fig. 12.20 *INL* of the ADC

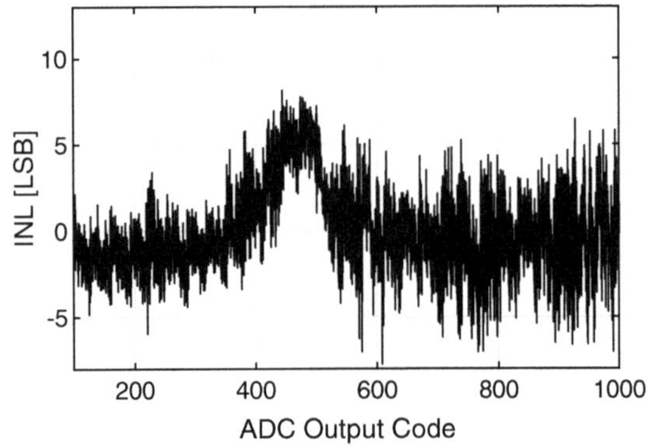

Fig. 12.21 *DNL* of the ADC

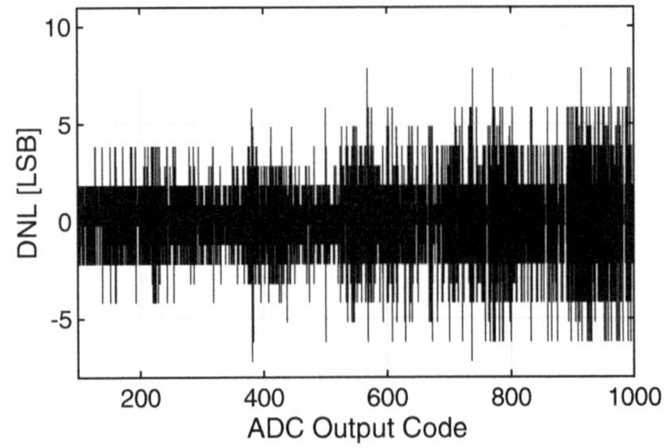

sized to manage an output load of about 200 fF, whereas the capacitive load in the measurement setup is actually much larger, due to the intrinsic large input capacitance of the available probes. This limits the possible clock frequency of the ADC to 1.1 MHz, corresponding to 100-kHz sampling rate (ten times lower than the nominal sampling rate of 1 MHz). However, this condition does not affect the performed linearity measurements.

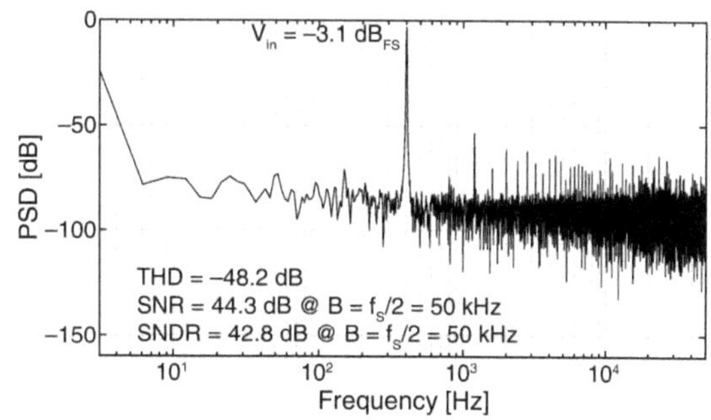

Fig. 12.22 Output spectrum of the ADC with a 400-Hz, -3.1-$\mathrm{dB_{FS}}$ sinusoidal input signal and $f_S = 100\,\mathrm{kHz}$

The ADC output spectrum, obtained with a 400-Hz, -3.1-$\mathrm{dB_{FS}}$ sinusoidal input signal and $f_S = 100\,\mathrm{kHz}$, is shown in Fig. 12.22. The achieved *THD* is lower than $-48\,\mathrm{dB}$ (i.e. 0.39 %), outperforming the specification, while the *SNR* and the signal-to-noise and distortion ratio (*SNDR*) over the Nyquist bandwidth ($f_S/2 = 50\,\mathrm{kHz}$) are 44.3 and 42.8 dB, respectively.

12.5.3 AFE Circuit Performance

The limited sampling frequency ($f_S = 100\,\mathrm{kHz}$), imposed by the measurement setup, prevents the effective exploitation of oversampling in the AFE circuit with a bandwidth $B = 4\,\mathrm{kHz}$. However, assuming, as reasonable, that the ADC noise does not change significantly when f_S is changed from 100 kHz to 1 MHz (i.e. the *SNR* over the Nyquist bandwidth remains the same as in Fig. 12.22), then we expect the ADC *SNR* under nominal operating conditions ($f_S = 1$ MHz) over the nominal bandwidth ($B = 4\,\mathrm{kHz}$) at full scale to be $SNR = 44.3\ \mathrm{dB} + 3.1\ \mathrm{dB} + 10\log\frac{f_S}{2B} = 68.3\ \mathrm{dB}$, corresponding to $ENOB = 11$ bits.

Combining the preamplifier (69 dB) and the ADC (68.3 dB) *SNR* values, the *SNR* of the complete AFE circuit becomes 65.1 dB, corresponding to $ENOB = 10.5$ bits, which is in line with the specifications. On the other hand, the *THD* is dominated by the preamplifier and, therefore, for the complete AFE circuit we obtain $THD = -43.4\,\mathrm{dB}$. A summary of the AFE circuit performance compared with the state-of-the-art is reported in Table 12.4.

12.6 Conclusions

A complete AFE circuit for MEMS capacitive accelerometers has been proposed. The main task of the AFE circuit is to read out the acceleration signal coming from the sensor and to convert it into digital domain. The proposed AFE circuit, designed in a 0.13-μm CMOS technology exploits a CT constant-charge topology

Table 12.4 AFE circuit performance summary

Parameter	[7]	[9]	[11]	This work
Sensitivity (κ_C)	0.4 fF/g	–	–	0.016 fF/g
Bandwidth (B)	2 kHz	30 Hz	20 kHz	4 kHz
Signal-to-noise ratio (SNR)	77 dB	69 dB	63 dB	65.1 dB
Total harmonic distortion (THD)	−65 dB	−40 dB	−70 dB	−43.4 dB
Effective number of bits ($ENOB$)	–	–	–	10.5 bits
Power supply voltage (V_{DD})	5 V	±2.5 V	1.8 V	1.2 V
Power consumption (P)	30 mW	0.9 μW	810 μW	90 μW
CMOS technology	0.5 μm	1.5 μm	0.35 μm	0.13 μm

for achieving 65.1 dB of *SNR* and −43.4 dB of *THD*, while consuming only 90 μW from a 1.2-V power supply. Thanks to a novel sensor biasing technique, the proposed AFE circuit achieves reliable operation even in the presence of strong sensor leakage currents and PVT variations, demonstrating that the CT constant-charge approach has indeed the potential for achieving lower power consumption than conventional SC DT solutions.

References

1. S.C. Ergen, A. Sangiovanni-Vincentelli, X. Sun, R. Tebano, S. Alalusi, G. Audisio, M. Sabatini, The tire as an intelligent sensor. IEEE Trans. Comput.-Aided Design Integr. Circuits Syst. **28**(7), 941–955 (2009)
2. G.K. Balachandran, V.P. Petkov, T. Mayer, T. Balslink, A 3-axis gyroscope for electronic stability control with continuous self-test, in *IEEE ISSCC Digest of Technical Papers* (2015), pp. 474–475
3. X. Wang, J. Zhao, Y. Zhao, G.M. Xia, A.P. Qiu, Y. Su, Y.P. Xu, A 1.2-μg/$\sqrt{Hz}$-resolution 0.4-μg-bias-instability MEMS silicon oscillating accelerometer with CMOS readout circuit, in *IEEE ISSCC Digest of Technical Papers* (2015), pp. 476–477
4. A. Donida, G. Di Dato, P. Cunzolo, M. Sala, F. Piffaretti, P. Orsatti, D. Barrettino, A 0.036-mbar circadian and cardiac intraocular pressure sensor for smart implantable lens, in *IEEE ISSCC Digest of Technical Papers* (2015), pp. 392–393
5. V. Viswam, J. Dragas, A. Shadmani, Y. Chen, A. Stettler, J. Müller, A. Hierlemann, Multi-functional microelectrode array system featuring 59,760 electrodes, 2048 electrophysiology channels, impedance and neurotransmitter measurement units, in *IEEE ISSCC Digest of Technical Papers* (2016), pp. 394–395
6. M. De Matteis, A. Pezzotta, M. Sabatini, M. Grassi, M. Croce, P. Malcovati, A. Baschirotto, A 0.13-μm-CMOS 90-μW 51-dB-SNR continuous-time accelerometer front-end with 10-bit SAR-ADC, in *Proceedings of IEEE Sensors* (2015), pp. 1393–1396
7. C.E. Furst, A low-noise/low-power preamplifier for capacitive microphones, in *Proceedings of IEEE ISCAS*, vol. 1 (1996), pp. 477–480
8. S.A. Jawed, D. Cattin, M. Gottardi, N. Massari, A. Baschirotto, A. Simoni, A 828-μW 1.8-V 80-dB dynamic-range readout interface for a MEMS capacitive microphone, in *Proceedings of ESSCIRC* (2008), pp. 442–445

9. J. Wu, G.K. Fedder, L.R. Carley, A low-noise low-offset capacitive sensing amplifier for a 50-$\mu g/\sqrt{Hz}$ monolithic CMOS MEMS accelerometer. IEEE J. Solid-State Circuits **39**(5), 722–730 (2004)
10. L. Picolli, M. Grassi, A. Fornasari, P. Malcovati, A 1.0-mW, 71-dB SNDR, fourth-order $\Sigma\Delta$ interface circuit for MEMS microphones. An. Int. Circ. Sign. Proc. **66**(2), 223–233 (2011)
11. R.R. Harrison, C. Charles, A low-power low-noise CMOS amplifier for neural recording applications. IEEE J. Solid-State Circuits **38**(6), 958–965 (2003)
12. A. Agnes, E. Bonizzoni, F. Maloberti, Design of an ultra-low power SA-ADC with medium/high resolution and speed, in *Proceedings of IEEE ISCAS* (2008), pp. 1–4
13. A. Agnes, E. Bonizzoni, P. Malcovati, F. Maloberti, An ultra-low power successive approximation A/D converter with time-domain comparator. An. Int. Circ. Sign. Proc. **64**(2), 183–190 (2010)
14. S. D'Amico, G. Cocciolo, M. De Matteis, A. Baschirotto, A 7.65-mW 5-bits 90-nm 1-GS/s ADC folded-interpolated without calibration, in *Proceedings of ESSCIRC* (2011), pp. 151–154

Part III
Power Management

The third part of this book is dedicated to recent developments in power management. The proliferation of low-power mobile devices has driven the need for efficiency optimization and on-chip or in-package integration. The first three chapters focus on three different DC/DC converter topologies, based on switched-capacitor, resonant, and inductive converters. The fourth chapter discusses precision issues due to packaging stress. The last two chapters present examples of power management for two emerging applications.

The first chapter, by Ravi Karadi, Gerard Villar Piqué, and Henk Jan Bergveld, describes switched-capacitor power converters, which are an interesting candidate for integrated converters but can also be used with discrete external capacitors. It discusses and compares alternative circuit topologies, and describes how these topologies can be synthesized. Further, two-phase and multi-phase topologies are studied and compared.

In the second chapter, by Jason T. Stauth, Christopher Schaef, and Kapil Kesarwani, the development of resonant and multimode switched-capacitor power converters is discussed. It is shown how these topologies compare with more classical power converter topologies, and theoretical and practical limitations are discussed as well. The chapter concludes with several design examples that were implemented and verified by measurements.

The third chapter, by Bruno Allard, Florian Neveu, and Christian Martin, gives an overview of inductive DC/DC converters. In order to minimize the values of the passive components, and thus enable in-package or ultimately on-chip integration, a focus of this work is to maximize the switching frequency. The chapter further describes the technology choices needed to minimize losses. Implemented prototypes are discussed using a special interposer with integrated capacitors as well as other advanced integration techniques.

The fourth chapter, by Mario Motz and Udo Ausserlechner, takes a different direction and gives an extensive overview of the problems caused by mechanical stress drift in precision analog circuits. It is shown how this stress occurs, how it can be modeled, and what the impact on various circuits, such as sensors, oscillators, and voltage references can be. Circuit techniques to measure and even compensate stress effects are described, and besides theory, many results from actual measurements are presented and analyzed.

In the fifth chapter, Stefan Dietrich and Stefan Heinen describe the application of LED drivers for illumination applications. Various topologies for these drivers are described and compared. Techniques for control and dimming are also discussed. After that, the design and implementation of a single-inductor multiple-output driver is elaborated which can drive multiple LED strings while using only a single inductor.

The sixth chapter, by Stefano Stanzione, Chris van Liempd, and Chris Van Hoof, describes an ultra-low-power DC/DC converter for energy harvesting applications. In such applications, the available power from harvesting can be very low, thus a key issue is to develop a converter that can maintain a decent efficiency when the power level is scaled down and to implement control loops that cause very little overhead. This work describes the theoretical background and shows the results of an implemented prototype.

Chapter 13
Switched-Capacitor Power-Converter Topology Overview and Performance Comparison

Ravi Karadi, Gerard Villar Piqué, and Henk Jan Bergveld

13.1 Introduction

In our modern society, the use of data-centric electronic devices has become ubiquitous. Due to the consumer demand of being able to use these devices anywhere, a multitude of battery-powered devices exists, including laptops, smart phones, smart watches and other so-called wearable electronic devices. Depending on the usage demands, energy storage in these devices can be implemented with a variety of different battery chemistries, where either primary (non-rechargeable) or secondary (rechargeable) batteries are used. In many cases, Li-ion batteries are used due to their superior energy density. Depending on the used battery chemistry for energy storage, the input voltage to the electronic device spans a relatively large range. At the same time, a large number of different load voltages for internal supply and for accommodating external interfaces need to be derived from the input voltage in the electronic device [1]. Bridging the gap between the input voltage range and the necessary load voltages in the electronic device can be accommodated by the power management block in Fig. 13.1 [2]. As can be seen in the figure, both voltage up and down conversion may be needed to bridge the voltage gap.

Voltage conversion in the power management block in Fig. 13.1 can be implemented in three different ways, i.e. with a continuous-time voltage regulator, an inductive or a switched-capacitor power converter (SCPC) [3]. In a continuous-time voltage regulator a dissipative pass device only allows implementing voltage down conversion from input voltage V_{in} to a lower output voltage V_{out} at a maximum efficiency of V_{out}/V_{in}. Despite their favorable Electro-Magnetic-Interference (EMI) behavior, in cases where V_{out} is much lower than V_{in}, the low efficiency may become a problem, especially at high power levels. Switched-mode circuits using

R. Karadi (✉) • G.V. Piqué • H.J. Bergveld
NXP Semiconductors, Eindhoven, The Netherlands
e-mail: Ravi.Karadi@nxp.com; Gerard.Villar.Pique@nxp.com; HenkJan.Bergveld@nxp.com

© Springer International Publishing Switzerland 2017

A. Baschirotto et al. (eds.), *Wideband Continuous-time ΣΔ ADCs, Automotive Electronics, and Power Management*, DOI 10.1007/978-3-319-41670-0_13

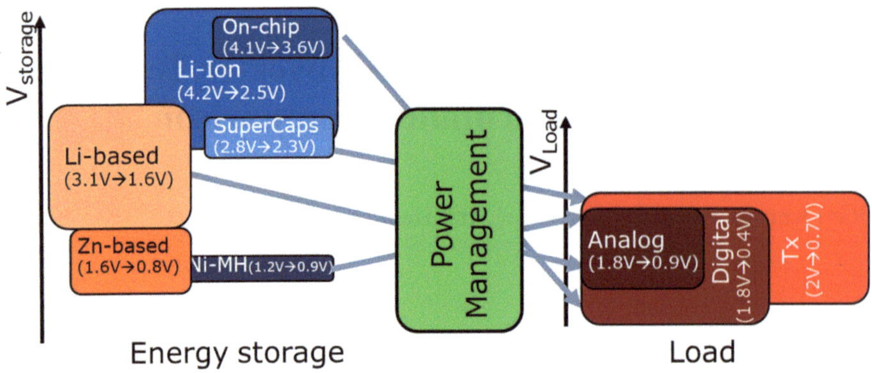

Fig. 13.1 Mismatch between energy storage voltage and load voltage [2]

reactive component(s) for temporal energy storage can implement both up and down DC/DC conversion at higher efficiency. If the primary reactive energy-storage component is an inductor, an inductive DC/DC converter can be implemented with two power switches and an inductor. Depending on how the switches and inductor are connected, either a down, an up, or an inverting up-down DC/DC converter can be constructed [4]. The ratio between input and output voltage is controlled by the duty cycle at which the power switches are operated. If the primary reactive energy-storage component is a capacitor, an SCPC can be implemented with a set of power switches and one or more so-called floating capacitors. As this paper will discuss in detail, the topology in which these switches and capacitor(s) are connected determines the ratio between input and output voltage. The output impedance of the converter is controlled to maintain a fixed output voltage when output current and/or input voltage vary.

Besides maximizing the run time of the battery-powered electronic devices by increasing the energy density of batteries and increasing the efficiency of voltage conversion, reducing the power consumption of the load is also important. For digital loads, increasing the power efficiency while maintaining the performance has led to the introduction of fine-grained power-management schemes, such as scaling the digital supply voltage with the activity of the connected block [5]. This need for local supply-voltage generation has sparked the development of DC/DC converters that are integrated with the load [1, 3].

Various methods exist to integrate the inductor with an inductive DC/DC converter [1, 3]. First, the inductor can be integrated on the same die as the converter power switches, converter control block and load, either as an air-core inductor [6, 7], or as post-processed thin-film inductor with magnetic core, e.g. based on CoZrTa alloys [8]. Alternatively, the inductor can be integrated elsewhere in the package, e.g. using bondwires [9] or package-trace air-core inductors [10], or realizing the inductor on a separate passive-integration die co-packaged with the active CMOS die [11]. In all cases, integrated inductive converters are used for relatively high output powers above 100 mW [3], because the efficiency performance of integrated inductive DC/DC converters decreases when realized for lower output power in smaller areas, in contrast to that of their integrated capacitive counterparts [12]. This makes SCPCs an attractive candidate for monolithic integration in the power range below 100 mW, but even when external SMD capacitors are used to realize higher output powers, competitive power densities compared to inductive DC/DC converters with external inductor can be shown [13].

Several monolithic capacitor integration possibilities exist for SCPCs, such as using the gate capacitance of MOSTs (MOSCAPs), using inter-metal capacitance in fringe capacitors (usually placed on top of MOSCAPs), or, when available, using Metal-Insulator-Metal (MIM) capacitors [1, 3]. Higher power densities can be obtained when non-standard capacitor realizations such as trench capacitors are used [14]. Besides purely inductive or switched-capacitor DC/DC converters, intermediate solutions are also appearing. On the one hand, floating capacitors are added to integrated inductive converters to create multiple voltage levels at the switching node [15, 16]. This reduces the ripple current in the inductor and the blocking voltage in the switches and therefore increases efficiency. On the other hand, inductors are added to integrated capacitive converters to reduce losses associated with (dis)charging the capacitors [17].

This paper deals with the fundamental aspect in SCPCs of how to realize the needed voltage conversion ratios in different ways, and how the resulting different topologies compare in performance. Section 13.2 gives a basic theoretical background on switched-capacitor power conversion. Sections 13.3 and 13.4 describe two-clock-phase and multi-clock-phase topology examples and how to synthesize these. A performance comparison of the various topologies in Sects. 13.3 and 13.4 is made in Sect. 13.5. Finally, conclusions are drawn in Sect. 13.6.

13.2 Theoretical Considerations for Switched-Capacitor Power Converters

The basic model of an SCPC is shown in Fig. 13.2a [3]. An ideal DC transformer multiplies the input voltage V_{in} with voltage conversion ratio M. As will be described in more detail in Sects. 13.3 and 13.4, this ratio M depends on the used circuit topology, i.e. how the capacitors are connected during the different clock phases of the converter. The average equivalent output impedance of the converter is modelled by R_{out}. As shown in Fig. 13.2b, R_{out} shows asymptotic behavior for both low and high values of switching frequency, f_s [18]. For low values of f_s, the asymptote is the Slow-Switching Limit (SSL), whereas for high f_s values the Fast-Switching Limit (FSL) defines the asymptote. In the SSL, the currents flowing into capacitors are impulsive and reach zero at the end of each clock phase, and R_{out} depends on f_s and the used floating capacitance. In the FSL, the voltage across the capacitors does not change during the conductance phase and R_{out} depends on the resistance in the charge and discharge path of the floating capacitors, i.e. mostly the on-resistance of

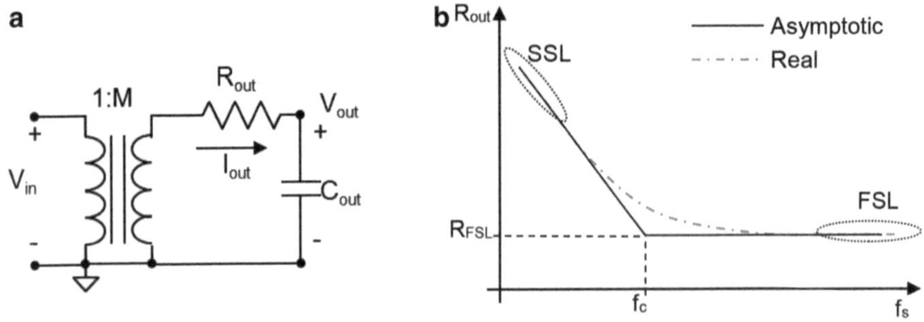

Fig. 13.2 (**a**) Averaged model of an SCPC, (**b**) equivalent output impedance R_{out} as function of switching frequency f_s

the power switches. The asymptotes cross at corner frequency, f_c. The overall curve for R_{out} can now be approximated by:

$$R_{out} = \sqrt{R_{FSL}^2 + R_{SSL}^2} \tag{13.1}$$

As can be seen from Fig. 13.2a, the maximum possible efficiency of an SCPC equals V_{out}/MV_{in}. However, since the averaged model does not consider the switching nature of the converter, in reality the achieved efficiency will be lower due to the occurrence of switching losses. These losses are caused by charging and discharging the parasitic capacitances associated with the plates of each floating capacitor in the converter, also known as bottom-plate losses, and by driver losses from turning the power switches on and off. Besides switching losses, conduction losses occur with a value of $(MV_{in} - V_{out})I_{out}$, as can be inferred from Fig. 13.2a.

In order to maintain a relatively high efficiency in cases where either V_{in} or V_{out} or both, have a large range, multiple values of M need to be used. In a multi-ratio SCPC, these multiple values of M can be generated using a single circuit [3]. A recent example of using eight values of M to maintain an efficiency above 90 % across the full range of a Li-ion battery as input source is given in [13]. Another important aspect in the design of any switched-mode power converter is the output voltage ripple. As with inductive DC/DC converters, interleaving can be applied in SCPCs as well to reduce output voltage ripple [3, 19]. This can be achieved by splitting up a large converter that delivers charge packets Q to the output at time instants $T_s = 1/f_s$ apart, in N identical parallel converters or modules, each delivering a charge packet Q/N to the output consecutively at time instants T_s/N apart. For integrated converters, splitting up a converter in parallel modules is not an issue. Scaling a converter for higher or lower output power then entails placing more or fewer of these modules in parallel. Finally, the control of output voltage V_{out} involves selecting the correct value of M (in case a multi-ratio converter is used), which usually is implemented open-loop using a look-up table, and modulating the value of R_{out} in a closed control loop [3]. Modulation of R_{out} is mostly done by

Table 13.1 All possible values of M in relation to number of floating capacitors

# floating capacitors	Possible conversion ratios M
1	$\frac{1}{2}; 1; 2$
2	$\frac{1}{3}; \frac{1}{2}; \frac{2}{3}; 1; \frac{3}{2}; 2; 3$
3	$\frac{1}{5}; \frac{1}{4}; \frac{1}{3}; \frac{2}{5}; \frac{1}{2}; \frac{3}{5}; \frac{2}{3}; \frac{3}{4}; \frac{4}{5}; 1; \frac{5}{4}; \frac{4}{3}; \frac{3}{2}; \frac{5}{3}; 2; \frac{5}{2}; 3; 4; 5$
4	$\frac{1}{8}; \frac{1}{7}; \frac{1}{6}; \frac{1}{5}; \frac{1}{4}; \frac{2}{7}; \frac{1}{3}; \frac{3}{8}; \frac{2}{5}; \frac{3}{7}; \frac{1}{2}; \frac{4}{7}; \frac{3}{5}; \frac{5}{8}; \frac{2}{3}; \frac{5}{7}; \frac{3}{4}; \frac{4}{5}; \frac{5}{6}; \frac{6}{7}; \frac{7}{8}; 1$ $\frac{8}{7}; \frac{7}{6}; \frac{6}{5}; \frac{5}{4}; \frac{4}{3}; \frac{7}{5}; \frac{3}{2}; \frac{8}{5}; \frac{5}{3}; \frac{7}{4}; 2; \frac{7}{3}; \frac{5}{2}; \frac{8}{3}; 3; \frac{7}{2}; 4; 5; 6; 7; 8$

varying f_s (in the SSL), but can also be achieved by varying the floating capacitance (in the SSL) or modulating the on-resistance or the duty cycle of the power switches (in the FSL).

For a given number of floating capacitors, all possible values of M that can be realized using these capacitors can be calculated based on the Fibonacci number series [3, 20]. This is illustrated in Table 13.1 [3].

Though Table 13.1 shows which values of M can be realized, the systematic synthesis of each of these values has not been addressed so far. In many cases, the ratios given in Table 13.1 can be realized in multiple ways, leading to different performance tradeoffs. This paper describes the different ways that exist to realize a desired value of M, using either two clock phases (Sect. 13.3) or more than two clock phases (Sect. 13.4), and their performances are compared in Sect. 13.5.

13.3 Two-Clock-Phase Topologies

In two-clock-phase SCPCs, the floating capacitors are charged and discharged in two consecutive clock phases, both clock phases fitting within one switching period, T_s. In steady state, the charging and discharging in the two clock phases are balanced. Section 13.3.1 gives an overview of the historically used two-clock-phase topologies, while Sect. 13.3.2 describes a synthesis method for two-clock-phase SCPC topologies.

13.3.1 Overview of Known Topologies

As described in Sect. 13.1, SCPCs can accommodate both voltage up and down conversion. In fact, a converter designed for a voltage conversion ratio M can also be used for the voltage conversion ratio $1/M$ by swapping input and output connections. As Table 13.1 shows, for each M that is realizable with a given number of floating capacitors, there is a corresponding $1/M$ value. Therefore, only voltage

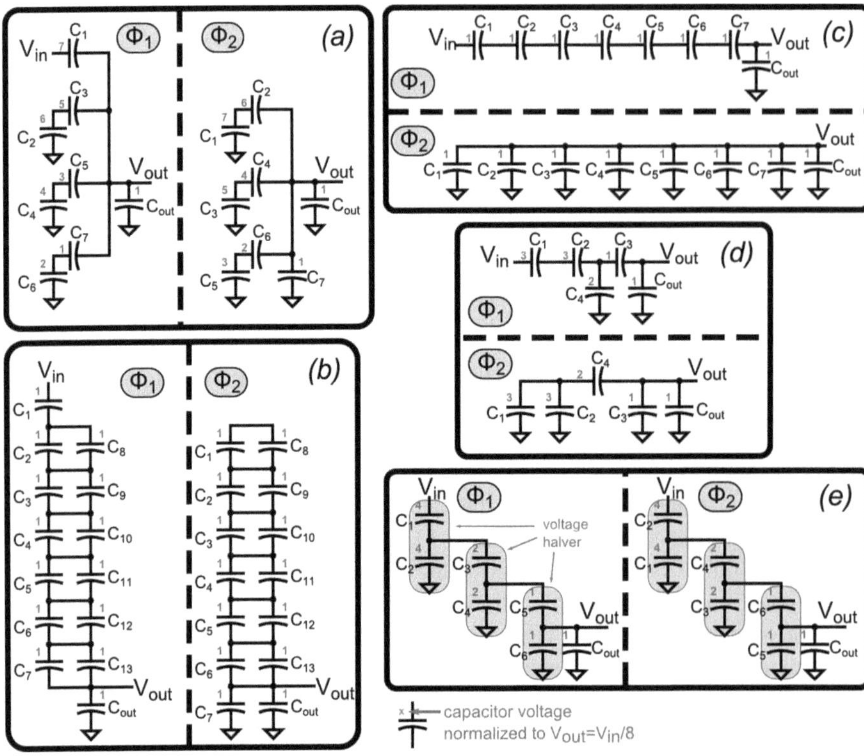

Fig. 13.3 Most well-known SCPC topologies for M = 1/8: (**a**) Dickson, (**b**) ladder, (**c**) series-parallel, (**d**) fractional-1, (**e**) recursive [24]

down conversion is considered in this section, but the five basic topologies shown also hold for up conversion [21, 22]. The five most well-known switched-capacitor power conversion topologies for $M = 1/8$ are shown in Fig. 13.3. For each topology, the circuit constellations in the two clock phases ϕ_1 and ϕ_2 are depicted, without showing the switch topology necessary to achieve these two states in a single circuit. The floating capacitors are denoted by C_i and the DC voltage across them is indicated as normalized to V_{out}, whereas the output buffer capacitor is C_{out} with voltage $V_{out} = V_{in}/8$ across it.

The Dickson topology is shown in Fig. 13.3a. This topology was inspired by the Greinacher or Cockroft-Walton topology, which effectively implements an AC/DC voltage conversion and therefore is not considered here. The Dickson topology is primarily applied in voltage up converters, since one of the switches in the charging path of each floating capacitor can be replaced with a diode, which simplifies converter control. To accommodate a voltage conversion ratio of M or $1/M$, $M - 1$ floating capacitors are needed, leading to seven floating capacitors in Fig. 13.3a. The ladder topology consists of two chains of series-connected floating capacitors that slide up and down along each other, which sets it apart from the other four topologies, since within each string the floating capacitors remain connected in the same order in both clock phases. To accommodate a voltage conversion ratio of M or $1/M$, $2M-3$ floating capacitors are needed, corresponding to thirteen floating capacitors in Fig. 13.3b. This inefficient use of the available capacitance makes the ladder topology less practical, but since all intermediate voltages are available at the ladder taps, it can be used as auxiliary converter to realize the needed drive voltages in a main SCPC [23].

The series-parallel topology shown in Fig. 13.3c uses $M-1$ floating capacitors to achieve a voltage ratio of M or $1/M$. In contrast to the Dickson topology, each floating capacitor has the same DC voltage across it in steady state. Figure 13.3d shows one example of a fractional topology using the minimum possible number of capacitors to realize $M = 1/8$ according to Table 13.1, i.e., four floating capacitors. Due to this most efficient use of capacitance compared to the other four topologies, many examples of fractional SCPCs can be found in the literature. However, though Table 13.1 shows what ratios are possible, in practice a fractional topology to realize a desired value of M is usually derived by trial and error.

A larger number of capacitors compared to the fractional topology of Fig. 13.3d can be used to achieve equally spaced voltage conversion ratios using a modular approach [24], as shown in Fig. 13.3e. With the smart reconfiguration of multiple replicas of the same 2:1 unit cell ($M = 1/2$) consisting of two floating capacitors, $2^N - 1$ voltage ratios can be achieved using N unit cells, so $2N$ floating capacitors. Though less voltage ratios are possible for a given number of floating capacitors compared to Table 13.1, the recursive approach of placing the same unit cell in series or parallel makes this topology easily reconfigurable. This can be an advantage compared to the fractional topology, especially when a multi-ratio converter is needed to maintain efficiency over a wider input or output voltage range.

13.3.2 Topology Synthesis

To break away from the trial and error aspect for the topology synthesis in Fig. 13.3d, a generic iterative algorithm for synthesizing SCPC topologies was developed [25]. A short summary of the method is given in this section. Figure 13.4 shows a conceptual block diagram of a two-clock-phase SCPC consisting of n_c floating capacitors with three external terminals.

Fig. 13.4 Conceptual SCPC with n_c floating capacitors

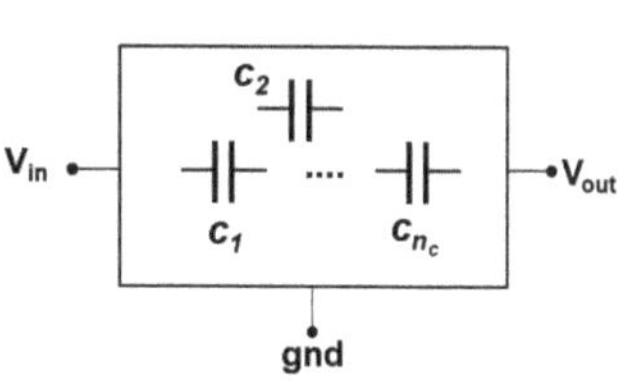

Table 13.2 Valid configuration pairs for $n_c = 2$, 3 and 4 (redundant pairs omitted)

Description	Config. identifier	Configuration pairs	$q_{\Phi1}, q_{\Phi2}$	q_{sum}
$n_c = 2$	2	$[c_1 \oplus c_2]$ & $[c_1 \| c_2]$	1,2	3
$n_c = 3$	3.1	$[(c_1 \oplus c_2) \oplus c_3]$ & $[(c_1 \| c_2) \| c_3]$	1,2+1	4
	3.2	$[(c_1 \oplus c_2) \| c_3]$ & $[(c_1 \| c_2) \oplus c_3]$	1+2,2	5
	3.3	$[(c_1 \oplus c_2) \| c_3]$ & $[c_1 \| (c_2 \oplus c_3)]$	1+1,1+1	4
$n_c = 4$	4.1	$[c_1 \oplus c_2 \oplus c_3 \oplus c_4]$ & $[c_1 \| c_2 \| c_3 \| c_4]$	1,3+1	5
	4.2	$[(c_1 \oplus c_2 \oplus c_3) \| c_4]$ & $[(c_1 \| c_2 \| c_3) \oplus c_4]$	1+3,3	7
	4.3	$[(c_1 \oplus c_2 \oplus c_3) \| c_4]$ & $[c_1 \| c_2 \| (c_3 \oplus c_4)]$	1+1,2+1	5
	4.4	$[((c_1 \oplus c_2) \| c_3) \oplus c_4]$ & $[((c_1 \| c_2) \oplus c_3) \| c_4]$	3,2+3	8
	4.5	$[(c_1 \oplus c_2) \| (c_3 \oplus c_4)]$ & $[((c_1 \| c_2) \oplus c_3) \| c_4]$	2,2+1	5
	4.6	$[(c_1 \oplus c_2 \oplus c_4) \| c_3]$ & $[((c_1 \| c_2) \oplus c_3) \| c_4]$	3,2+1	6
	4.7	$[(c_1 \oplus c_2) \| c_3 \| c_4]$ & $[(c_1 \| c_2) \oplus c_3 \oplus c_4]$	3+2,2	7
	4.8	$[((c_1 \oplus c_2) \| c_3) \oplus c_4]$ & $[c_1 \| (c_2 \oplus c_3) \| c_4]$	2,2+2	6
	4.9	$[(c_1 . c_2) \| (c_3 . c_4)]$ & $[c_1 \| (c_2 . c_3) \| c_4]$	2,2+1	5
	4.10	$[((c_1 . c_2) \| c_3) \| c_4]$ & $[c_1 \| (c_2 . c_3 . c_4)]$	1+1,2+1	5
	4.11	$[((c_1 . c_2) \| c_3) \| c_4]$ & $[(c_1 . c_4) \| (c_2 . c_3)]$	2+2,2	6

ci⊕cj implies the same charge flows through ci and cj, ci||cj implies charge is shared between ci and cj

The Fibonacci limit for the achievable conversion gains of SCPCs was introduced in Table 13.1. According to this limit, with n_c floating capacitors all conversion ratios given by Eq. (13.2) can be realized:

$$M(n_c) = \frac{V_{out}}{V_{in}} = \frac{1 \leq P \leq F_{n_c+1}}{1 \leq Q \leq F_{n_c+1}} \tag{13.2}$$

where P and Q are integers and F_n is the n_cth Fibonacci number. The synthesis algorithm for a target conversion ratio, $M = P/Q$, is divided into two steps:

I) *Iterative generation of configuration pairs*: In this step, all valid configuration pairs (without considering connections to the external terminals) with the required number of floating capacitors, as given by the Fibonacci limit of Eq. (13.2), are generated through iteration. For each configuration pair, n_c floating capacitor voltages and the output voltage V_{out} need to be resolved, so valid configuration pairs are the ones that yield $n_c + 1$ valid KVL(Kirchoff Voltage Law) equations while maintaining charge balance for each floating capacitor. These valid configuration pairs are depicted in Table 13.2, where c_i denote the floating capacitors, and $q_{\Phi1}$ and $q_{\Phi2}$ are the effective charge coefficients during the two clock phases, and $q_{sum} = q_{\Phi1} + q_{\Phi2}$. A configuration pair describes charge flow through the floating capacitors and corresponds to the set of KVL equations for each clock phase.

II) *Terminal assignment*: A converter with a desired voltage conversion ratio, $M = V_{out}/V_{in} = P/Q$ has charge conversion ratio $q_{out}/q_{in} = 1/M = Q/P$. After normalization, $q_{out} = Q$ and $q_{in} = P$. With terminal assignment, the three exter-

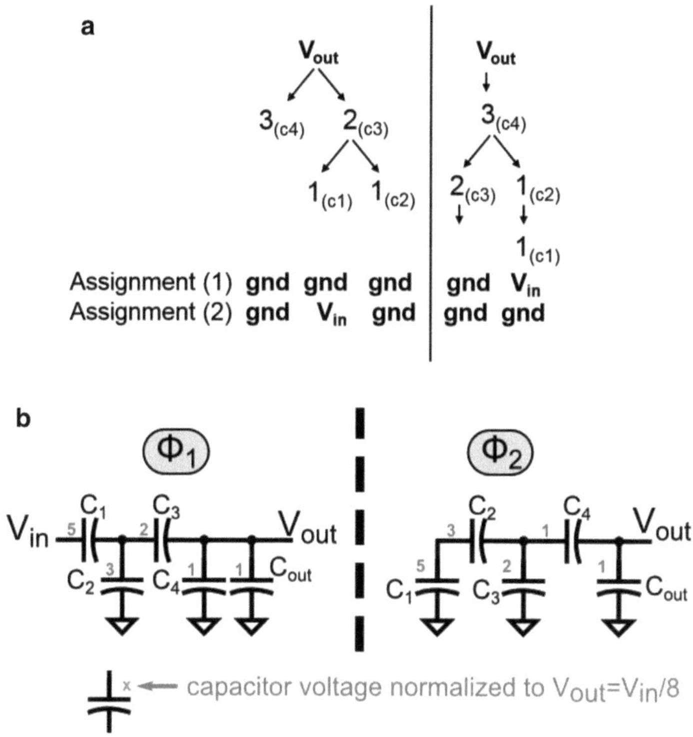

Fig. 13.5 (**a**) Charge flow diagram for configuration pair (4.4) in Table 13.2, (**b**) synthesized alternative fractional-2 SCPC topology to realize M = 1/8

nal terminals are connected to the configuration pairs with $q_{sum} \geq max(P,Q)$ such that the charge conversion ratio is satisfied over the two phases Φ_1 and Φ_2.

A more detailed example of applying the two synthesis steps is given in [25]. To further illustrate the application of this algorithm, consider $M = 1/8$, which implies $P = 1$, and $Q = 8$. Following the steps of the algorithm:

1) $n_c = 4$ from the Fibonacci limit on the possible conversion ratios, see Table 13.1.
2) Table 13.2 gives all possible configuration pairs for $n_c = 4$. The exact explanation of this table can be found in [25].
3) Out of 11 possible configuration pairs, only the configuration pair (4.4) satisfies $q_{sum} \geq max(P,Q)$, since it has a value $q_{sum} = 8$. Its charge flow diagram is shown in Fig. 13.5a. The terminal assignment with $P = 1$ and $Q = 8$ can be done in two different ways resulting in two different topologies with the same M. Figure 13.3d shows the results of polarity assignment to the capacitors corresponding to assignment (1) and Fig. 13.5b shows it for assignment (2).

It should be noted that the floating capacitors of the two topologies with $M = 1/8$ have identical charge coefficients as defined in [18], hence the same SSL and FSL

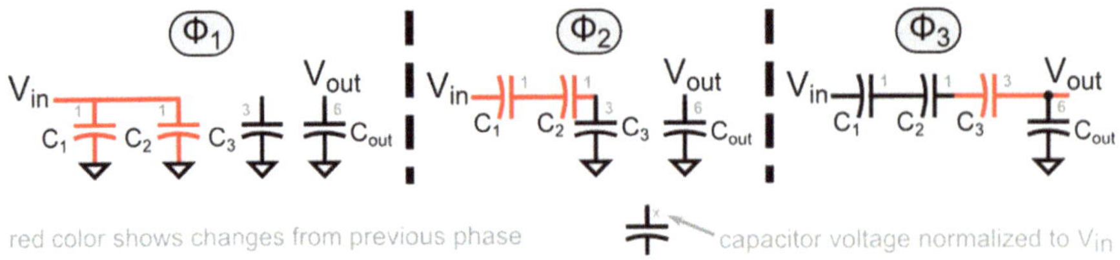

Fig. 13.6 SCPC topology using three clock phases for M = 6 [26]

impedances. However, they differ in the capacitor and switch blocking voltages, hence one topology can be better suited over the other in a given application. This point will be developed further in Sect. 13.5.

13.4 Multi-Clock-Phase Topologies

The Fibonacci constraint on the minimum required number of floating capacitors can be extended by employing additional clock phases within each switching period T_s. In these multi-clock-phase topologies, the charge balance of the capacitors can occur over more than two clock phases. Two recent examples of multi-clock-phase SCPCs are given in Sect. 13.4.1, while Sect. 13.4.2 shows examples of alternative synthesized multi-clock-phase topologies.

13.4.1 Overview of Known Topologies

An example of a topology using three clock phases achieving a voltage up-conversion ratio $M = 6$ using three floating capacitors is shown in Fig. 13.6 [26]. By exploiting an additional clock phases, this topology uses one less floating capacitor to realize $M = 6$. The operation of the converter is as follows:

- Φ_1: C_1 and C_2 are charged to V_{in} while C_3 is unconnected: normalized charge coefficients $q_{c1} = 2$, $q_{c2} = 2$, $q_{c3} = 0$.
- Φ_2: C_1 and C_2 are placed in series with V_{in} and C_3 is charged to $3V_{in}$: charge coefficients $q_{c1} = -1$, $q_{c2} = -1$, $q_{c3} = 1$.
- Φ_3: C_1, C_2 and C_3 are in series with V_{in} to charge C_{out} to $6V_{in}$: charge coefficients $q_{c1} = -1$, $q_{c2} = -1$, $q_{c3} = -1$.

The charge coefficients in the three clock phases reveal an important difference between multi-clock-phase and two-clock-phase topologies: in multi-clock-phase topologies, the amount of (dis)charge supplied to a floating capacitor does not need to be the same during each clock phase unlike the two-clock-phase topologies. The charge balancing of the floating capacitors occurs over more than two clock

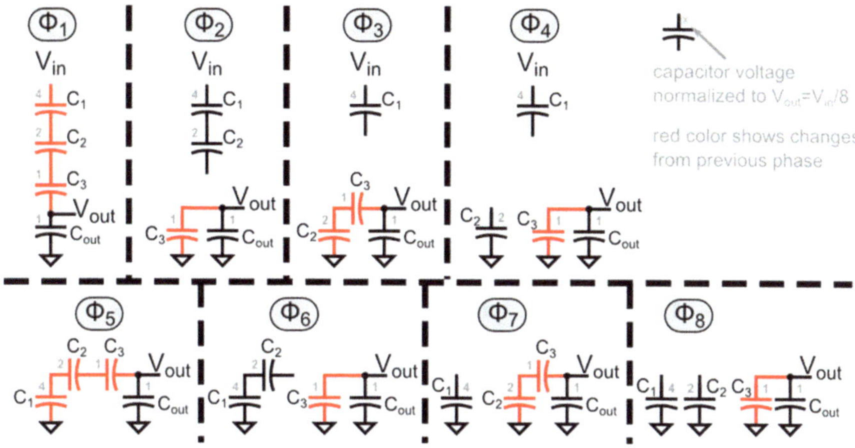

Fig. 13.7 Binary eight-clock-phase topology providing M = 1/8 [27]

phases, with some capacitors potentially being inactive during certain phase(s). This feature will result in different output impedance characteristics for a multi-clock-phase topology compared to a two-clock-phase topology. In the realization of the switched-capacitor up converter using this topology, the capacitance values of C_2 and C_3 were increased to enable the reduction of the capacitance of C_1 and hence its full integration, leaving only two SMD floating capacitors. Therefore, the potential of this multi-clock-phase topology to realize the desired M with a lower number of capacitors than prescribed by Table 13.1 was fully exploited to yield a high-efficiency, low-volume DC/DC converter.

Another approach to realizing multi-clock-phase converters is based on binary topologies [27]. The binary topologies employ 2^{n_c} clock phases with n_c floating capacitors to provide the binary conversion ratios given by:

$$M = \frac{1 \leq P \leq 2^{n_c} - 1}{2^{n_c}} \tag{13.3}$$

Figure 13.7 shows an example with $M = 1/2^3 = 1/8$ that uses three floating capacitors with eight clock phases. A unit charge is supplied to V_{out} during each clock phase, resulting in a total of eight units of charge over a complete switching period. Therefore, the charge coefficient of each floating capacitor, when it is used, is always 1/8, the minimum possible. This minimum charge coefficient will result in the lowest SSL impedance as will be discussed in Sect. 13.5 (given the same total amount of floating capacitance and the same switching frequency). In [27], this topology has been implemented using a cascade of N 2:1 unit cells ($M = 1/2$) operating in successively doubling frequencies, which is the same as having a single switching frequency with 2^N clock phases. This approach has the advantage of modularity, similar to the two-clock-phase variant in Fig. 13.3e.

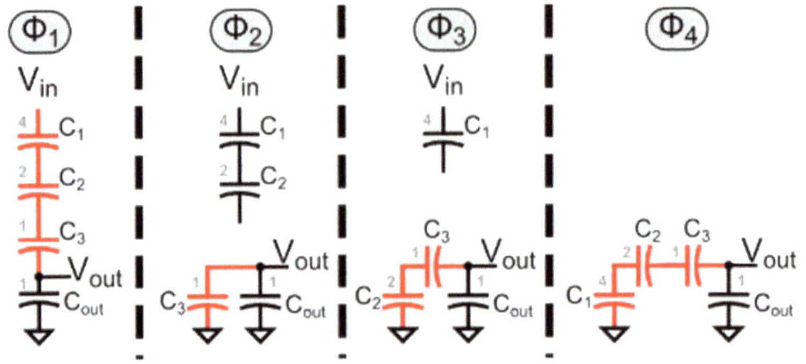

Capacitor	Φ1	Φ2	Φ3	Φ4
C1	1	0	0	-1
C2	1	0	-2	1
C3	1	-4	2	1
Vout	1	4	2	1

Fig. 13.8 Alternative M = 1/8 topology with four clock phases, including charging behaviour

Two observations can be made about this topology: (1) from the charge-flow point of view, the configurations in the phases Φ_2, Φ_4, Φ_6, and Φ_8 are identical to each other, just as the configurations in phases Φ_3, and Φ_7 are identical to each other, (2) each configuration has only one KVL loop. This means that there are actually only four unique configurations in this topology. This is consistent with the requirement of four linearly independent loops to uniquely resolve the voltages across the capacitors and the output voltage V_{out} [25].

13.4.2 Topology Synthesis

By combining the configurations with identical charge flow in Fig. 13.7 in a single phase, a multi-clock-phase fractional topology using four clock phases instead of eight clock phases can be made as shown in Fig. 13.8. The (dis)charging behavior of this topology is depicted in the table shown in the inset, normalized to $q_{in} = q_{out}/8$.

By manipulating the table in Fig. 13.8, other multi-clock-phase topologies can be generated. As an example, consider a target conversion ratio of $M = 1/7$. This converter delivers a normalized total charge of $7q_{in}$ to V_{out} over one switching period. To achieve this, the last two rows of the table in Fig. 13.8 can be adapted, resulting in the (dis)charging behavior of the table in Fig. 13.9. The (dis)charging of the floating capacitors has also been changed to maintain charge balance. The resulting topology is shown in Fig. 13.9 resulting in a four-clock-phase topology with $M = 1/7$. In general, with n_c floating capacitors, topologies with a conversion

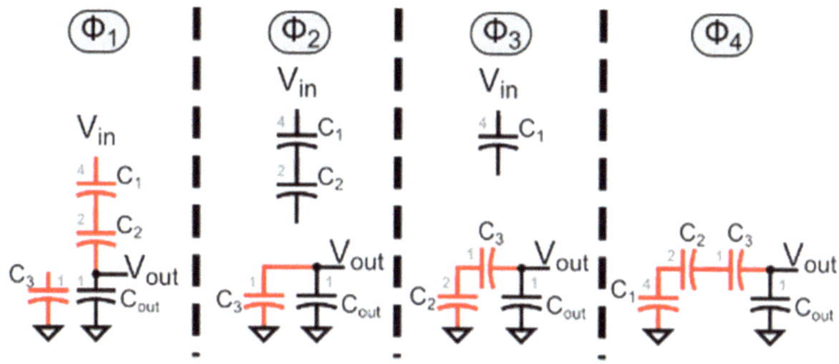

Capacitor	Φ_1	Φ_2	Φ_3	Φ_4
C_1	1	0	0	-1
C_2	1	0	-2	1
C_3	0	-3	2	1
V_{out}	1	3	2	1

Fig. 13.9 Four-clock-phase topology to generate M = 1/7 including charging behavior

ratio given by:

$$M = \frac{1 \le P \le 2^{n_c}}{1 \le Q \le 2^{n_c}} \tag{13.4}$$

can be realized using a maximum of $n_c + 1$ clock phases (discounting repetition of configurations with identical charge flow, as in the case of binary multi-clock-phase topologies).

13.5 Topology Performance Comparison

13.5.1 Comparison Between Alternative Two-Clock-Phase Topologies

As described in previous sections, there are many different topologies that can provide the same conversion ratio M. However, they might present different advantages or disadvantages that makes them more or less suitable for different applications. From the functionality point of view of an SCPC, the most important parameter besides M, is the amount of output impedance R_{out} (see Fig. 13.2). In the SSL region, the output impedance R_{SSL} can be computed as follows:

$$R_{SSL} = \frac{m}{f_s C_T} \tag{13.5}$$

where f_s is the switching frequency, C_T is the total amount of floating capacitance and m is a topology-dependent parameter that is determined by circuit analysis [18].

In the FSL region, the output impedance R_{FSL} only depends on the on-resistance of the switches (neglecting the interconnect parasitic resistances) and the topology, and can be expressed as follows:

$$R_{FSL} = p\frac{k_{sw}}{A_{swT}} \tag{13.6}$$

where p is a topology-dependent parameter obtained by circuit analysis [18], A_{swT} is the total area of the switches in [mm^2], and k_{sw} is a technology-related parameter that states the on-resistance of the switches per unit of silicon area (assuming the same voltage rating for all the devices), in [Ω·mm^2]. Therefore, the inverse ratio A_{swT}/k_{sw} describes the amount of conductance that is used by the design.

Table 13.3 shows the values of the p and m parameters for a number of different topologies that provide the same conversion ratio ($M = 1/8$) with a different number of floating capacitors n_c. As shown, all the different two-clock-phase topologies (A–E) have the same m and p values, which means that actually all the considered topologies will present the same output impedance provided that they operate at the same switching frequency, use the same amount of total floating capacitance and the same total amount of area for the switches.

Topologies F and G in Table 13.3 represent the two different multi-clock-phase topologies that were introduced in Sect. 13.4, also providing $M = 1/8$. However, because of their multi-clock-phase nature, conclusions from the comparison with the two-clock-phase topologies cannot be derived directly. This will be commented in Sect. 13.5.2, along with the content of the last two columns.

In more detail, the parameters shown in Table 13.3 can be described as follows:

- $n_c \rightarrow$ the number of required floating capacitors.
- $n_{sw} \rightarrow$ the number of power switches.
- $m \rightarrow$ product of R_{SSL}, the switching frequency and the total amount of floating capacitance (given the optimum relative sizing of all the floating capacitors), see Eq. (13.5).
- $p \rightarrow$ ratio between R_{FSL} and the amount of conductance, given the optimum relative sizing of all the power switches, see Eq. (13.6).
- $l \rightarrow$ this parameter can be used as an indication of the bottom-plate power losses that a topology suffers from. If all the energy spent to charge and discharge the parasitic bottom-plate capacitors is considered lost (pessimistic approach), the l parameter is calculated as:

$$l = \sum_{i=1}^{n_C} C_{Ri} V_{Ri}^2 \tag{13.7}$$

Table 13.3 Performance parameter comparison of various M = 1/8 topologies

Topology (M = 1/8)	n_c	n_{sw}	m	p	l	$\sum_{i=1}^{n_{sw}} g_i V_{offi}^2$	$\sum_{i=1}^{n_c} V_{Ci} q_{Ci}$	V_{Ci} (norm. to V_{out})							q_{Ci} (norm. to $q_{out}/8$)							ml	$m \sum_{i=1}^{sw\_events} g_i V_{offi}^2$
								C_1	C_2	C_3	C_4	C_5	C_6	C_7	C_1	C_2	C_3	C_4	C_5	C_6	C_7		
A: Fig. 13.3a	7	22	0.766	15.13	0.031	0.028	28	7	6	5	4	3	2	1	1	1	1	1	1	1	1	0.024	0.021
B: Fig. 13.3c	7	22	0.766	15.13	0.625	0.239	7	1	1	1	1	1	1	1	1	1	1	1	1	1	1	0.479	0.183
C: Fig. 13.3e	6	20	0.766	15.13	0.125	0.08	12	4	4	2	2	1	1		½	½	1	1	2	2		0.096	0.061
D: Fig. 13.3d	4	13	0.766	15.13	0.152	0.092	13	3	3	1	2				1	1	3	2				0.116	0.070
E: Fig. 13.5b	4	13	0.766	15.13	0.067	0.058	15	5	3	2	1				1	1	2	3				0.051	0.044
F: Fig. 13.7	3	10	0.328	27.5	0.292	0.163	7	4	2	1					1	1	1					0.096	0.067
G: Fig. 13.8	3	10	0.571	24.66	0.135	0.099	16	4	2	1					1[a]	2[a]	4[a]					0.077	0.062

Topologies: A: Two-clock-phase Dickson, B: Two-clock-phase series-parallel, C: Two-clock-phase recursive [24], D: Two-clock-phase fractional-1, E: Two-clock-phase fractional-2, F: Eight-clock-phase binary [27], G: Four-clock-phase alternative

[a]There is difference in charge flow per phase. The shown value is the maximum one

where C_{Ri} is the value of each floating capacitor normalized to the total amount of floating capacitance C_T, and V_{Ri} is the voltage across each floating capacitor, normalized to the input voltage V_{in}. From the l parameter definition in Eq. (13.7), it becomes straightforward to compute the total amount of bottom-plate losses:

$$P_{bot} = V_{in}^2 f_s \alpha C_T l \tag{13.8}$$

where α is a technology-dependent ratio representing the average of both top-plate and bottom-plate parasitics of the floating capacitors relative to their capacitance.

- The summation of the $g \cdot V_{off}^2$ (conductance g and blocking voltage V_{off}) values of all the power switches. The blocking voltage is squared because conceptually, the on-resistance of a power switch increases with the square of its blocking voltage. This parameter provides an intuitive indication about the size of the switches, and as a result, also about the energy that needs to be spent to drive their gates.
- The summation of $V_{ci} \cdot q_{ci}$ for all the floating capacitors. This parameter becomes especially relevant in fully integrated solutions, since the relative sizes of the floating capacitors is proportional to the amount of charge flowing through them and the operating voltage across them.
- $V_{Ci} \rightarrow$ The voltage across each of the floating capacitors C_i (normalized to the output voltage $V_{out} = V_{in}/8$), as shown in Sects. 13.3 and 13.4.
- $q_{Ci} \rightarrow$ The amount of charge flowing through each of the floating capacitors C_i (normalized to the amount of input charge per switching period $q_{in} = q_{out}/8 = I_{out} \cdot T_s/8$).

Based on comparing only the p and m parameters of two-clock-phase topologies A–E, it could be concluded that all the different topologies would provide the same performance. However, looking at the rest of the table content, substantial advantages could be obtained by making a proper topology choice for a particular application-technology combination. Below, different possible scenarios will be discussed.

13.5.1.1 Scenario 1: Floating Capacitors Realized with External SMD Capacitors

For this scenario, the most interesting options would be the fractional topologies (D and E), because they require the smallest number of capacitors and pins. Since the amount of bottom-plate parasitics of SMD capacitors is normally negligible, in case of a high-voltage application, the fractional-1 topology (D) would represent a better fit since the operating voltage across the floating capacitors is smaller. On the other hand, if this becomes irrelevant, e.g. because of the relatively low voltage of the application, then the fractional-2 topology (E) would be preferred because of the lower combination of conductance and blocking voltage of the power switches, which would likely result in lower switching losses.

13.5.1.2 Scenario 2: Fully Integrated Switched-Capacitor Power Converter

In the fully-integrated scenario, the relative importance of the different parameters changes quite significantly. In the integrated domain, the number of floating capacitors is not a problem by itself, but their total size is. If the application is sufficiently low-voltage enough, so that the voltage across the capacitors does not impact their implementation towards a lower capacitance density, then a Dickson topology (A) would clearly be the best choice due to its extremely low bottom-plate losses, and low area required for the switches. On the other hand, when a particular application runs at a relatively high voltage, a series-parallel topology (B) could become interesting due to the minimization of the voltage across all the floating capacitors, despite the significantly larger amount of bottom-plate and switching losses. The fractional-2 topology (E) shows a good trade-off between blocking voltage and conductance of the required switches, as well as very low bottom-plate power losses and applied voltage across the floating capacitors. Despite not standing out in any of the main converter parameters, the recursive topology (C) provides a very interesting modular approach to implement a multi-ratio SCPC by means of smart interconnection of unit cells that implement $M = 1/2$. This implies that a single cell could be designed, replicated and smartly used to easily create a large range of conversion ratios (though not all the possible ones). Unfortunately, in cases where extremely high or low ratios are needed, the number of needed unit cells and therefore floating capacitors would make a design based on external components (scenario 1) unpractical.

It is important to take into account that all parameter values shown in Table 13.3 are merely indicative of a trend, and would only apply to cases where the voltage rating of devices (both floating capacitors and power switches) could be gradually adjusted in a continuous way. Unfortunately, most technology processes provide devices with very specific voltage ratings that change in discrete steps. This clearly precludes the existence of an absolute optimum choice for every different application-technology combination.

13.5.2 Comparison Between Two-Clock-Phase and Multi-Clock-Phase Topologies

In this section, the more conventional two-clock-phase topologies will be compared to multi-clock-phase topologies, in terms of the achievable gains and performance considering similar parameters to the ones used in Sect. 13.5.1.

Although parameters related to output impedance (p and m) and the possible conversion ratios are common to all the two-clock-phase topologies (besides the ladder topology of Fig. 13.3b), this is not the case with topologies operating with multiple clock phases. As a result, the two-clock-phase topologies will be compared to the approach presented in [27] (as illustrated by the binary multi-clock-phase

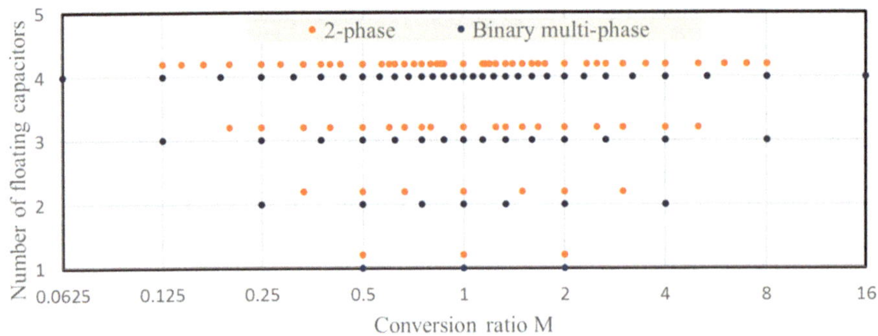

Fig. 13.10 Comparison between achievable voltage conversion ratios M for fractional two-clock-phase and binary multi-clock-phase topologies

topology F, Fig. 13.7), since this is the finest example of a multi-ratio multi-clock-phase topology that the authors are aware of. This particular approach allows for a systematic way to synthesize a different number of binary-scaled conversion ratios that operate in multi-clock-phase, with a small number of floating capacitors. As introduced in Sect. 13.4.1, a smart connection of voltage-halver stages is used that each operate at twice the frequency of its precedent, resulting in a particular case of multi-clock-phase design (with 2^{n_c} clock phases for n_c floating capacitors).

Figure 13.10 shows the achievable number of conversion ratios M (up/down) that can be obtained from a fractional two-clock-phase (see Table 13.1) and a binary multi-clock-phase topology, as a function of the number of floating capacitors. As expected from Sect. 13.4, multi-clock-phase topologies are able to achieve a higher maximum ratio for a given number of floating capacitors than fractional two-clock-phase ones. However, it is the density of available ratios in a particular gain range that allows multi-ratio designs to keep a high power efficiency over a certain range of input or output voltage.

As observed from Fig. 13.10, a fractional two-clock-phase design could provide more conversion ratios in a particular gain range than the binary multi-phase approach, eventually resulting in a higher average power efficiency. As an example of that, Fig. 13.11 compares the theoretical maximum power efficiency (V_{out}/MV_{in}) available from multi-ratio designs using four floating capacitors based on either a two-clock-phase or a binary multi-clock-phase topology. In this example an output voltage of 5 V has been considered for an input voltage range corresponding to Li-ion batteries (2.6–4.2 V), as in [13]. In this example, the two-clock-phase fractional topology delivers an average efficiency across the input voltage range of 96.4 %, compared to 95.4 % for the binary multi-clock-phase topology. However, the more evenly distributed ratios from the binary multi-clock-phase topology makes the ratio control easier by increasing the linearity.

Besides the conversion ratios that can be generated from a particular approach (either using two or multiple clock phases), the performance of an SCPC is also strongly determined by its output impedance and switching losses. Figure 13.12a

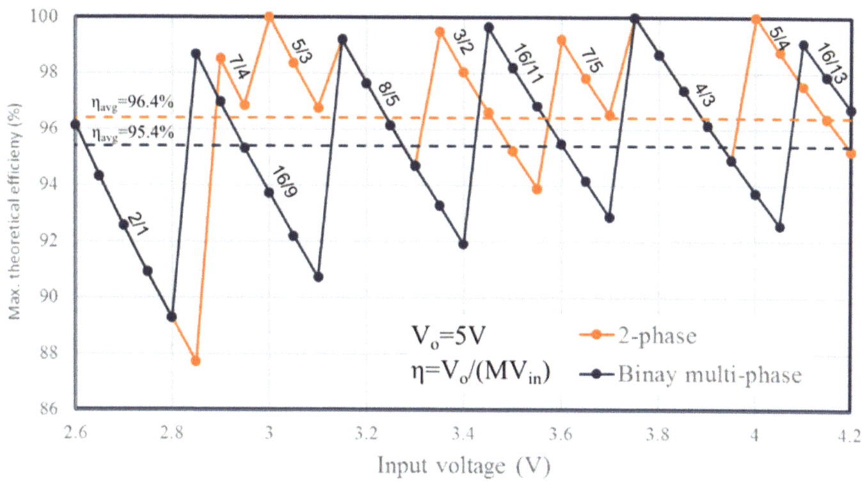

Fig. 13.11 Impact of achievable voltage conversion ratios with fractional two-clock-phase and binary multi-clock-phase topologies in practical multi-ratio converters

shows, as a function of the step-down conversion ratio M, the ratio between the switching frequency of a multi-clock-phase design (f_{mph}) and a two-clock-phase (f_{2ph}) design. Both designs provide the same R_{SSL} with the same total amount of floating capacitance C_T. It is clearly noticeable that multi-clock-phase topologies require a lower f_s to provide the same amount of impedance, although they have a higher number of switching events due to their higher number of clock phases within each switching period T_s. From Eq. (13.5), for a given value of R_{SSL}, the value of m varies with that of f_s. Therefore, the ratio of switching frequency coincides with the ratio of m values of each family of topologies.

Both binary multi-clock-phase and two-clock-phase topologies can also be compared based on their R_{FSL} value. Figure 13.12b shows the ratio of the total amount of area spent on implementing the power switches for the multi-clock-phase case (A_{mph}) and for the two-clock-phase designs (A_{2ph}), as a function of the implemented conversion ratio M. From this graph it becomes clear that the two-clock-phase topologies will require less area spent in the switches, thus making them more appealing when implementing an SCPC with external floating capacitors, especially when high output power is needed and a lower switch area leads to a lower cost [13]. Similar to the m ratio corresponding to the f_s ratio in Fig. 13.12a, the switch-area ratio in Fig. 13.12b coincides with the ratio of p values according to Eq. (13.6).

Considering that multi-clock-phase topologies require larger power switches (for the same R_{FSL}) but require lower f_s (for the same R_{SSL} and C_T) albeit with having more switching events per cycle, it is relevant to compare their expected amount of switching losses. An indication of the amount of switching losses that can be expected from either design approach could be computed as the product of the

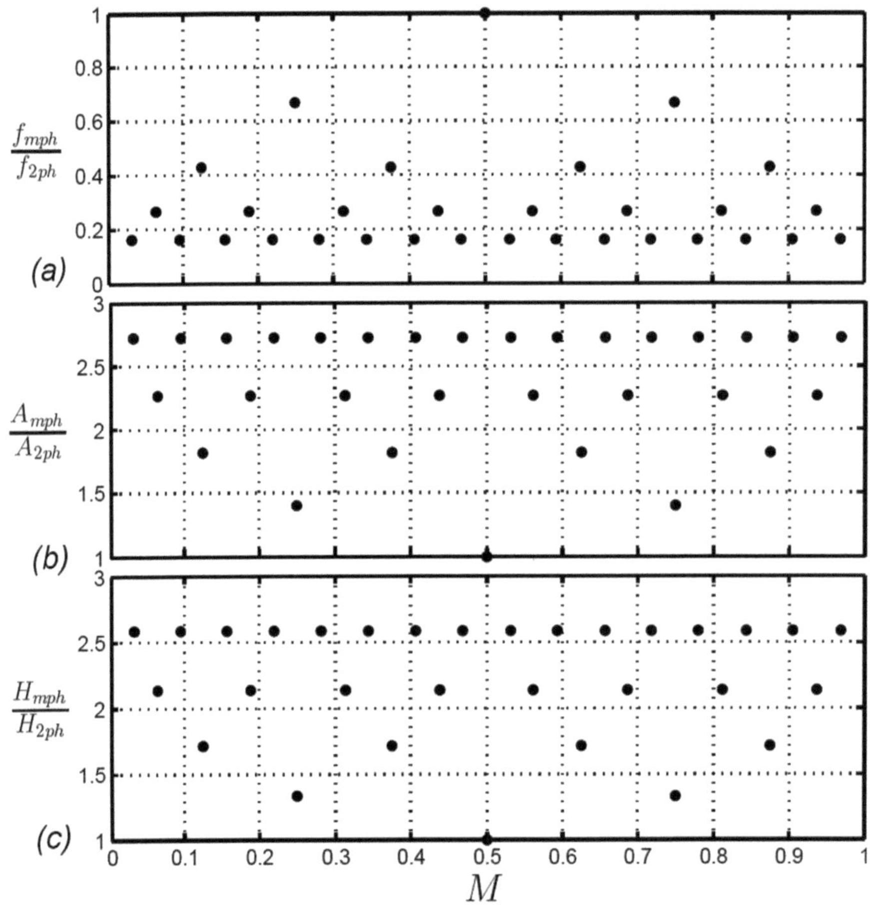

Fig. 13.12 Comparison of two-clock-phase vs. binary multi-clock-phase topologies in terms of (**a**) switching frequency, (**b**) switch area and (**c**) switching losses indication, as a function of the step-down conversion ratio M

switching frequency and the amount of area that is switched per cycle. In case of two-clock-phase topologies, the amount of area that switches per cycle is the same as the total amount of area spent in switches (A_{swT}). However, in multi-clock-phase designs, some switches might be turned-on/off more than once per switching period T_s, meaning that the amount of area that switches is bigger than A_{swT} by a factor $switching\_events/n_{sw}$ (if all the switches are equally sized, as in the binary multi-clock-phase topology). To address this issue, the following parameter is defined as an indicative parameter of the amount of switching losses:

$$H = f_s A_{swT} \frac{switching\_events}{n_{sw}} \begin{cases} 2\ phases \rightarrow switching\_events = n_{sw} \\ mult.ph \rightarrow switching\_events > n_{sw} \end{cases} \tag{13.9}$$

Figure 13.12c shows the ratio of this parameter for both the multi-clock-phase and two-clock-phase topology (H_{mph}/H_{2ph}), as a function of the conversion ratio M. It is observed that in general higher switching losses should be expected from multi-clock-phase designs providing the same conversion ratio and output impedance, despite the lower switching frequency. As in previous results, both approaches coincide for $M = 1/2$, for which a single floating capacitor is used and only two clock phases are possible.

Unfortunately, the amount of bottom-plate losses and the blocking voltage of the different devices could not be included in the results of Fig. 13.12b, c without loss of generality, due the wide variety of two-clock-phase topologies that could be implemented, providing the same conversion ratio. In case that not all the devices of a particular design could have the same voltage rating, this should also be taken into account when considering the different design strategies.

For completeness, the multi-clock-phase designs of Figs. 13.7 and 13.8, providing $M = 1/8$, have also been included in Table 13.3 (topologies F and G, respectively). For the particular case of $M = 1/8$, the different two-clock-phase topologies A–E can be compared to multi-clock-phase designs F and G, including the blocking voltages of the devices as well as bottom-plate losses. The relative amount of bottom-plate losses is obtained from the ml product which takes into account the difference in switching frequency. To compute the relative amount of switching losses taking into account the blocking voltage of the power switches (last column of the table), which switch is used more than once needs to be carefully considered for the multi-clock-phase topologies, as well as their relatively lower f_s (represented by the m value). From the Table 13.3, the following is observed:

- Multi-clock-phase topologies are clearly an interesting solution for designs with discrete floating capacitors, where reducing the number of components and pins is important.
- Multi-clock-phase topologies operate at lower f_s, although their bottom-plate losses are not necessarily smaller. However, when using discrete floating capacitors, this may be irrelevant.
- Multi-clock-phase designs require a larger amount of area for the switches, which is relevant for high-power designs [13, 27].
- Despite operating at lower f_s, no lower switching losses result from multi-clock-phase designs since they have more switching events per cycle.

13.6 Conclusions

Various SCPC topologies are possible to realize a desired voltage conversion ratio M. A major distinction between possible topologies is the number of clock phases that is used within a switching cycle. Within the group of two-clock-phase SCPCs, different topologies are possible that use different numbers of floating capacitors to realize a desired value of M. The fractional topology uses the minimum number of floating capacitors to realize a given M value according to the Fibonacci limit, but the synthesis of such a fractional topology is usually performed by trial and error. Instead, this paper gives an example of systematic derivation of two-clock-phase topologies. Comparing the performance of various two-clock-phase topologies to each other, the output impedance is the same given the same switching frequency, total floating capacitance and total switch area. However, this paper has shown that depending on the application, performance differences e.g. in the amount of bottom-plate losses, the voltages appearing across the floating capacitors or the blocking voltages of power switches, different topologies come out favorably for different applications.

Using more than two clock phases allows the extension of the achievable range of M values of two-clock-phase converters. This allows realizing a target M with a lower number of floating capacitors, which can be advantageous in applications using external floating capacitors. A comparison between two-clock-phase and binary multi-clock-phase topologies shows that though the range of M values is smaller, the density of available M values in a certain needed range is higher in two-clock-phase topologies. This increases the achievable average efficiency across e.g. the desired input voltage range compared to binary multi-clock-phase topologies. Moreover, two-clock-phase topologies use less area to implement the power switches, which may be favorable for high-power applications. However, the modular approach to generate different conversion ratios with the binary multi-clock-phase reduces complexity in generating a larger number of conversion ratios.

References

1. H.J. Bergveld, M. Rose, Integration trends in monolithic power ICs: application and technology challenges, in *Proceedings of the IEEE Bipolar/BiCMOS Circuits and Technology Meeting (BCTM)*, 2015, pp. 13–19
2. G. Villar Piqué, Embedded voltage regulation and energy management for IoT/IoE, in *Proceedings of the IEEE International Solid-State Circuits Conference Forum (ISSCC Forum)*, Feb 26, 2015
3. G. Villar Piqué, H.J. Bergveld, E. Alarcon, Survey and benchmark of fully integrated switching power converters: switched-capacitor versus inductive approach. IEEE Trans. Power Electron. **28**(9), 4156–4167 (2013)
4. R.W. Erickson, D. Maksimovic, *Fundamentals of Power Electronics* (Springer Science & Business Media, 2001)
5. V. De, Energy efficient computing in nanoscale CMOS: challenges and opportunities, in *Proceedings of the IEEE Asian Solid-State Circuits Conference (A-SSCC)*, 2014, pp. 121–124
6. J. Wibben, R. Harjani, A high-efficiency DC–DC converter using 2 nH integrated inductors. IEEE J. Solid-State Circuits **43**(4), 844–854 (2008)
7. M. Wens, M. Steyaert, A fully integrated CMOS 800 mW four-phase semiconstant ON/OFF-time step-down converter. IEEE Trans. Power Electron. **26**(2), 326–333 (2011)
8. D.S. Gardner, G. Schrom, F. Paillet, B. Jamieson, T. Karnik, S. Borkar, Review of on-chip inductor structures with magnetic films. IEEE Trans. Magn. **45**(10), 4760–4766 (2009)
9. C. Huang, P.K. Mok, A 100 MHz 82.4% efficiency package-bondwire based four-phase fully-integrated buck converter with floating capacitor for area reduction. IEEE J. Solid-State Circuits **48**(12), 2977–2988 (2013)

10. E.A. Burton, G. Schrom, F. Paillet, J. Douglas, W.J. Lambert, K. Radhakrishnan, M.J. Hill, FIVR—fully integrated voltage regulators on 4th generation Intel® Core™ SoCs, in *Proceedings of the IEEE Applied Power Electronics Conference and Exposition (APEC)*, 2014, pp. 432–439
11. H.J. Bergveld, K. Nowak, R. Karadi, S. Iochem, J. Ferreira, S. Ledain, E. Pieraerts, M. Pommier, A 65 nm-CMOS 100 MHz 87% efficient DC-DC down converter based on dual-die system-in-package integration, in *Proceedings of the IEEE Energy Conversion Congress and Exposition (ECCE)*, 2009, pp. 3698–3705
12. G. Villar Piqué, Potential benefits of integrated switching power converters: inductive vs. switched-capacitor, in *International Workshop on Power Supply on Chip (PowerSoc)*, 2012
13. G. Villar Piqué, H.J. Bergveld, R. Karadi, A 1 W 8-ratio switched-capacitor boost power converter in 140 nm CMOS with 94.5% efficiency, 0.5 mm thickness and 8.1 mm^2 PCB area, in *Proceedings of the IEEE Symposium on VLSI Circuits (VLSIC)*, 2015, pp. 338–339
14. Paul, D. Jiao, S. Sapatnekar, C.H. Kim, Deep trench capacitor based step-up and step-down DC/DC converters in 32 nm SOI with opportunistic current borrowing and fast DVFS capabilities, in *Proceedings of the IEEE Asian Solid-State Circuits Conference (A-SSCC)*, 2013, pp. 49–52
15. G. Villar Piqué, E. Alarcon, *CMOS Integrated Switching Power Converters: A Structured Design Approach* (Springer, 2011)
16. W. Kim, D. Brooks, G. Wei, A fully integrated 3-level DC-DC converter for nanosecond-scale DVFS. IEEE J. Solid-State Circuits **47**(1), 206–219 (2012)
17. C. Schaef, J. Stauth, A 3-phase resonant switched capacitor converter delivering 7.7 W at 85% efficiency using 1.1 nH PCB trace inductors. IEEE J. Solid-State Circuits **50**(12), 2861–2869 (2015)
18. M.D. Seeman, S.R. Sanders, Analysis and optimization of switched-capacitor DC-DC converters. IEEE Trans. Power Electron. **23**(2), 841–851 (2008)
19. Y. Li, J. Jiang, W. Ki, C. Yue, S. Sin, U. Seng-Pan, R. Martins, A 123-phase DC-DC converter ring with fast DVS for microprocessors, in *Proceedings of the IEEE International Solid-State Circuits Conference (ISSCC)*, 2015, pp. 364–365
20. M. Makowski, D. Maksimovic, Performance limits of switched-capacitor DC-DC converters, in *Proceedings of the IEEE Power Electronics Specialists Conference (PESC)*, 1995, pp. 1215–1221
21. T. Van Breussegem, Monolithic capacitive CMOS DC-DC converters. Ph.D. Thesis, Katholieke Universiteit Leuven, 2012
22. J. Tsai, S. Ko, C. Wang, Y. Yen, H. Wang, P. Huang, P. Lan, M. Shen, A 1 V input, 3 V-to-6 V output, 58%-efficient integrated charge pump with a hybrid topology for area reduction and an improved efficiency by using parasitics. IEEE J. Solid-State Circuits **50**(11), 2533–2548 (2015)
23. H. Meyvaert, G. Villar Piqué, R. Karadi, H.J. Bergveld, M. Steyaert, A light-load-efficient 11/1 switched-capacitor DC-DC converter with 94.7% efficiency while delivering 100 mW at 3.3 V. IEEE J. Solid-State Circuits **50**(12), 2849–2860 (2015)
24. L. Salem, P. Mercier, A recursive switched-capacitor DC-DC converter achieving 2^N-1 ratios with high efficiency over a wide output voltage range. IEEE J. Solid-State Circuits **49**(12), 2773–2787 (2014)
25. R. Karadi, Synthesis of switched-capacitor power converters: an iterative algorithm, in *Proceedings of the IEEE Workshop on Control and Modeling for Power Electronics (COMPEL)*, 2015, pp. 1–4
26. R. Karadi, G. Villar Piqué, 3-phase 6/1 switched-capacitor DC-DC boost converter providing 16 V at 7 mA and 70.3% efficiency in 1.1 mm^3, in *Proceedings of the IEEE International Solid-State Circuits Conference (ISSCC)*, 2014, pp. 92–93
27. L. Salem, P. Mercier, A battery-connected 24-ratio switched capacitor PMIC achieving 95.5%-efficiency, in *Proceedings of the IEEE Symposium on VLSI Circuits (VLSIC)*, 2015, pp. 340–341

Chapter 14
Resonant and Multimode Switched Capacitor Converters for High-Density Power Delivery

Jason T. Stauth, Christopher Schaef, and Kapil Kesarwani

14.1 Introduction

Modern power electronics are driven by a constant need to improve efficiency in the effort to extend battery life of portable electronics, relax thermal constraints in power intensive applications, and help mitigate CO_2 emissions by reducing wasted energy. However, in a variety of applications, *size, form-factor, and ultimately the power-density of the energy-management subcomponents are becoming critical* as these constrain both cost and size of the packaged system [1–3]. By far, the most common power delivery architectures are based on inductive energy storage. For example, buck, boost, and transformer-coupled DC-DC converters currently dominate the market for typical low-moderate voltage and power applications. This places substantial pressure on inductive components to achieve both high volumetric energy- and power-density and also low loss at increasing switching frequencies [2].

To illustrate this, consider the well-known relationships for current and voltage ripple (Δi_L and Δv_C) in a conventional buck converter:

$$\Delta i_L \propto \frac{V_{IN}}{f_{SW} L}, \tag{14.1}$$

and

$$\Delta v_C \propto \frac{\Delta i_L}{f_{SW} C}. \tag{14.2}$$

J.T. Stauth (✉) • C. Schaef • K. Kesarwani
Thayer School of Engineering at Dartmouth, 14 Engineering Drive, Hanover, NH, USA
e-mail: jason.t.stauth@dartmouth.edu

© Springer International Publishing Switzerland 2017

A. Baschirotto et al. (eds.), *Wideband Continuous-time ΣΔ ADCs, Automotive Electronics, and Power Management*, DOI 10.1007/978-3-319-41670-0_14

In Eqs. (14.1) and (14.2), V_{IN} is the supply voltage, f_{SW} is the switching frequency, L and C are the inductance and capacitance respectively. Also note that the units of (1) are [V-s/H] and units of (2) are [A-s/F]. Here, the tradeoffs among voltage, switching frequency, and size of passive components are readily apparent. For a fixed input voltage, the only way to improve the [V-s/H] FOM is by utilizing higher frequency or larger inductance. To improve the [A-s/F] FOM for a fixed Δi_L requires utilizing higher frequency or larger capacitance. To ultimately achieve higher power-density, these trends drive up switching frequencies, but demand also that active and passive components remain efficient [2–4].

Fortunately, trends in active devices are generally in support of higher switching frequencies. Scaled CMOS technologies are approaching f_{max} levels in the near-THz regime [5], providing increasingly efficient operation at switching frequencies appropriate for even the fastest power converters, although notably also trending towards lower breakdown voltages. At higher voltages, wide-bandgap technologies are gaining higher levels of integration while relaxing tradeoffs among breakdown voltage, frequency, and efficiency [6]. Capacitor technologies are also improving, one example being deep-trench technologies with densities approaching many 100 s of nF/mm^2—orders of magnitude higher than traditional MIM and MOS capacitors [7–9]. However, inductor components remain a key limitation in traditional buck and boost topologies, especially given the need for higher inductance-density while maintaining low-loss at high frequency. Available magnetic materials scale poorly to frequencies above 10 MHz, and air-core topologies remain practically constrained to (at most) 10's of nH for mm-scale geometries [2].

In the past decade, switched-capacitor DC-DC converters have gained academic and commercial interest due to a number of favorable architectural advantages and scaling properties [9–11]. In [12], it was shown that SC converters can outperform conventional magnetic-based topologies (buck and/or boost) purely in terms of conduction loss across a wide range of power and voltage levels. This is related to the active device figure of merit, the V-A product, which relates the breakdown voltage and peak current rating needed for all active devices in the circuit. More generally, the SC approach eliminates bulky and expensive magnetic components which often dominate size and cost, and limit operating frequencies due to core and winding loss. Another advantage that has emerged in favor of SC topologies is the ability to segment designs into *large numbers of interleaved phases* [9, 10, 13]. This capability enables two main advantages: in principle, interleaving can reduce input and output voltage ripple to near-negligible levels, permitting elimination of bypass capacitance and therefore full utilization of the available capacitance density for energy transfer. Interleaving also opens up the possibility of soft-charging the parasitic bottom-plate component, significantly reducing its impact on efficiency [8, 14].

However, the switched capacitor approach also has several limitations. First, it relies heavily on the capacitance density available in a given technology. While capacitive energy storage is often touted as superior to inductive energy storage due to the relatively higher energy density of capacitors as compared to inductors, [3, 15], *to fully realize this advantage, a given converter must be able to access and utilize the available energy-density*. In the case of SC converters, accessing the available energy density comes with a penalty of higher charge-sharing losses. To mitigate charge sharing losses, SC converters remain constrained by a fundamental relationship similar to (2) and necessitate either high frequency or high capacitance to minimize conduction loss in the circuit. Therefore, an argument can be made that some of the recent performance gains in the SC area are derived largely from nascent component technologies (e.g. deep-trench capacitors) or advanced process options (e.g. deep submicron silicon-on-insulator technologies). Finally, while variable regulation can be achieved with SC designs through linear modulation of the output impedance, architectural reconfiguration [8–10], and recursive topologies [16], these approaches tend to result in lower efficiency and/or power-density. This is caused by the additional stray junction capacitance of reconfigurable architectures and higher series resistance in recursive approaches. It can also be appreciated that techniques which vary switching frequency to modulate output impedance trade off power loss for regulation capability and also cannot boost the output voltage above the nominal level.

Hybrid, soft-charging, and resonant operation of SC converters is increasingly promising to address some of the above-mentioned limitations [17–22]. In the resonant switched capacitor (ReSC) approach, the efficiency and power density of SC architectures are improved through the use of small magnetic components that resonate with flying capacitors and tune out their reactive impedance [23]. This approach also enables soft- or zero-current-switching (ZCS) which reduces power loss and stresses in the semiconductor components. Other developments in the ReSC area include the use of multi-mode operation to achieve variable conversion ratios and voltage regulation, and also to increase efficiency across the output power range [19–22]. Importantly, the ReSC concept extends and merges with a variety of SC architectures spanning Dickson, series-parallel, and flying-capacitor multilevel (FCML) topologies [19, 24]. While an inductive impedance is still needed, compared to traditional buck and boost topologies, the hybrid approach can operate with much smaller inductance as the switched-capacitor (SC) front end can reduce the energy storage requirements (volt-second product) in the inductor by a significant margin (potentially 10–100×).

This paper will provide an overview or resonant and soft charging switched-capacitor DC-DC converters. Section 14.2 will discuss nominal resonant operation and a generalized comparison of SC and ReSC topologies. Section 14.3 will present candidate circuit architectures and their advantages and disadvantages. Section 14.4 will discuss an expanded suite of operating modes for hybrid and resonant SC converters. Section 14.5 will present implementation examples and future work in the space.

14.2 Resonant Switched Capacitor Converters

Consider Fig. 14.1a, b which shows simplified switching cells for nominally 1:1 SC and resonant switched capacitor (ReSC) converters. Here a flying impedance is switched between two voltage sources with a net voltage difference of ΔV. From the perspective of the flying impedance, this process presents a square wave of voltage, which in turn drives a respective current waveform. In the switched capacitor case, the current waveform is impulsive and is governed by the $R_{ESR} \cdot C_X$ time constant, where R_{ESR} represents the total resistance in the flying impedance loop and C_X is the flying capacitance. In the ReSC case, assuming the frequency of the voltage waveform is at the fundamental resonant frequency, f_O, of the impedance, the current waveform is sinusoidal with amplitude $\frac{2}{\pi}\frac{\Delta V}{R}$. These representative waveforms are shown in Fig. 14.1d. By integrating to get the charge transferred between voltage ports per cycle, it can be shown that in both cases, the process can be modelled as using an ideal transformer model with output resistance, R_{EFF} (Fig. 14.1c) [12, 18].

The equations that describe the parameter, R_{EFF}, for the 1:1 switching cells in Fig. 14.1a, b are shown as Eqs. (14.3) and (14.4). Here it should be noted that while the SC case is accurate for any switching frequency (f_{SW}), the ReSC case assumes operation at the fundamental resonant frequency ($f_{SW} \approx f_O$). Another distinction here can be made for the SC converter class: the 'coth' relationship in (3) governs the transition of the converter between the 'slow-switching limit' (SSL) and 'fast switching limit' (FSL). For example at low switching frequencies, (3) will approximate to $R_{EFF\_SC} \approx (f_{SW}C_X)^{-1}$, while at high frequencies to $R_{EFF\_SC} \approx 4R_{ESR}$. The boundary between the FSL and SSL is therefore defined here as $f_{FSL-SSL} \equiv (4R_{ESR}C_X)^{-1}$.

The ReSC relationship can be appreciated by examining the argument of the tanh in (4), which depends on the quality factor, Q_X, of the resonant network.

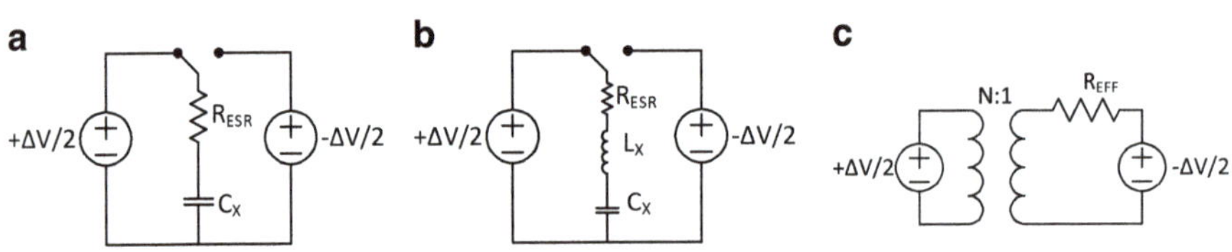

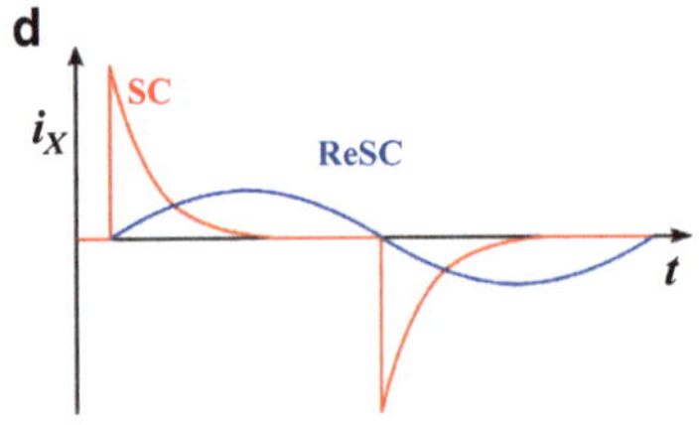

$$R_{EFF\_SC} = \frac{1}{f_{SW}C_X}\coth\left(\frac{1}{4f_{SW}R_{ESR}C_X}\right) \tag{14.3}$$

$$R_{EFF\_ReSC} = \frac{1}{f_{SW}C_X}\tanh\left(\frac{\pi}{4Q_X}\right) \approx \frac{\pi^2}{2}R_{ESR} \tag{14.4}$$

Fig. 14.1 Ideal switching cells and behavioral transformer model. (**a**) SC cell; (**b**) ReSC cell; (**c**) behavioral; (**d**) respective current waveforms

For example, at high-Q, the relationship simplifies to $R_{EFF\_\text{ReSC}} \approx \frac{\pi^2}{2} R_{ESR}$. Then consider the ratio of $R_{EFF\_SC}/R_{EFF\_\text{ReSC}}$, while maintaining the assumption that $f_{SW} = f_0 << f_{FSL-SSL}$, and that C_X is the same for both. The result follows as:

$$\frac{R_{EFF\_SC}}{R_{EFF\_\text{ReSC}}} \approx \frac{4}{\pi} \frac{1}{2\pi f_0 C_X R_{ESR}} = \frac{4}{\pi} Q_X \tag{14.5}$$

Essentially, this indicates that at a given frequency, assumed resonant for the ReSC, and where $f_0 \ll f_{FSL\text{-}SSL}$ (operation below the FSL boundary), the effective resistance of the ReSC topology is $\sim Q_X$ times lower than the equivalent SC topology. The following conclusions can therefore be made:

1. For SC and ReSC topologies with the same C_X and R_{ESR}, at the same frequency, the ReSC topology can operate with Q times lower R_{EFF} (or conduction loss).
2. Given the same C_X and R_{ESR}, but now assuming the converters operate with the same R_{EFF}, the ReSC topology can operate at ~Q times lower frequency (or switching loss)

On the one hand, this seems like a compelling argument for the resonant topology as it provides a degree of freedom beyond what is available in the SC topology to minimize total power loss (or maximize efficiency). However, a limitation to the above analysis is that it neglects the additional losses associated with the inductor, which ultimately limits the achievable performance benefit. Another way to phrase this is that for the ReSC topology, it is *essential to consider the ESR of the additional inductor component* and its effect on power loss and efficiency.

In order to provide a more accurate assessment of the ReSC topology as compared to a comparable SC converter, a realistic model for available magnetic components must be included. Figure 14.2 provides one such comparison. Here, the power loss of a representative SC converter and ReSC converter is shown versus switching frequency. For the comparison, a fixed flying capacitance of

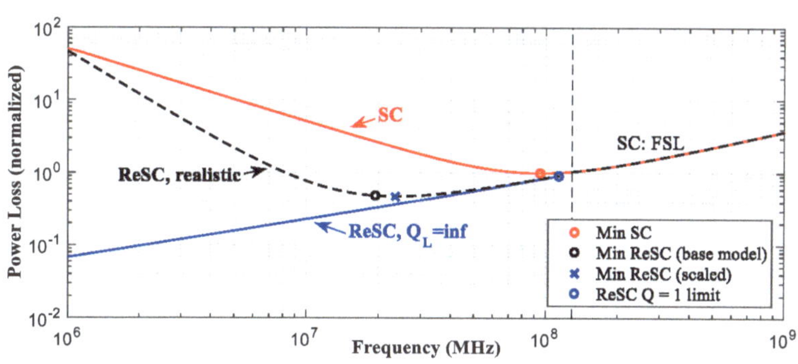

Fig. 14.2 Comparison of optimized SC and ReSC designs (nominally 2:1) given ideal and realistic assumptions about inductor ESR

10 nF with 1 % bottom plate ratio and load current of 1A was used. Also, all designs are assumed to be implemented in the same, representative CMOS process technology. It is also the case that all points on the curves are optimized at the given switching frequency (e.g. CMOS switches are sized to minimize total switching and conduction loss).

In-line with the analysis in [10], it can be appreciated that the SC converter optimization requires both a consideration of device sizes and switching frequency. In Fig. 14.2, the curve for the SC converter shows a minimum power loss at a switching frequency ~100 MHz. This happens to occur near the SSL-FSL boundary, as is found to be common in high power-density designs [18]. All power loss levels in Fig. 14.2 are normalized to the minimum SC power loss.

The curves for the ReSC case consider two scenarios: one where the inductor is assumed to add no parasitic resistance ($Q_L = \infty$, such that the effective quality factor, Q_X, is dominated by switch resistance only), and another case where a realistic mm-scale air-core inductor is used. For the $Q_L = \infty$ case, the conclusion is simply that the ReSC converter power loss scales in proportion to $\sqrt{f_{SW}}$. However, it can be appreciated that this is oversimplified because at fixed capacitance, lower switching frequency implies a larger inductor. Assuming the volume of the inductor is fixed, the ESR will scale in approximate proportion with inductance [2]. To second order, however, we assume that AC resistance in the inductor scales with $R_{AC} \approx k\sqrt{f_{SW}}$. Therefore, for a given base inductor model, ESR scaling depends both on inductance and frequency.

A representative curve for realistic ReSC performance is also shown in Fig. 14.2. Here it can be seen that, similar to SC designs, there is a minimum power loss associated with a particular operating frequency range. The predicted performance for the exact inductor base model, in this case coilcraft 0806SQ 6.9 nH with added 40 mΩ DC contact resistance, is able to achieve a factor of two reduction in loss compared to the best case SC design. This also occurs at ~5× lower frequency. The design could be very slightly improved by scaling the base inductance to an incrementally smaller level as indicated by the 'scaled' Min ReSC point. In all cases, the curves converge in the SC FSL. This is to be expected, as in the FSL, switch resistance dominates any reactive impedance in the circuit. Also note that an expanded set of 'realistic' ReSC curves can be generated by considering a range of inductor base models and physical size or volume [18].

14.2.1 ReSC Topologies

The simplified circuit schematics for two representative, nominally 2:1, ReSC topologies are shown in Fig. 14.3. Figure 14.3a shows the schematic of what we will term the 'indirect' ReSC. Here, four transistors, $M_{1\text{-}4}$ (assumed n-channel for simplicity), are controlled by switching signals $S_{1\text{-}4}$. In resonant operation, S_1 and S_3 are operated in phase, with S_3 shifted to the common mode of M_3. Switch signals S_2 and S_4 are operated complementary to S_1 and S_3. Assuming a nominal load current,

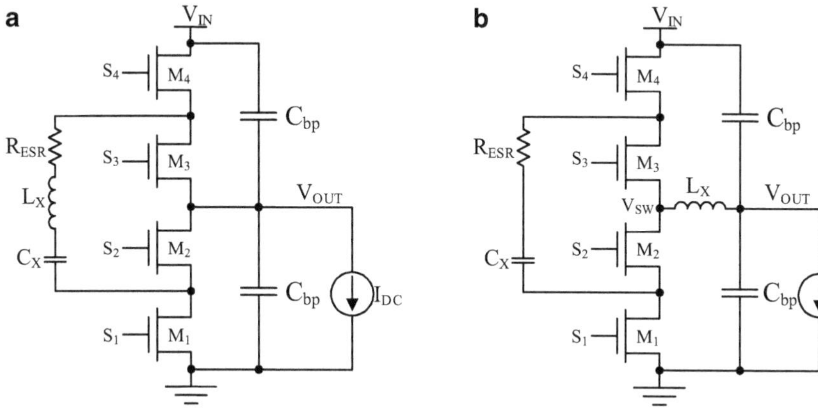

Fig. 14.3 ReSC topologies. (**a**) Indirect ReSC topology; (**b**) direct ReSC topology (3-level)

I_{DC}, V_{OUT} will be slightly lower than $V_{IN}/2$, driving a resonant current excitation in the flying impedance, assuming operation at the resonant frequency. Similar to the discussion of Fig. 14.1, the current in the inductor, L_X, will be sinusoidal with mean value of zero (C_X blocks DC current).

Figure 14.3b shows the schematic of what we will term the 'direct' ReSC. In this topology, the inductor directly connected to the output terminal. The switching process now modulates only the flying capacitance, C_X, similar to a nominal 2:1 SC converter. Also, R_{ESR} is shown as lumped with C_X, but in principle it should capture all resistance in the full resonant loop which includes switches, C_X, L_X, and one of C_{bp}. It should be apparent that the schematic in Fig. 14.3b is equivalent to the more well known 3-level buck converter, [25–27], which is a version of the general class of flying-capacitor multilevel DC-DC converters [28]. However, when we discuss this as the 'direct ReSC converter,' we are supposing operation at a resonant frequency. Also, the 'direct' vs 'indirect' distinction is similar to the distinction made in buck-boost DC-DC converters where in the 'direct' conversion case, the inductor current is unidirectional and couples directly to the output terminal in all phases.

Figure 14.4 shows representative current waveforms for the indirect and direct ReSC converters. In the indirect topology, it can be appreciated that all of the power in the inductor current waveform is concentrated at the fundamental resonant frequency. However, because the current waveform in the indirect topology appears as full-wave rectified, it has an appreciable DC component. It has been shown that approximately 80 % of the power in the indirect waveform is concentrated at DC [18]. The remaining spectral content appears at even harmonics of the fundamental resonant frequency.

Figure 14.5 overlays the power spectral densities of these waveforms (normalized to the total power) with the quality factor and ESR of a representative mm-scale air-core inductor versus frequency. In the modestly simplified inductor profile, the ESR

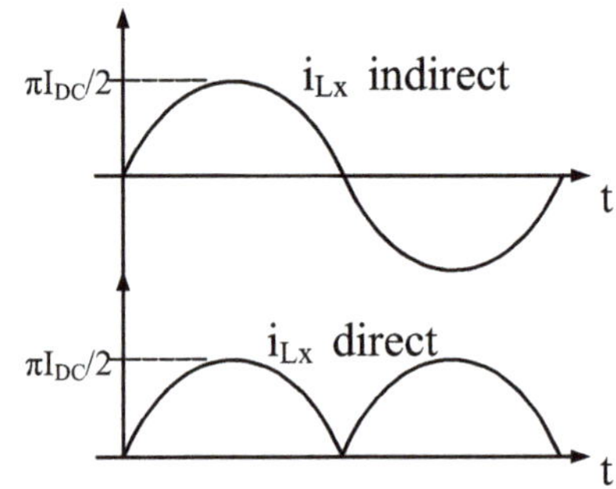

Fig. 14.4 Current waveforms

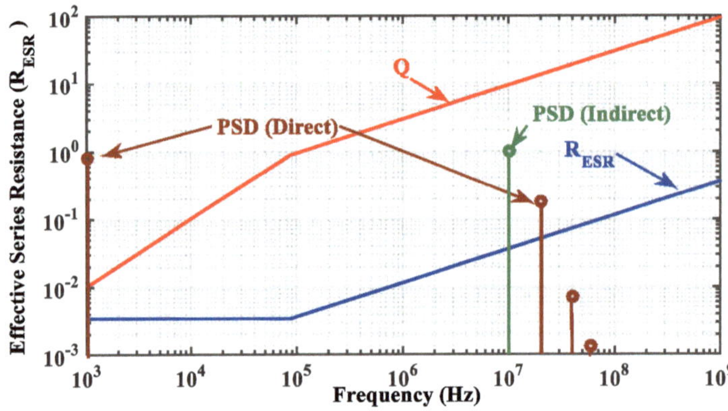

Fig. 14.5 PSD comparison: direct vs indirect

follow the DC resistance up until a frequency where the skin depth is approximately equal to the conductor radius. At higher frequencies ESR is dominated by AC resistance: $R_{AC} \approx k\sqrt{f_{SW}}$. Here, the advantages of the 'direct' approach are more apparent. Because of the large DC term in the PSD of inductor current, the direct topology has the advantage of lower effective resistance in the inductor component. As AC resistance may be much higher than DC resistance, this can provide a substantial reduction in power loss in the direct ReSC topology. A similar argument can be made for core loss, if cored inductors are considered, [19].

However, there are tradeoffs and cases where it may be more favorable to use the indirect approach. One disadvantage of the 'direct' topology is more complicated gate drive circuits and potentially higher voltage stresses on the CMOS switches. For example, the switches are exposed to the full voltage swing on the flying capacitor, C_X. The peak-peak voltage swing on the flying capacitor for the 2:1 converter in Fig. 14.2b. can be written as:

$$\Delta V_{Cs} = \pi Q_X I_{DC} R_{ESR} \approx \frac{8}{\pi} Q_X \Delta V, \tag{14.6}$$

where ΔV is again the net voltage difference between V_{OUT} in its nominal and loaded states (Fig. 14.1). Therefore, switches may be exposed to a voltage much greater than just $V_{IN}/2$ plus the loaded voltage difference, ΔV. Also, it is seen in (14.6) that this is a function of both the load and quality factor of the converter. However, in the indirect topology, switches are only exposed to ΔV itself because they are always coupled to common voltages constrained by bypass capacitors. Therefore the indirect topology may be able to achieve higher power or power-density levels for a given peak voltage stress on switching components, although with potentially lower efficiency.

14.3 Hybrid and Multimode Operation

Another opportunity in the hybrid SC converter area is the potential to leverage an expanded set of operating modes to affect variable regulation and higher efficiency across the output power range. For example, consider the direct (three-level) converter in Fig. 14.3b. In addition to nominally resonant operation, this converter can use additional switching states to couple the inductor to either ground or V_{IN}. The converter can also use high impedance (deadtime) states to provide discontinuous conduction mode (DCM-like) behavior. Similar operating modes have also been used for the indirect topology [21, 22].

Figure 14.6 shows an expanded set of operating states for the converter in Fig. 14.3b. As is well known in the 3-level buck converter literature, in scenarios with output voltages higher than $V_{DD}/2$, the converter can configure the inductor to V_{DD} for a portion of the switching cycle to increase the average voltage at the switching node, V_{SW}. Similarly, for scenarios with output voltages lower than $V_{DD}/2$, the converter can configure the inductor to *ground* for a portion of the switching cycle to decrease the average voltage at the switching node. This process results in equivalent 3-level operation because the voltage on C_X should attain a nominal level of $V_{DD}/2$.

The benefits and challenges of 3-level operation are well known. A primary benefit is the ability to reduce the inductance by a factor of four. A first factor of two reduction is achieved due to the additional mid-rail voltage level (reduces V in the [V-s/H] FOM). A second factor of 2 is achieved because the switching frequency is effectively double as there are now 4 switching states per cycle (reduces s in the [V-s/H] FOM). This advantage scales with higher conversion ratio (N-level) FCML topologies: in general, inductance scales inversely with N^2 for a constant [V-s/H] FOM or equivalently, constant current ripple.

However a central challenge of the 3-level (or N-level FCML) converter is maintenance of voltage balance on the flying capacitor(s) [28]. Without some mechanism to measure and correct for variation in the nominal voltage on C_X, this voltage can drift due to minor asymmetries in switch timing and load transients. Variation in the voltage on C_X can result in a reduction of efficiency or at worst, higher switching stresses and potentially breakdown failures of the power devices.

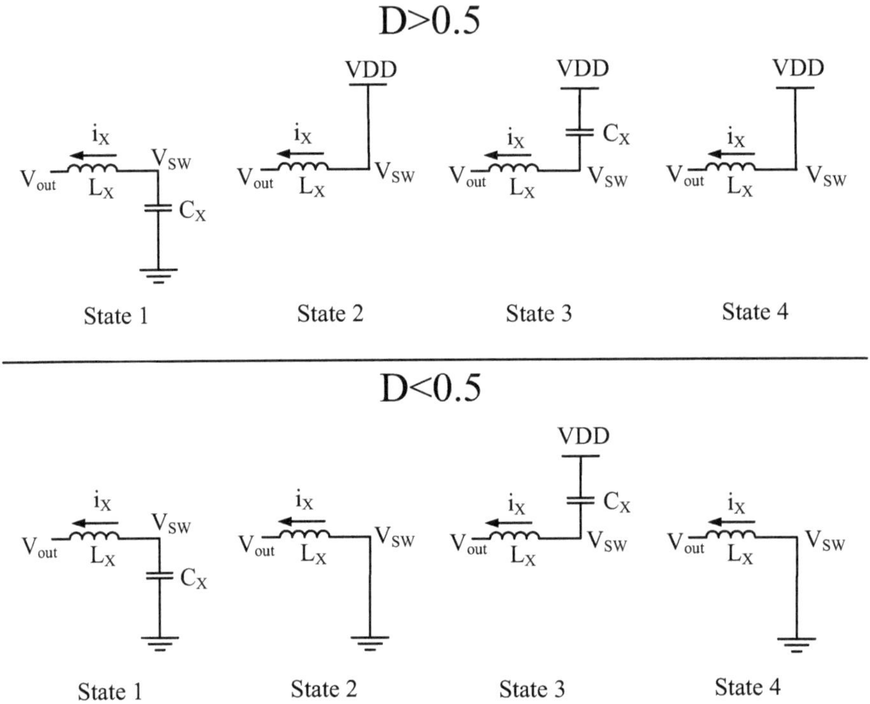

Fig. 14.6 Operating states for variable regulation modes

Here it should be noted, that while this problem is known in the 3-level and FCML converter classes, in [19], it was shown that the same problem affects all hybrid inductive SC converters.

While the 3-level converter is most well known as an augmented form of the buck converter, it merges into the ReSC circuit class. For example, depending on the switching frequency and voltage conversion ratio, the inductor current may take on a variety of forms spanning resonant, quasi-resonant, continuous, and discontinuous conduction. Figure 14.7 shows these possible operating modes segmented by voltage ratio and conduction angle. For the 2:1 3-level converter, the factor, *k*, is ½, but it can be appreciated that similar operating modes are possible in a wide range of converter topologies with different nominal conversion ratios.

Here, the inductive mode is identical to the 3-level buck converter. Also, similar to the standard treatment of the buck converter, the switching frequency is well beyond the LC corner and the inductor current approximates a triangle wave. The duty cycle can be used to affect variable output regulation both above and below the nominal conversion ratio 1:k (note that only step-down modes are shown in Fig. 14.7). Also, depending on the load current level, the converter may operate

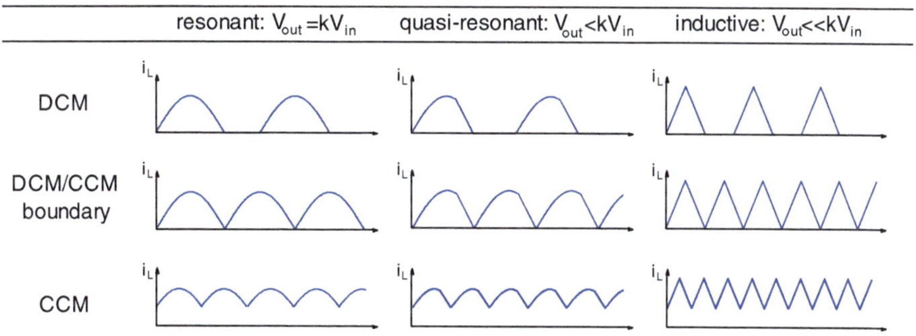

Fig. 14.7 Multimode operating scenarios: inductor current

more efficiently in continuous conduction mode, CCM (at high load current), or DCM (at low current). Pulse frequency modulation (PFM) control is also possible, and is treated here as a form of DCM.

As switching frequencies are reduced, however, the hybrid inductive-SC topology enters a quasi-resonant mode of operation. For example, in the switch states that couple L_X and C_X directly, resonant operation happens when $f_{SW} \approx f_0$. The quasi-resonant mode is therefore used when both lower frequency operation is most efficient and when variable regulation is needed. In quasi-resonance and when the output voltage is less than the nominal conversion ratio, the inductor current follows a partial resonant cycle (during *State-1* and *State-3* in Fig. 14.6) and then is linear during a hard switched state (*State-2* and *State-4* in Fig. 14.6). For the case where output voltage is greater than the nominal conversion ratio, the linear and resonant states are switched to affect a higher average voltage at the switching node, V_{SW}, but the same principle applies.

In pure resonant mode, the inductor current waveform can take several trajectories depending only on the switching frequency. For example, when $f_{SW} = f_0$, the converter operates in resonance with zero-current switching transitions. If a long dead-time is inserted between these transitions, the converter can operate in a DCM mode. This mode is the same as the dynamic off-time modulation (DOTM) mode discussed in [22]. DCM is useful to affect variable output regulation by tuning the loadline of the converter, much like in the SC case. It is also useful to provide higher efficiency in light load as it can be shown that both conduction and switching loss scale linearly with switching frequency. At higher switching frequencies, the converter enters CCM where zero-current-switching no longer occurs. CCM is useful at high load currents as this can be used (much like in conventional buck converters) to reduce the ratio of RMS/DC current in the inductor, providing higher efficiency [19].

Figure 14.2 is a plot of output impedance, R_{EFF}, normalized to the total (parasitic) series resistance, R_{ESR}. The x-axis is switching frequency, normalized to the SSL-FSL boundary frequency for an equivalent SC converter. The plots include the cases for pure (direct) ReSC converter, SC converter, and the hybrid approach discussed above. Each of the circuits operate with the same total flying capacitance, C_X, and assume the same parasitic resistance, R_{ESR}. While the resonant and hybrid cases must accommodate the additional ESR of the inductor component, for simplicity this is neglected here.

Across the operating frequency range, it can be seen that the SC output impedance (curve 1) follows an expected relationship, i.e. (3). The pure ReSC converter (which assumes the 'direct' topology) also follows the expected behavior, quantitatively detailed in [18, 19]. However, the hybrid multimode approach traces out a more favorable curve than both previous topologies. For example, below the resonant frequency, the hybrid approach maintains the advantage of Q times lower R_{EFF} than the SC topology, while still permitting a linear tradeoff between switching and conduction loss. Above the resonant frequency, the hybrid approach enters CCM and the effective resistance approaches R_{ESR}, but at approximately Q times lower frequency than the equivalent SC converter. While derivation of the equation for R_{EFF} versus frequency for the hybrid topology is complicated, it can be captured by a behavioral model somewhat similar to the 'root-sum-squared' model [12], often used for the SC approach:

$$R_{EFF\_hybrid} = \sqrt[\alpha]{1 + \left(\frac{\pi^2 f_0}{8 f_{SW}}\right)^{\alpha}} * R_{ESR}. \tag{14.7}$$

In Eq. (14.7), f_0 is the resonant frequency, f_{SW} is the switching frequency, and α is a behavioral term that can be fit in a regression model for the given converter. It can be shown that a value of $\alpha \approx 9.2$ provides a behavioral fit to the actual model with a worst case error of approximately 2 %.

Importantly, what is shown in Fig. 14.8 and Eq. (14.7) is that the behavior of the hybrid inductive-SC DC-DC converter family is quite similar to the behavior of the underlying SC circuit family. The advantage can be quantified by the quality factor, Q_X, of the converter and the resonant frequency. While the control, gate driving, and design procedure is somewhat more complicated than a pure SC design, the

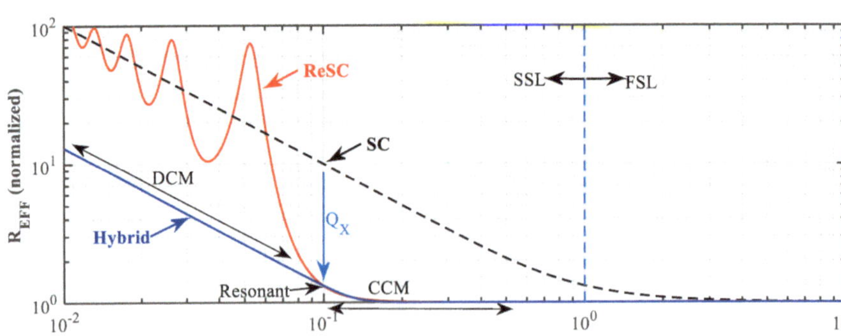

Fig. 14.8 Normalized output impedance: R_{EFF} vs switching frequency

macro-behavior is similar and can be generalized into a unified relationship, (14.7). However, while the variety of operating modes permit loss minimization across the output power spectrum, the hybrid converter class has the advantage that it can also provide variable voltage regulation without reconfiguring the native converter architecture.

14.4 Implementation Examples

To justify some of the above discussion, here we present several recent implementations of hybrid-multimode, nominally resonant switched capacitor designs. In 2014 [22], presented a nominally 2:1 ReSC that utilized DOTM or the DCM concept highlighted in Fig. 14.7 to regulate the output voltage below the nominal level. This design used die-attached mm-scale air-core inductors and on-chip flying and bypass capacitors to achieve 85 % efficiency at 0.64 W/mm^2 in 180 nm bulk CMOS. In 2015 [21], presented a nominally 2:1 ReSC with 3-phase interleaving and an all N-channel power train to further improve efficiency and power-density to 89 % at 0.75 W/mm^2. The three-phase design used mm-scale printed circuit board (PCB) trace inductors with inductance values of 1.1 and 4.5 nH. Regulation was achieved using quasi-resonant modes similar to those in Fig. 14.7 to boost the output voltage above the nominal 2:1 level, and reduce it below this level as well. An overview of these designs and specifications is shown in Table 14.1.

More recently, several discrete implementations were built to explore further aspects of multimode operation of nominally resonant SC converters. The work in [19] explored hybrid multimode operation of 3-level and 4-level flying capacitor multilevel (FCML) DC-DC converters. The 3-level prototype operated exclusively at nominally 2:1 conversion ratios. By varying the frequency, as shown in Fig. 14.7, the converter was able to seamlessly transition from CCM operation (high frequency), to resonant operation, and finally to DCM by inserting a variable deadtime in between resonant transitions. Figure 14.9 shows measured efficiency for the 24 to 12 V step-down converter for load currents ranging between 100 mA and 10 A (~1.2 W to 120 W output power). While peak efficiency in the resonant mode was over 99 %, the converter was able to seamlessly transition to CCM at higher current and DCM at low current to maintain high efficiency (over 98 %) for most of the load range.

In another implementation, operation across the full suite of possible operating modes was explored. Figure 14.10 shows the printed circuit board design that can be operated as a 3-level or 4-level FCML. Ceramic flying capacitors (4.7 μF, X7R) and an inductance of 600 nH were used with a resulting resonant frequency of approximately 100 kHz. All different operating regimes described in Sect. 14.3 were tested in 3-level operation and measured inductor current waveforms are shown in Fig. 14.11. These can be compared to the theoretical waveforms in Fig. 14.7. It can be seen that based on the required conversion ratio and resulting duty cycle

Table 14.1 Highlights of previous designs using resonant-multimode operation

Source	Kesarwani et al. [22]	Schaef et al. [21]
Die photo	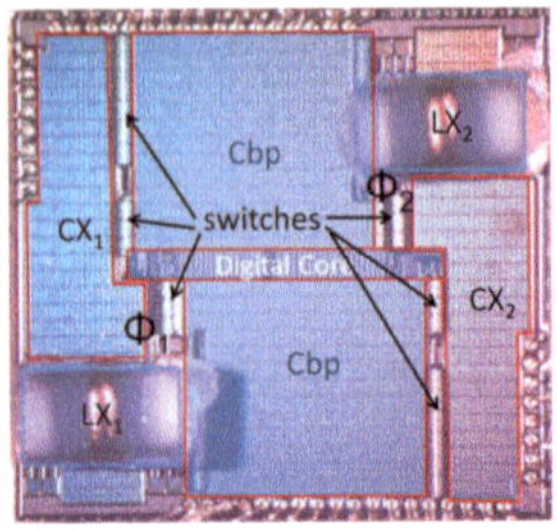	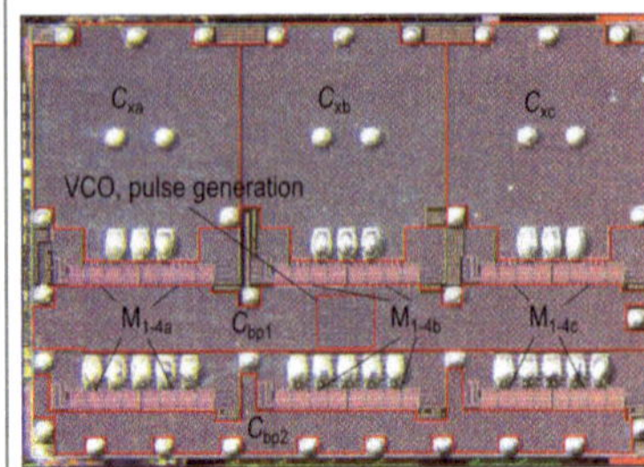
Converter	2-Phase nominally 2:1 ReSC	3-Phase nominally 2:1 ReSC
Regulation	Dynamic off-time modulation (DOTM or DCM)	Variable quasi-resonant buck and boost modes
Power train	Complementary	N-channel with bootstrapped gate drive
Inductor(s)	Die attached mm-scale air-core (1.9–5.5 nH)	Printed circuit board trace inductors (1.1–4.5 nH)
Capacitor(s)	Flying ∼18 nF MIMBypass ∼22 nF MIM	Flying ∼25 nF MIMBypass ∼12 nF MIM
Process	180 nm Bulk CMOS	180 nm Bulk CMOS
Size	∼3 × 3 mm	∼3.8 × 2.6 mm
Frequency	∼20–30 MHz	∼20–50 MHz
Efficiency @ power density	∼85 % @ 0.64 W/mm^2	∼85 % @ 0.91 W/mm^2 (1.1 nH)∼89 % @ 0.75 W/mm^2 (4.5 nH)

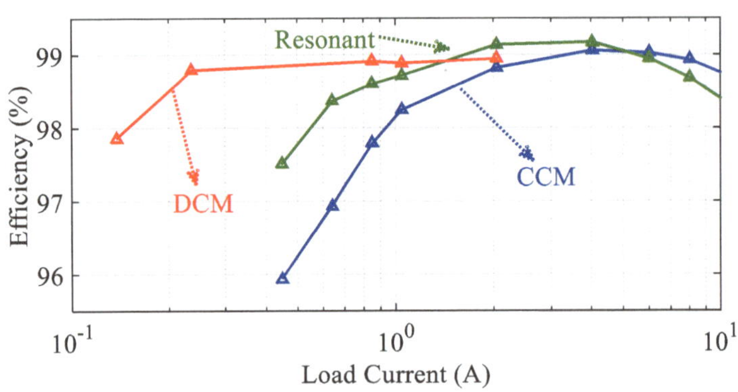

Fig. 14.9 Measured efficiency vs load current: discrete, nominally 2:1 (24 V:12 V) ReSC converter using multimode operation [19]

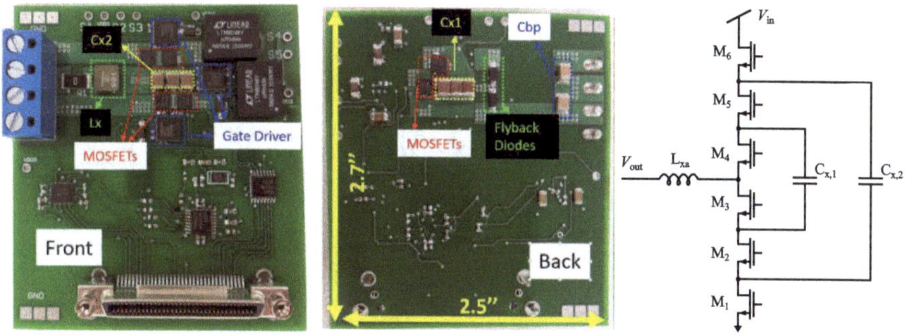

Fig. 14.10 Printed circuit board prototype: 3-level and 4-level FCML DC-DC converter

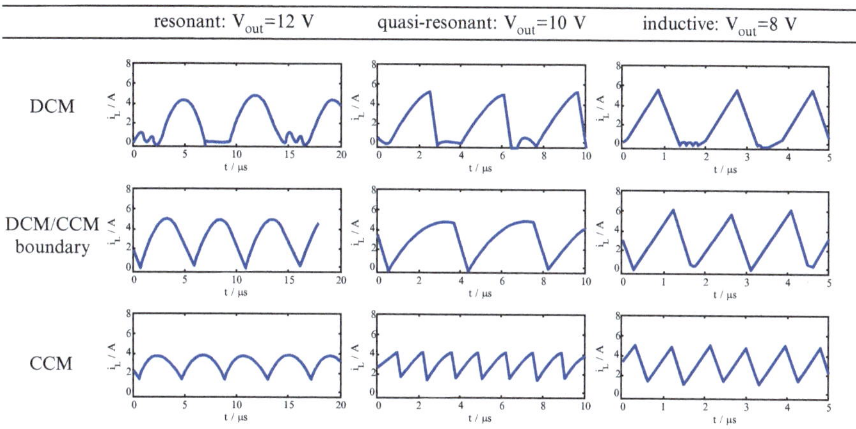

Fig. 14.11 Measured operating waveforms for discrete converter prototype shown in Fig. 14.10. These can be compared to the theoretical curves in Fig. 14.7

that the converter shows resonant, quasi-resonant or inductive behavior. Moreover, the frequency can be modulated to achieve DCM or CCM behavior to maximize efficiency for various load currents.

These operating modes can be demonstrated for 4-level operation and generally apply to all resonant or hybrid switched capacitor converters. However, the resonant frequency in different phases of operation can vary and careful tuning of the individual periods is required to achieve efficient operation as shown in Fig. 14.12 for the 4-level converter. As different flying capacitor configurations are connected in series with the inductor, the pulse widths in resonant operation need to be tuned to the resulting resonant periods in each phase to maintain the required voltage balance on the flying capacitors. Moreover, the pulse-widths need to be adjusted across the different operating regimes, e.g. in CCM operation the pulse-widths need to be of equal length to maintain charge-balance. Efficient multi-mode operation is therefore not only a design challenge but also requires careful consideration of the control strategy.

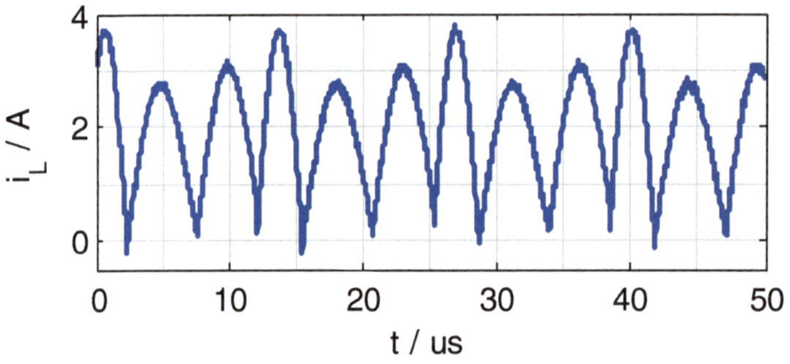

Fig. 14.12 Measured operation of the 4-level FCML in resonant mode: note that resonant frequencies in different phases is variable and requires careful tuning

14.5 Conclusions

In order to meet the challenges of future high-density power management applications, there is a need to explore architectures that break from the fundamental constraints that limit conventional DC-DC architectures. While the switched capacitor approach is favorable for many of these applications, it also has limitations that will continue to constrain efficiency, power-density, and the capability for efficient voltage regulation. Hybrid and resonant SC converters show tremendous promise to address some of these challenges. While the limitations of current inductor technologies must be factored in, the hybrid approach has many of the same scaling features and promising attributes of the SC circuit class. In this paper, we have shown also that hybrid-inductive SC converters have many of the same capabilities as more conventional DC-DC converters: the ability to regulate output voltage efficiently, and the capability for multi-mode operation to maintain high efficiency across the output power range. However these topologies benefit greatly from merging the advantages of both SC and inductive converters: the volume and inductance of magnetic components are greatly reduced while the utilization of the capacitor energy density is increased.

Going forward, there is a great opportunity for future work on hybrid, soft-switching, and resonant switched capacitor converters as these topologies will undoubtedly find a strong niche in a range of future applications.

References

1. D. Andersen, Waiting for PSOC... applications of PSIP/PSOC products, in *IEEE Workshop on Power Systems on Chip (PowerSOC)*, 2012
2. D.J. Perreault, J. Hu, J.M. Rivas, Y. Han, O. Leitermann, R.C.N. Pilawa-podgurski, A.D. Sagneri, C.R. Sullivan, Opportunities and challenges in very high frequency power conversion, in *IEEE Applied Power Electronics Conference (APEC)*, 2009, pp. 1–14

3. S.R. Sanders, E. Alon, H. Le, M.D. Seeman, M. John, V.W. Ng, The road to fully integrated DC-DC conversion via the switched-capacitor approach. IEEE Trans. Power Electron. **28**(9), 4146–4155 (2013)
4. D.J. Perreault, A.D. Sagneri, Design of miniaturized, isolated DC-DC converters operating at radio frequencies, in *IEEE Workshop on Power Systems on Chip (PowerSOC)*, 2012
5. International Technology Roadmap for Semiconductors, 2011
6. H.A. Mantooth, M.D. Glover, P. Shepherd, Wide Bandgap technologies and their implications on miniaturizing power electronic systems. IEEE J. Emerg. Sel. Top. Power Electron. **2**(3), 374–385 (2014)
7. L. Chang, R.K. Montoye, B.L. Ji, A.J. Weger, K.G. Stawiasz, R.H. Dennard, A fully-integrated switched-capacitor 2:1 voltage converter with regulation capability and 90 % efficiency at 2.3A/mm^2, in *2010 IEEE Symposium on VLSI Circuits (VLSIC)*, 2010, pp. 0–1
8. T.M. Andersen, F. Krismer, J.W. Kolar, T. Toifl, C. Menolfi, L. Kull, T. Morf, M. Kossel, M. Brandli, P. Buchmann, P.A. Francese, 4.7 A sub-ns response on-chip switched-capacitor DC-DC voltage regulator delivering 3.7 W/mm^2 at 90 % efficiency using deep-trench capacitors in 32 nm SOI CMOS, in *2014 IEEE International Solid-State Circuits Conference Digest of Technical Papers (ISSCC)*, 2014, pp. 90–91
9. T.M. Andersen, F. Krismer, J.W. Kolar, T. Toifl, C. Menolfi, L. Kuli, T. Morf, M. Kossel, M. Brandii, P.A. Francese, 20.3 A feedforward controlled on-chip switched-capacitor voltage regulator delivering 10 W in 32 nm SOI CMOS, in *2015 IEEE International Solid-State Circuits Conference (ISSCC)*, 2015, pp. 1–3
10. H.-P. Le, S.R. Sanders, E. Alon, Design techniques for fully integrated switched-capacitor DC-DC converters. IEEE J. Solid-State Circuits **46**(9), 2120–2131 (2011)
11. M.S. Makowski, D. Maksimovic, Performance limits of switched-capacitor DC-DC converters, in *IEEE Power Electronics Specialists Conference (PESC)*, 1995, pp. 1215–1221
12. M.D. Seeman, S.R. Sanders, Analysis and optimization of switched-capacitor DC–DC converters. IEEE Trans. Power Electron. **23**(2), 841–851 (2008)
13. G.V. Pique, A 41-phase switched-capacitor power converter with 3.8 mW output ripple and 81 % efficiency in baseline 90 nm CMOS, in *International Solid State Circuits Conference (ISSCC)*, **23**(2), 98–100 (2012)
14. N. Butzen, M. Steyaert, 12.2 A 94.6 %-efficiency fully integrated switched-capacitor DC-DC converter in baseline 40 nm CMOS using scalable parasitic charge redistribution, in *IEEE International Solid State Circuits Conference (ISSCC), Digest of Technical Papers*, 2016
15. J.J. Cooley, S.B. Leeb, Per panel photovoltaic energy extraction with multilevel output DC-DC switched capacitor converters, in *IEEE Applied Power Electronics Conference (APEC)*, 2011, pp. 419–428
16. L.G. Salem, P.P. Mercier, 4.6 An 85 %-efficiency fully integrated 15-ratio recursive switched-capacitor DC-DC converter with 0.1-to-2.2 V output voltage range, in *2014 IEEE International Solid-State Circuits Conference Digest of Technical Papers (ISSCC)*, 2014, pp. 88–89
17. R.C.N. Pilawa-podgurski, D.J. Perreault, Merged two-stage power converter with soft charging switched-capacitor stage in 180 nm CMOS. IEEE J. Solid-State Circuits **47**(7), 1557–1567 (2012)
18. K. Kesarwani, R. Sangwan, J.T. Stauth, Resonant switched-capacitor converters for chip-scale power delivery: design & implementation, in *Power Electronics, IEEE Transations*, 2014, pp. 1–1
19. K. Kesarwani, J.T. Stauth, Resonant and multi-mode operation of flying capacitor multi-level DC-DC converters. *IEEE Control Model. Power Electron.* (2015)
20. K. Kesarwani, J.T. Stauth, The direct-conversion resonant switched capacitor architecture with merged multiphase interleaving: Cost and performance comparison, in *2015 IEEE Applied Power Electronics Conference and Exposition (APEC)*, 2015, pp. 952–959
21. C. Schaef, K. Kesarwani, J.T. Stauth, 20.2 A variable-conversion-ratio 3-phase resonant switched capacitor converter with 85 % efficiency at 0.91 W/mm^2 using 1.1 nH PCB-trace inductors, in *2015 IEEE International Solid-State Circuits Conference - (ISSCC) Digest of Technical Papers*, 2015, pp. 1–3

22. K. Kesarwani, R. Sangwan, J.T. Stauth, 4.5 A 2-phase resonant switched-capacitor converter delivering 4.3 W at 0.6 W/mm^2 with 85 % efficiency, in *2014 IEEE International Solid-State Circuits Conference Digest of Technical Papers (ISSCC)*, 2014, pp. 86–87
23. J.T. Stauth, M.D. Seeman, K. Kesarwani, Resonant switched-capacitor converters for sub-module distributed photovoltaic power management. IEEE Trans. Power Electron. **28**(3), 1189–1198 (2013)
24. Y. Lei, R. May, R.C.N. Pilawa-podgurski, Split-phase control: achieving complete soft-charging operation of a Dickson switched-capacitor converter, in *IEEE Control and Modeling for Power Electronics (COMPEL)*, 2014
25. G.V. Pique, E. Alarcon, Monolithic integration of a 3-level DCM-operated low-floating-capacitor buck converter for DC-DC step-down conversion in standard CMOS, in *IEEE Power Electronics Specialists Conference (PESC)*, 2008, pp. 4229–4235
26. W. Kim, D. Brooks, G. Wei, A fully-integrated 3-level DC-DC converter for nanosecond-scale DVFS. IEEE J. Solid-State Circuits **47**(1), 206–219 (2012)
27. V. Yousefzadeh, E. Alarcon, D. Maksimovic, Three-level buck converter for envelope tracking in rf power amplifiers, in *IEEE Applied Power Electronics Conference (APEC)*, 2005, pp. 1588–1594
28. T.A. Meynard, H. Foch, Multi-level conversion: high voltage choppers and voltage-source inverters, in *IEEE Power Electronics Specialists Conference*, pp. 397–403, 1992

Chapter 15
Heterogeneous Integration of High-Switching Frequency Inductive DC/DC Converters

Bruno Allard, Florian Neveu, and Christian Martin

15.1 Introduction

Many electronic products are designed for compactness as shown in Fig. 15.1. Each sub-block is designed accordingly. The camera example in Fig. 15.1 features a tiny board crowded with a lens, a connector and a dedicated power supply (bottom right zone). This power supply occupies a significant part of the board and passive components are the major contributors.

It is a common knowledge that an industrial trend is to improve dramatically the power density of the power supply. Innovations cover the following fields: converter architectures, passive devices, control, design and packaging [1]. Recent literature covers many demonstrations of switched-capacitor converters either on-chip or in-package. Beyond the impressive results [2], switched-capacitor converters face a limitation due to the available capacitance density. This translates into a limitation in power density [3] where inductive converters offer a better trade-off at similar level of efficiency. This article will now on focus on high power density inductive buck converters for embedded products: this is called Power-System-on-Chip [1, 4]. Reviews of high switching frequency buck converters are available in [5, 6]. Following statements may be derived:

B. Allard (✉)
University of Lyon, Ampere, UMR CNRS 5005, Lyon, France
e-mail: bruno.allard@insa-lyon.fr

F. Neveu
INSA Lyon, Villeurbanne, France

C. Martin
University Claude Bernard Lyon 1, Villeurbanne, France

© Springer International Publishing Switzerland 2017

A. Baschirotto et al. (eds.), *Wideband Continuous-time ΣΔ ADCs, Automotive Electronics, and Power Management*, DOI 10.1007/978-3-319-41670-0_15

Fig. 15.1 Teardown of a commercial electronic product (*left*), zoom on the embedded camera (*right*) [on-line: www.ifixit.com/Teardown/Xbox+One+Kinect+Teardown/19725]

- The natural corner frequency of the output filter of the converter correlates with the switching frequency: the high switching frequency is used to enhance the converter transient performances but this is not a driving design indicator so far. Higher switching frequency is intended to reduce the values of passive components.
- The efficiency drops with increasing frequency. The efficiency figure must also be related to the output-to-input voltage conversion ratio. Selecting the literature results for given ratios, the different peak efficiency values drop in a similar manner. One result seems to be above the general trend (ratio of 0.5, [7, 8]).
- Efficiency profits from thin silicon technologies whatever the switching frequency.
- A penalizing effect of thin technologies is the limitation in input voltage range.

There is then a real challenge to handle a standard input voltage range (3.3 V) with a thin technology (CMOS 40 nm, 1.2 V). Efficiency in the vicinity of 90 % is reported in literature but for a high ratio of output-to-input voltage (larger than 0.85) [9]. A challenge is evidently to achieve similar efficiency but for a ratio below 0.5.

Various strategies are reported in literature to fight power losses inside a non-isolated DC/DC converter, namely a buck converter as considered here.

- Soft switching, either zero-voltage or zero-current schemes, has been applied to integrated buck converters [10]. Variability and parasitic devices are limitations to the optimization of the architecture. Moreover a shift in the actual switching frequency complicates the filtering of generated noise.
- Multi-phase architecture: the current is shared upon several phases and the Joule losses are reduced [10]. Independent inductors are considered generally. A phase shedding scheme enables to adjust the number of operating phase according to the actual output current, i.e. operates the exact amount of silicon necessary at one moment with respect to the output current.

- Multiple inductors may be coupled [11]. One magnetic device will then be affected by a current waveform at twice the switching frequency. At similar inductance values, the ripple is reduced. At similar ripple and inductance values, the switching frequency can be lowered, i.e. the switching losses in the active devices are reduced.
- Resonant gate drivers have been experimented [12]. The gain on losses is rather limited and does not compensate the penalty on silicon area to accommodate the air-core inductance needed for each single driver.
- Transistor width segmentation is an ultimate solution to adapt on the fly the silicon area to the actual output current [12]. This necessitates an evaluation of the output current. The scheme superimposes a modulation to the pulse-width modulation generally considered at high switching frequency. A risk exists for the stability of the converter.
- High-voltage MOSFETs create more switching and conduction losses than low voltage counterparts. Cascode association of low-voltage MOSFETs has been experimented to recreate a high-voltage MOSFET but with lower losses [13]. If a thin CMOS technology is to considered, a cascode power stage is a solid candidate to stand high input voltage [7, 14].

With respect to the challenges considered here-above, it is reasonable to select hard switching operation at +100 MHz switching frequency. A cascode power stage fits the objective to rely on standard MOFSETs of the thin CMOS technology. Using coupled inductors inside a multi-phase architecture enables to reduce the values of passive components for a targeted output current ripple. The targeted system is pictured in Fig. 15.2 (left). Other specifications are listed in Table 15.1. Transient performances will not be considered here. In [15] is detailed a large-bandwidth voltage converter. The analogue voltage-mode control loop accounts for only 1.5 point of efficiency. In [15] is reported 78 % peak efficiency at the output power considered here with similar voltage conditions at input and output. The article details a proposal of a power stage that will limit power losses at similar switching frequency value. The power stage is detailed in Sect. 15.2.

As explained later, interconnections are crucial for high-switching frequency operation. Moreover the converter should benefit from a high-quality decoupling at the input. A 2.5 D approach is introduced with a capacitive interposer intended to receive the active part of the converter and the passive devices (Fig. 15.2, right). A high quality decoupling is obtained as detailed in Sect. 15.3. Measurements are given in Sect. 15.4.

15.2 Power Stage

The choice for a cascode configuration power stage stands as an alternative to the standard power stage, either with complementary transistors or with dual N-transistors, when the input voltage is of higher value than the transistor nominal

Table 15.1 Specifications of the target converter

Name	Min	Typ.	Max	Unit
V_{IN}	3.0	3.3	3.6	V
V_{OUT}	–	1.2	–	V
I_{LOAD}	50	280	500	mA
I_{SLEEP}	–	15	–	mA
$I_{STANDBY}$	–	150	–	uA
Static voltage accuracy	−2.5	–	+2.5	%
Dynamic voltage accuracy	−6	–	+6	%
Efficiency at $I_{LOAD,TYP}$	90	–	–	%
Efficiency at $I_{LOAD,MIN}$	80	–	–	%
Efficiency at I_{SLEEP}	70	–	–	%
F_{SW}	100	–	–	MHz
Ambient temperature	−40	55	125	Celsius
Package height	–	1.2	4	mm
Total area	–	–	10	mm^2
Technology platform	40 nm bulk CMOS by Infineon			

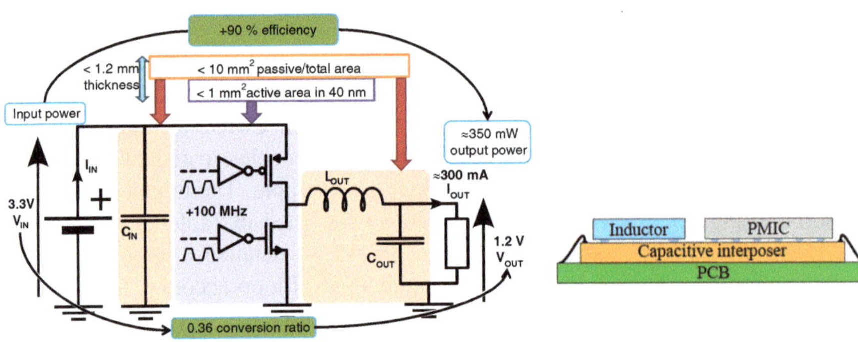

Fig. 15.2 Schematic of the targeted power stage (*left*) and interconnection proposal (*right*)

drain-to-source voltage. Figure 15.3 shows three options. Details on the limitations of the standard power stage at high switching frequency have been reported in [15]. The 2-transistor cascode configuration has been reported in [7, 14]. Good results have been experimentally demonstrated and the conclusion is favorable to the cascode behavior. The configuration is generic and can theoretically be extended to a large number of transistors, due to the systematic arrangements [16]. The 3-transistor cascode power stage is pictured in Fig. 15.3 (right): physical implementation was first reported in literature by B. Serneels in [19, 20] in the context of high-voltage driver. Recently it was demonstrated in the context of a class-D amplifier [21]. Besides the capability of the 3-transistor cascode power stage was not demonstrated at high switching frequency.

The selected technology for silicon implementation (Table 15.1) must be considered here as a case where the nominal technology voltage (1.2 V) is not in range

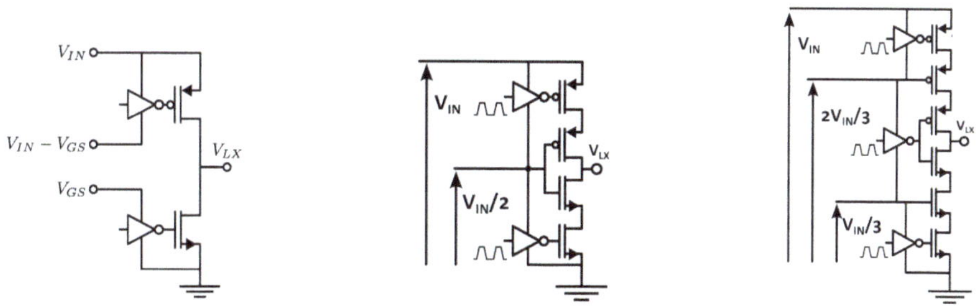

Fig. 15.3 Typical power stage configurations: standard (*left*), 2-stage cascode (*center*), 3-stage cascode (*right*)

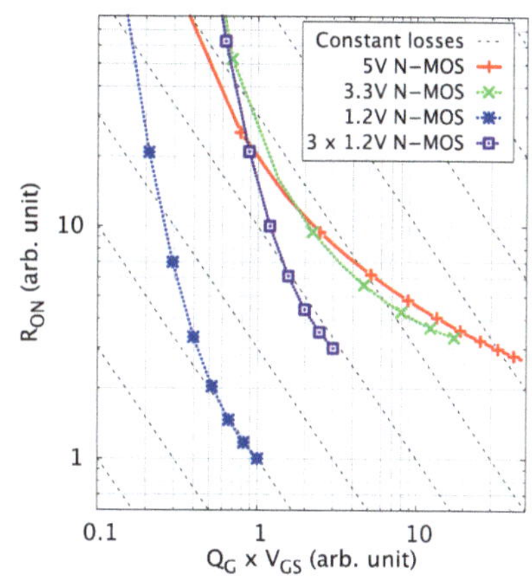

Fig. 15.4 Figure-of-merit for transistor options in a standard CMOS 40 nm node

with input voltage (3.3 V). High-voltage transistors may be designed as Lateral-Drain (LDMOS) or Extended-Drain (EDMOS) structures. Transistors impact the power stage configuration in terms of losses and silicon area. Second-order issues are safe-operating-area, sensitivity to temperature or electro-magnetic-interference. Losses are related to the static on-state resistance and parasitic capacitances for given operating conditions (voltage, current, frequency, temperature). Depending on the channel width of a transistor, it is possible to evaluate the losses for given operating conditions. In nominal operating conditions for the targeted power stage, it is then easy to compare the impact of various transistor options depending on channel width or silicon area. Figure 15.4 pictures iso-value curves of losses in the $(R_{on}, Q_G.V_{GS})$ plane as dashed lines. Lower losses are toward the bottom left corner. The figure-of-merit of several transistor options are then plotted. As expected the standard transistor is better in terms of losses and silicon area but can not stand the targeted input voltage. The 5 and 3.3 V transistor options are readily available to stand the input voltage but will not help reach the efficiency target. Fortunately stacking 3 standard transistors offers a better trade-off, hence the interest for the cascode power stage.

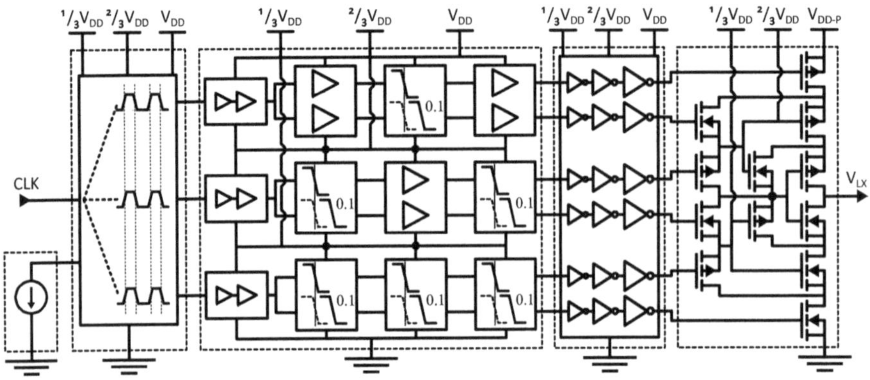

Fig. 15.5 Schematic of the power stage

As no transistors in the power stage is to stand more than 1.2 V, a systematic arrangement in the design leads to the schematic in Fig. 15.5. To take care of the Safe Operating Area of each power MOSFETs inside the cascode power stage, polarization MOSFETs have been added to the power stage.

The 3-floating drivers in Fig. 15.3 (right) are distributed on two-rows of local drivers, shared between the 3 transistors in the stack. This design is scalable to any odd number of transistors in the stack. Local intermediate voltage levels are required as 1/n of the transistor number in the stack. The generation of these local voltage levels as well as the associated efficiency are discussed at the end of the article. The power stage requires 3 driving signals with perfect synchronization. Skew within these latter signals is the key figure for robustness of the power stage. At 100 MHz (10 ns period), an accuracy of less than 100 ps has to be considered. Otherwise a transistor in the arrangement may have to stand a voltage higher than the nominal rating, hence inducing stress and aging. The layout of the power stage must comply to this objective: propagate synchronous signals from the global control input to the drivers' inputs through dedicated level-shifters to take care of the local change in voltage swings.

The power stage has been designed in a first step for continuous conduction mode (CCM) as the objective is to demonstrate the capability of a cascode configuration and the benefit of the 2.5 D assembly. In Fig. 15.6, a color code represents the voltage levels from input voltage to ground respectively in red, blue, green and gray. In CCM operation Fig. 15.6 gives the mapping of voltages when the output voltage is respectively high (left) or low (right).

Active-well transistor is mandatory to cancel polarization of parasitic junctions in the cascode stage (Fig. 15.7, left and center). The final layout is pictured in Fig. 15.7 (right) where the power stage has been optimized for 150 mA output current and 100 MHz operation. A standard power stage with 3.3 V transistors was also designed for similar specifications and with similar care for purpose of comparison.

The full power stage requires 4 decoupling capacitors for the various input voltages. These capacitors, as well as the output filter capacitor, are removed off-chip to the passive interposer where the active die and the inductor will also take

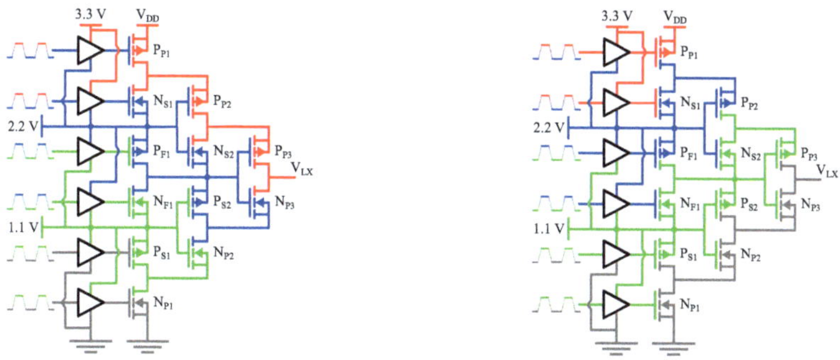

Fig. 15.6 Operation of the 2-stage cascode in CCM: high voltage on output (*left*), low voltage on output (*right*)

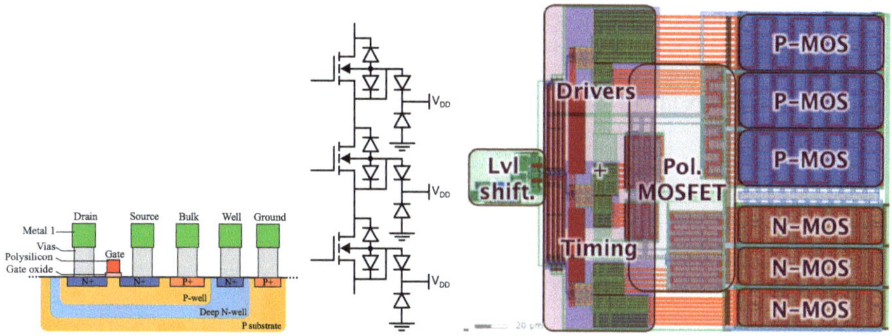

Fig. 15.7 Cross-section of the active-well transistor (*left*), 3-transistor cascode parasitic junctions (*center*), Layout of the power stage (*right*)

place. Decoupling capacitors' values and the inductance value have been optimized with respect to output current ripple and output voltage accuracy constraints. Decoupling of main input voltage requires 33 nF with ESR accounted for 100 mΩ. Other decoupling capacitors are set to 11 nF with an accounted ESR of 150 mΩ. Finally the output filter capacitor is set to 16 nF (150 mΩ maximum ESR value and accompanied by an inductor of 60 nH (reference PFL1005-60NMRU).

15.3 High Quality Decoupling

In [12] is demonstrated a single stage of a buck converter for 200 MHz operation. Passive components are placed off-chip. Results show that dumping resistors had to be implemented on-chip to kill oscillations introduced by parasitic devices (L and C) between the off-chip devices and the on-chip power stage. Authors report a penalty of at least 2 points of efficiency. It is of utmost importance to reduce the length of interconnection between the decoupling capacitors and the power stage. The active

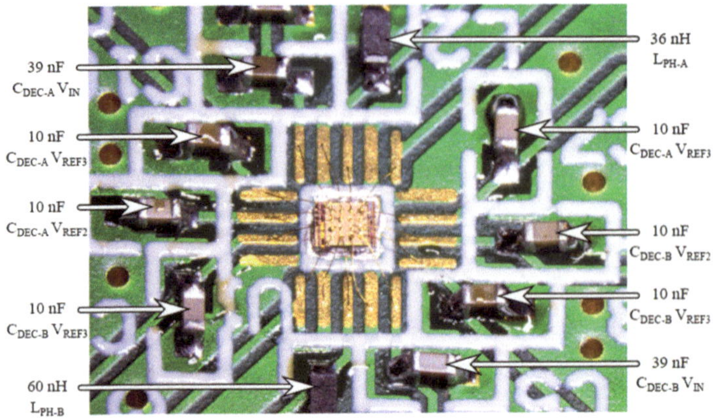

Fig. 15.8 Active die with off-chip passive devices on a board

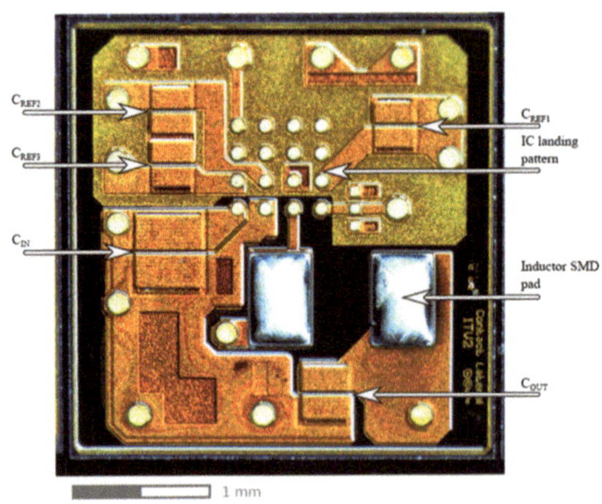

Fig. 15.9 Capacitive interposer manufactured by IPDia [17]

Table 15.2 Measured parameters of designed capacitor structures

	C_{MEAS} (nF)	ESR (mΩ)	SRF (MHz)
C_{IN}	33	70	330
C_{OUT}	16	110	490
$C_{REF1,2,3}$	11	150	630

die has been tested in a chip-on-board configuration with on-board passive devices. Figure 15.8 pictures the lengthy connections accounting for large parasitic devices.

In the present case, an interposer was manufactured by IPDiA to accommodate all decoupling capacitors, the output filter capacitor and all landing pads necessary to the flip-chipping connection of the active die and the inductor [17], Fig. 15.9. Table 15.2 lists the experimental performances of the embedded capacitors.

The various steps of converter manufacturing are summarized in Fig. 15.10. The obtained converter is pictured in Fig. 15.11 (left). It is ready to be connected to a system. For test purpose, samples are connected to a board through wire bonding

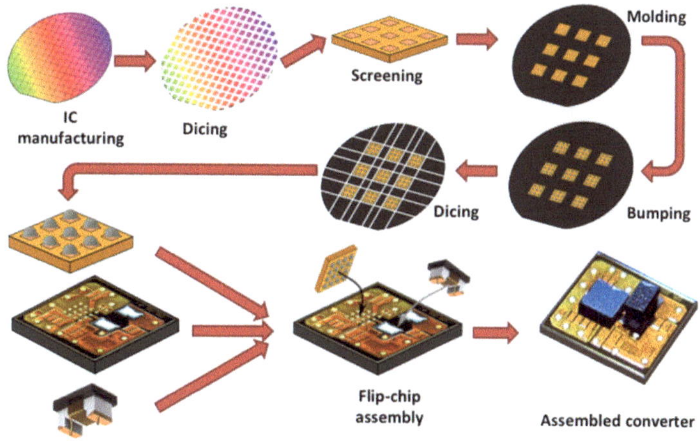

Fig. 15.10 Main steps of the 2.5 D converter manufacturing

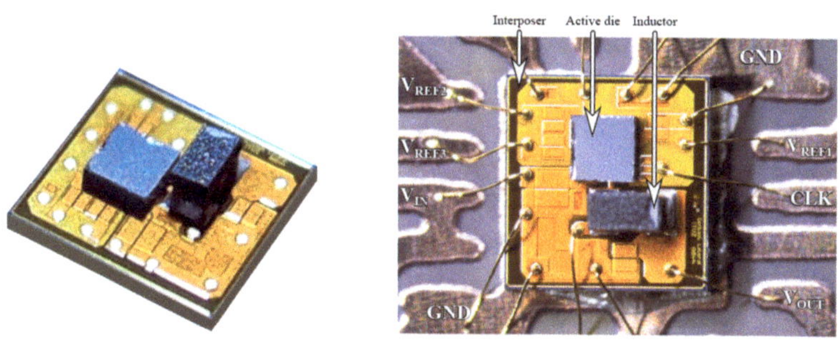

Fig. 15.11 Converter sample (*left*) and assembly on board (*right*)

(Fig. 15.11, right). No extra passive devices are added. Standard power stage and cascode power stage have been experimentally compared to interpret the benefit of interposer on the one hand, and the benefit of a cascode stage on the other hand. Measurements are presented in Sect. 15.4.

Measurements are not easy due to the compactness of the assembly. For the sake of internal signal sensing, the interposer in Fig. 15.9 is not optimal in geometry. Figure 15.12 gives the main parasitic devices awaited in the assembly when probing the power stage output voltage (V_{LX}) and the converter output voltage (V_{OUT}). Unfortunately this system of parasitic devices creates ringings and interferences in several configurations. Not all signals could be measured. As an example, Fig. 15.13 pictures the V_{OUT} node voltage when a probe is connected or not to the V_{LX} node.

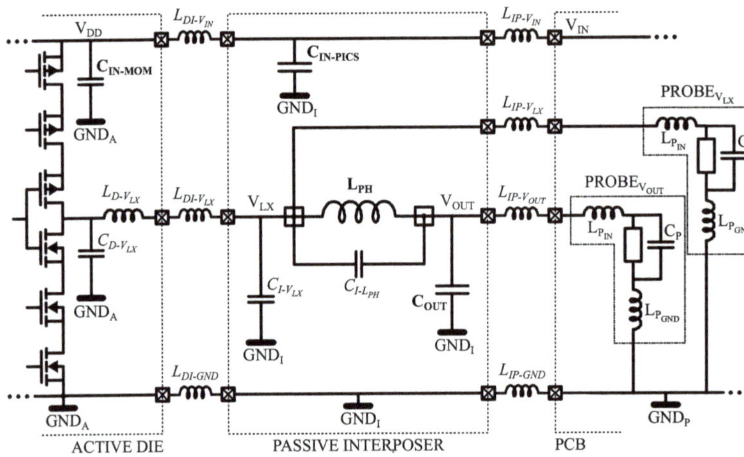

Fig. 15.12 Equivalent circuit to the 2.5 D converter from parasitic device point of view

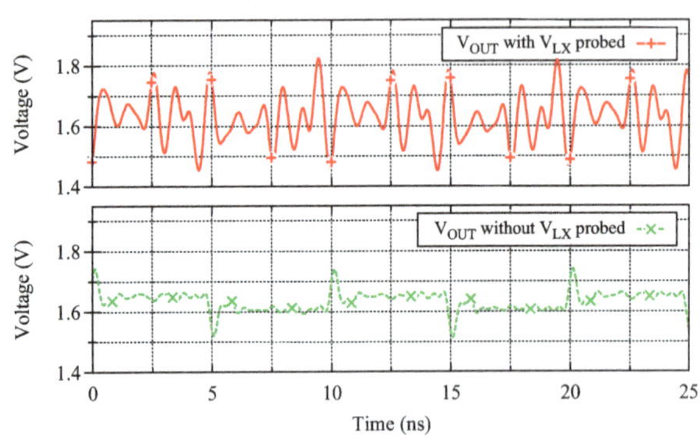

Fig. 15.13 Measurements of output voltage in two sensing configurations

15.4 Measurements

One important thing is to asses the behavior of the cascode power stage. Synchronicity or skew are key parameters for the power stage robustness. So far experimental observation is carried out on observable signals. As said previously not all of them may be sensed without detrimental interference. The voltage at V_{OUT} node has been monitored in various operating conditions for both the standard and cascode power stage on the interposer with the same inductor value (60 nH) and same switching frequency (100 MHz). As shows Fig. 15.14, it is impossible to detect a difference between the switching waveforms.

Efficiency is measured from total input and output power. The standard power stage has also an input voltage (3.3 V) and separate input voltages for the drivers. A mechanism of charge recycling between the drivers of the cascode power stage has not been explained but the consequence is that the currents drawn from the intermediate input voltages are very low. Figure 15.15 details some major results.

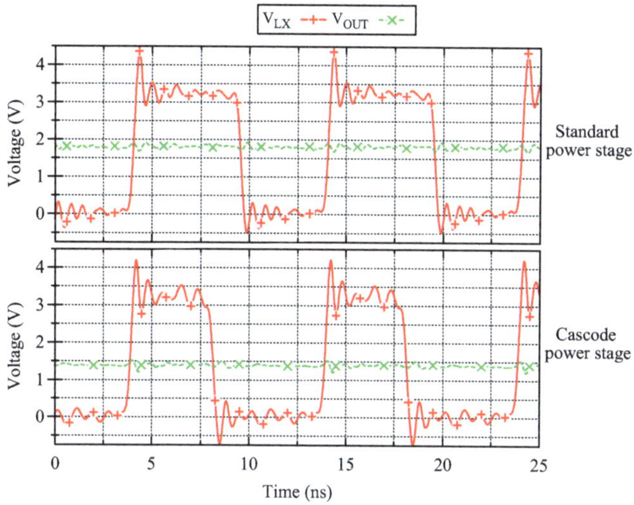

Fig. 15.14 Standard and cascode power stage behavior on interposer ($V_{IN} = 3.3$ V, $V_{OUT} = 1.8$ V, $I_{OUT} = 150$ mA, $f_{sw} = 100$ MHz)

First column compares the performance of the standard power stage on board and on interposer. It is obvious that in similar operating conditions, the interposer offers a benefit on efficiency of at least 4 points. This is mainly due to the reduction in interconnection parasitics. Voltage ringing is indeed very low and no particular precautions is to be taken inside the active part. Otherwise a technique is to insert resistance to dump ringings at the cost of additional losses. The first row in Fig. 15.15 shows that the cascode power stage offers superior performances than the standard counterpart. Somewhere 10 points of efficiency have been gained. They are repaid by some complexity in the design but the silicon penalty is affordable (+ 0.006 mm^2, i.e. + 14 %). This efficiency budget can be used to operate the cascode power stage at 200 MHz with performances still comparable to that of the standard power stage at half the switching frequency. The converter footprint would be smaller if smaller passive component values are selected to keep the output current and voltage ripples equal.

Results in Fig. 15.15 are given in terms of losses in Fig. 15.16. Losses in the drivers and total losses from input voltages are separated for both power stages. Post-layout simulation (PLS) results are added as well. Distribution of losses over the standard and cascode power stages reflect the trend as given globally by efficiency figures. PLS results are in good agreement with measurements for drivers but not for other losses. In this case PLS results do not take into account the interconnection parasitics.

Finally it is interesting to look at the cascode power stage sensitivity to operating conditions around values considered for optimization (nominal current, nominal output reference voltage, nominal frequency). Figure 15.17 (left) indicates that the switching frequency does not significantly affect the efficiency in contrary to the

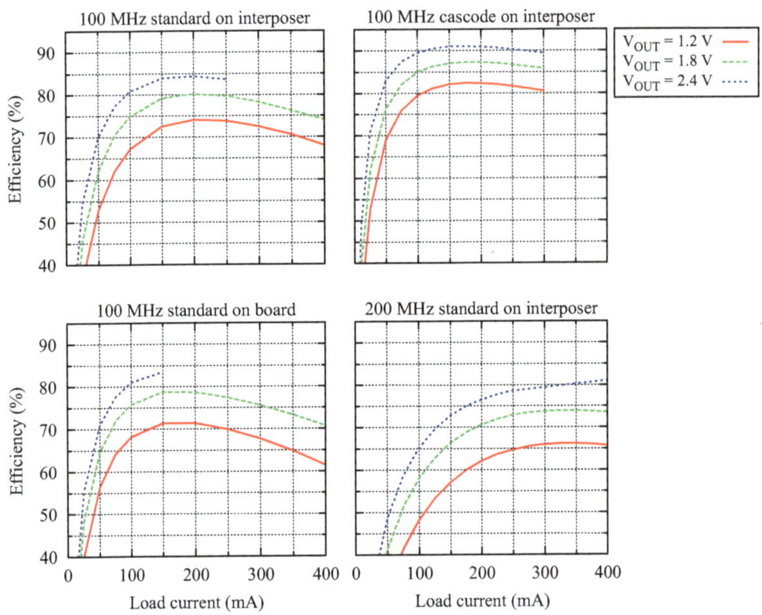

Fig. 15.15 Efficiency of standard and cascode power stages, on board or on interposer

input-to-output voltage ratio. Figure 15.17 (right) indicates that power transistor silicon area optimization is very sensitive. Drivers' losses remain insensitive to the output current and efficiency drops heavily apart from the optimization point. This corroborates the interest for multiphase architectures where the maximum output current is handled over the full number of phases. Lower output current values open the possibility of phase shedding.

15.5 Perspectives

The 3-transistor cascode power stage has been demonstrated in CCM and open-loop operation. One improvement will be to integrate the generation of the intermediate voltage level from the main input voltage through a switched-capacitor DC/DC. A redesign of the power stage is necessary to introduce two global input signals so as to be able to turn-off both the top and bottom side of the output stage. Voltage-mode control has already been demonstrated at 200 MHz for a standard power stage. This control option can be applied without any changes for the cascode power stage. More interestingly sliding-mode control could benefit from the rather insensitivity of efficiency to the switching frequency if the frequency range is limited. The fabricated interposer possesses landing pads for one inductor. A new version of the interposer must be fabricated to accommodate coupled inductors. Such a magnetic

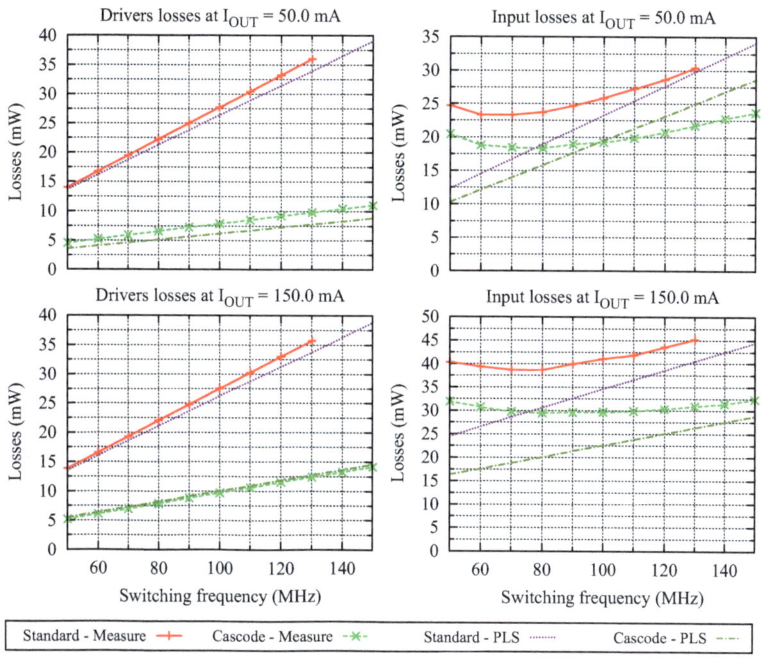

Fig. 15.16 Simulated and measured losses for standard and cascode power stage

Fig. 15.17 Efficiency of cascode power stage: sensitivity to operating conditions

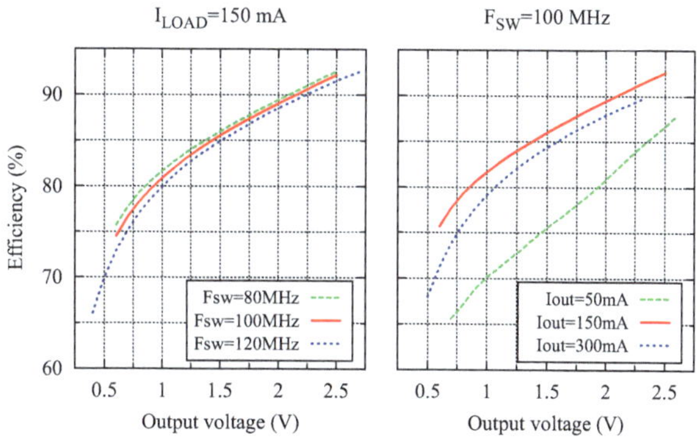

device has been manufactured by Tyndall Institute upon specifications as pictured in Fig. 15.18 [18]. A rather discrete test board as pictured in Fig. 15.19 is unfortunately not sufficient as the magnetic device no more benefits from the proximity of the capacitive interposer.

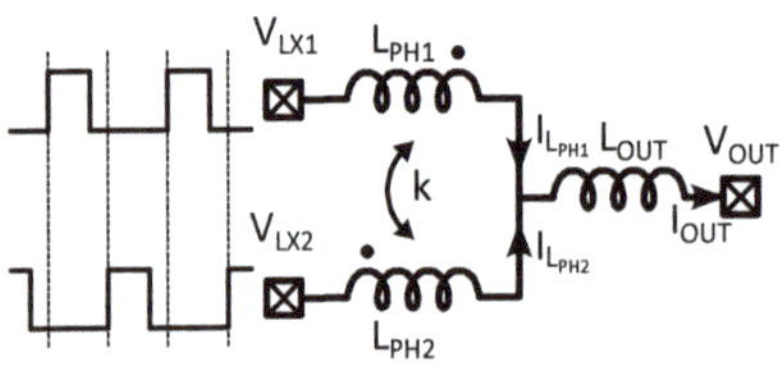

Fig. 15.18 Magnetic device suitable for a 2-phase buck converter (*left*) and integrated sample (*right*)

Fig. 15.19 Test board ready for functional verification of the two-phase buck converter

15.6 Conclusion

Low footprint DC/DC converters and/or low form factor DC/DC converters require further integration. When considered for System-on-Chip, the switched-capacitor architectures are preferable but high efficiency is related to the quality of embedded capacitors and on-chip deep trench capacitors are quite expensive. When the targeted power density leads to select an inductive architecture, the switching frequency is the degree of freedom to push the converter in the era of air-core inductors, compatible with on-chip manufacturing. However output power capability is limited. An intermediate solution to accommodate magnetic-core inductors is a 2.5 D assembly using a capacitive interposer. Each part of the system is to be manufactured with the cost-optimal industrial process compatible with the awaited performances. The article has shown the benefit of an interposer as interconnections between devices are shortened, hence lower parasitic effects at high switching frequency. On the other hand this article has reported the performances of a 3-transistor cascode power stage as a better trade-off between high input voltage and high efficiency in advanced CMOS technology nodes.

Acknowledgements This work is supported by the European Commission through the Seventh Framework Programme (FP7), under the project grant PowerSWIPE No. 318529.

References

1. F. Waldron, R. Foley, J. Slowey, A. Alderman, B. Narveson, S. Mathuna, Technology roadmapping for power supply in package (PSiP) and power supply on chip (PwrSoC). IEEE Trans. Power Electron. **28**(9), 4137–4145 (2013)
2. Y. Pascal, G. Pillonnet, Efficiency comparison of inductor-, capacitor-, and resonant-based converters fully integrated in CMOS technology. IEEE J. Emerging Sel. Top. Circuits Syst. **PP**(99), 1–9 (2015)
3. G. Villar-Pique, H. Bergveld, E. Alarcon, Survey and benchmark of fully integrated switching power converters: switched-capacitor versus inductive approach. IEEE Trans. Power Electron. **28**(9), 4156–4167 (2013)
4. C.O. Mathuna, N. Wang, S. Kulkarni, R. Anthony, N. Cordero, J. Oliver, P. Alou, V. Svikovic, J.A. Cobos, J. Cortes, F. Neveu, C. Martin, B. Allard, F. Voiron, B. Knott, C. Sandner, G. Maderbacher, J. Pichler, M. Agostinelli, A. Anca, M. Breig, C.O. Mathuna, R. Anthony, Power supply with integrated passives - the EU FP7 powerswipe project, in *2014 8th International Conference on Integrated Power Systems (CIPS)* (2014), pp. 1–7
5. F. Neveu, C. Martin, B. Allard, Review of high frequency, highly integrated inductive DC-DC converters, in *2014 8th International Conference on Integrated Power Systems (CIPS)* (2014), pp. 1–7
6. F. Neveu, B. Allard, C. Martin, A review of state-of-the-art and proposal for high frequency inductive step-down DC-DC converter in advanced CMOS. J. Analog Integr. Circ. Sig. Process **87**(2), 1–11 (2016)
7. E. Burton, G. Schrom, F. Paillet, J. Douglas, W. Lambert, K. Radhakrishnan, M. Hill, FIVR - Fully integrated voltage regulators on 4th generation intel core SoCs, in *2014 Twenty-Ninth Annual IEEE Applied Power Electronics Conference and Exposition (APEC)* (2014), pp. 432–439
8. N. Kurd, M. Chowdhury, E. Burton, T. Thomas, C. Mozak, B. Boswell, P. Mosalikanti, M. Neidengard, A. Deval, A. Khanna, N. Chowdhury, R. Rajwar, T. Wilson, R. Kumar, Haswell: a family of IA 22 nm processors. IEEE J. Solid State Circuits **50**(1), 49–58 (2015)
9. M.K. Song, M. Dehghanpour, J. Sankman, D. Ma, A VHF-level fully integrated multi-phase switching converter using bond-wire inductors, on-chip decoupling capacitors and DLL Phase synchronization, in *2014 Twenty-Ninth Annual IEEE Applied Power Electronics Conference and Exposition (APEC)* (2014), pp. 1422–1425
10. S. Abedinpour, B. Bakkaloglu, S. Kiaei, A multistage interleaved synchronous buck converter with integrated output filter in 0.18 um SiGe process. IEEE Trans. Power Electron. **22**(6), 2164–2175 (2007)
11. J. Wibben, R. Harjani, A high-efficiency DC-DC converter using 2 nH integrated inductors. IEEE J. Solid-State Circuits **43**(4), 844–854 (2008)
12. M. Bathily, B. Allard, F. Hasbani, A 200 MHz integrated buck converter with resonant gate drivers for an RF power amplifier. IEEE Trans. Power Electron. **27**(2), 610 –613 (2012)
13. M. Wens, M. Steyaert, A fully-integrated 0.18 m CMOS DC-DC step-down converter, using a bondwire spiral inductor, in *Custom Integrated Circuits Conference, 2008 (CICC, 2008) (IEEE,* 2008), pp. 17–20
14. K. Ostman, J. Jarvenhaara, S. Broussev, I. Viitaniemi, A 3.6-to-1.8 V cascode buck converter with a stacked LC filter in 65-nm CMOS. IEEE Trans. Circuits Syst. Express Briefs **PP**(99), 1–5 (2014)
15. M. Bathily, B. Allard, F. Hasbani, V. Pinon, J. Verdier, Design flow for high switching frequency and large-bandwidth analog DC/DC step-down converters for a polar transmitter. IEEE Trans. Power Electron. **27**(2), 838–847 (2012)
16. G. Schrom, P. Hazucha, J.-H. Hahn, V. Kursun, D. Gardner, S. Narendra, T. Karnik, V. De, Feasibility of monolithic and 3D-stacked DC-DC converters for microprocessors in 90nm technology generation, in *Proceedings of the 2004 International Symposium on Low Power Electronics and Design (ISLPED)* (2004), pp. 263–268

17. F. Lallemand, F. Voiron, Silicon interposers with integrated passive devices, an excellent alternative to discrete components, in *European Microelectronics Packaging Conference (EMPC)* (2013), pp. 1–6
18. C. Mathuna, N. Wang, S. Kulkarni, S. Roy, Review of integrated magnetics for power supply on chip (PwrSoC). IEEE Trans. Power Electron. **27**(11), 4799–4816 (2012)
19. B. Serneels and T. Piessens and M. Steyaert and W. Dehaene, A high-voltage output driver in a 2.5-V 0.25- mu;m CMOS technology, *IEEE Journal of Solid-State Circuits*, **40**(3), 576–583 (2005) doi: http://dx.doi.org/10.1109/JSSC.2005.843599
20. B. Serneels and M. Steyaert and W. Dehaene, A 237mW aDSL2+ CO Line Driver in Standard 1.2V 0.13μm CMOS, *IEEE International Solid-State Circuits Conference. Digest of Technical Papers*, 524–619 (2007) doi: http://dx.doi.org/10.1109/ISSCC.2007.373525
21. B. Serneels and E. Geukens and B. De Muer and T. Piessens, A 1.5W 10V-output Class-D amplifier using a boosted supply from a single 3.3V input in standard 1.8V/3.3V 0.18μm CMOS, *IEEE International Solid-State Circuits Conference*, 94–96 (2012) doi: http://dx.doi.org/10.1109/ISSCC.2012.6176890

Chapter 16
Electrical Compensation of Mechanical Stress Drift in Precision Analog Circuits

Mario Motz and Udo Ausserlechner

16.1 Introduction: Why Do We Need Stress Compensation Circuits?

It has been known for a long time that (i) mechanical stress alters the parameters of electronic devices [1], and (ii) that integrated circuits are subject to large mechanical stress caused by wafer manufacturing and package assembling processes [2, 3]. In practice, these effects are often negligible, because the principles of circuit design engineering heavily rely on matching of paired devices instead of absolute values of individual devices. For example, the gain of amplifiers is defined by ratios of resistances and the offset depends on the matching of differential input transistor pairs and current mirrors. Although some parameters of single devices are notably altered by package stress (roughly by 10 %), both elements of the pairs vary synchronously, so that their ratio is constant. Of course, this holds only for homogeneous stress and adequate layout techniques like common centroids or interdigitated arrangements [4, 5].

Nonetheless, occasionally circuit designers are bound to rely on absolute values of parameters of single devices, because they define voltages, currents, frequencies, or sensitivities of transducers (temperature, pressure, or magnetic field sensors). In these cases, it is straightforward to trim the circuits after packaging and store the calibration data in on-board memories. Yet, this strategy fails to cope with lifetime drifts of mechanical stress. Sensor designers seem to have been the first ones to note that these drifts are indeed relevant.

M. Motz (✉) • U. Ausserlechner
Infineon Technologies, Siemensstrasse 2, Villach, Austria
e-mail: mario.motz@infineon.com

© Springer International Publishing Switzerland 2017

A. Baschirotto et al. (eds.), *Wideband Continuous-time ΣΔ ADCs, Automotive Electronics, and Power Management*, DOI 10.1007/978-3-319-41670-0_16

Originally, there were many candidates that might cause these stress drifts [6], but in the course of time, it became apparent that moisture ingress from ambient air into the mold compound of plastic encapsulated packages is the major contributor to mechanical stress drift in the field [7]. In 2002 we conducted an in-house investigation with a stress-sensor testchip after the principles of [8] (Fig. 16.1): MOS-current mirrors with current flow directions in orthogonal directions were used as stress-sensitive elements. The advantage of such a system is that the signal is in the current domain and so it is simple to multiplex an array of elements onto a single terminal in a package like PG-SSO-4-1, which offers only four terminals (two supply terminals, one input and one output current terminal). A moderately "wet" condition of the mold compound was provoked by storing the package in boiling water for 3 h and a moderately "dry" condition was established by storing the package three hours at 180 °C in an oven, which was flushed with dry air. Due to the principle of stress measurement, only the difference of in-plane normal stress components σ_{xx}–σ_{yy} and the in-plane shear stress component σ_{xy} were accessible [9]. Thereby the long edges of the chip are parallel to the x-direction and the short edges are parallel to the y-direction. The patterns of both stress components in dry state agree with [10, 11]. These stress components show an increase in the dry state over wet state by roughly a factor of two. Near the middle of the long edge of the chip, the difference σ_{yy}–σ_{xx} increases by about 20 MPa.

We assembled a Hall plate with biasing circuit and amplifiers for the output signal in the same package [12] and stored it at ambient moisture for several days. Then we measured the magnetic sensitivity while forcing the following sequence of temperatures: start at ambient temperature, then cool down to −40 °C and ramp up to 160 °C, wait there for three hours, then ramp down to −40 °C, and finally ramp up to room temperature. The measurement results in Fig. 16.2 reveal a hysteretic behavior: prior to bake out at 160 °C the mold compound of the package was soaked up with ambient moisture, which causes less compressive stress on the chip and thus more magnetic sensitivity of the Hall plate than after bake out. Between start and stop the magnetic sensitivity exhibits a drift of nearly 2 %. This drift can be reduced to about 0.75 % with optimized mold compounds, but it cannot be avoided altogether for packages, which have to comply with automotive quality standards. Interestingly, when the sample remains in the test fixture over week-end at constant ambient moisture, the magnetic sensitivity drifts back to its original value. So, the package did not disintegrate irreversibly but the stress changes reversible with the water content of the package. This has also been verified by other investigations [7, 13].

These findings led us to the idea to measure the relevant stress components on-chip and feed them into an electronic compensation circuit, which counteracts the systematic effects of the now known mechanical stress on the main circuit (Fig. 16.3). Such a method is versatile: it works for arbitrary packages and silicon technologies. Although the general idea is straightforward, there are challenges: which stress components are relevant, and how can we measure them accurately despite process tolerances and over a wide temperature range.

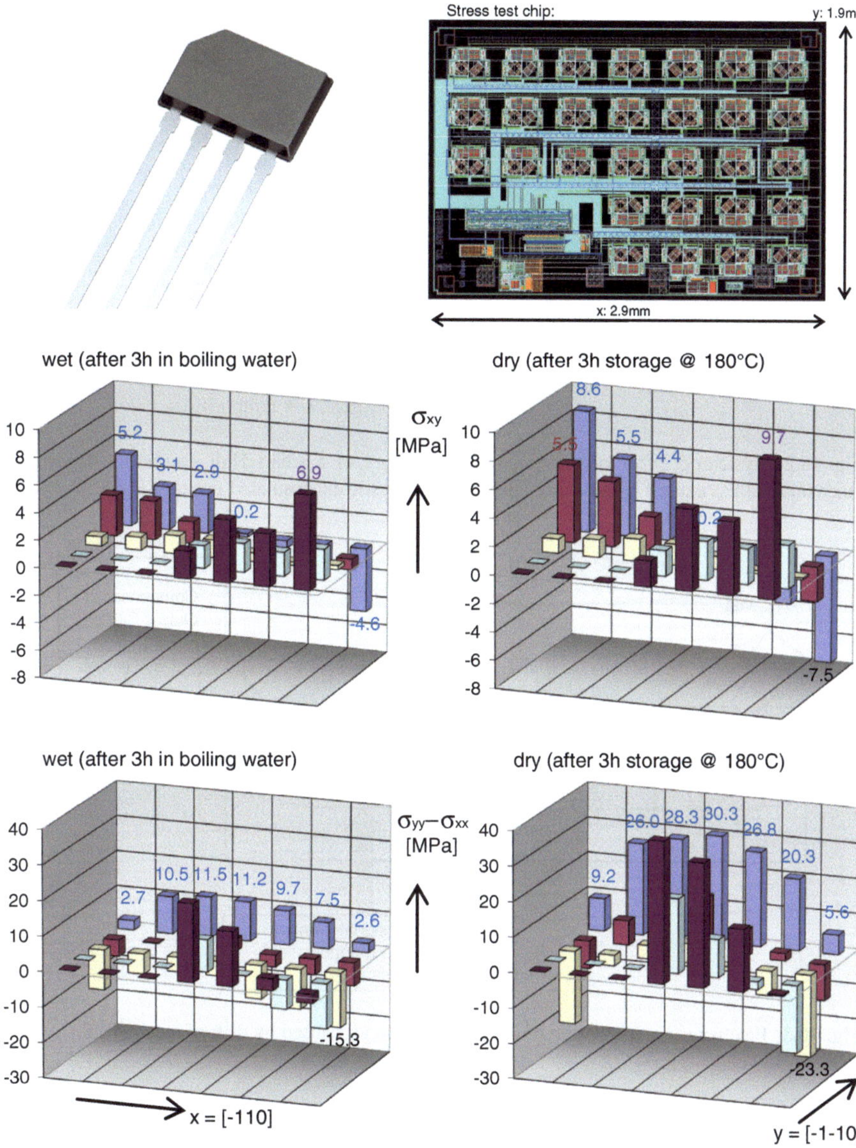

Fig. 16.1 Stress in small plastic encapsulated package (PG-SSO-4-1) as measured with a stress test chip: comparison of wet vs. dry condition at room temperature. The x-axis is parallel to the long edge of the chip and points in [-110] direction of the silicon crystal. The y-axis is parallel to the short edge of the chip and points in [-1-10] direction of the silicon crystal (cf. Fig. 16.7). The silicon chip is 0.22 mm thick, the copper leadframe is 0.2 mm thick, and the overall package is 1 mm thick. Plotted values are averages over 15 devices

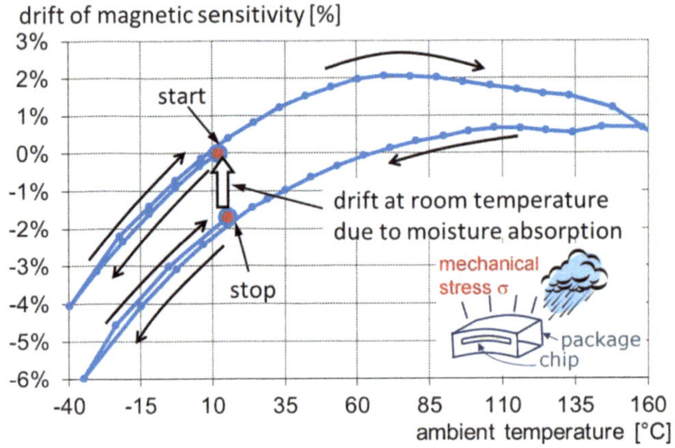

Fig. 16.2 Hysteretic effect of the magnetic sensitivity of a smart Hall sensor due to drift of mechanical stress caused by moisture absorption in plastic encapsulated package PG-SSO-4-1

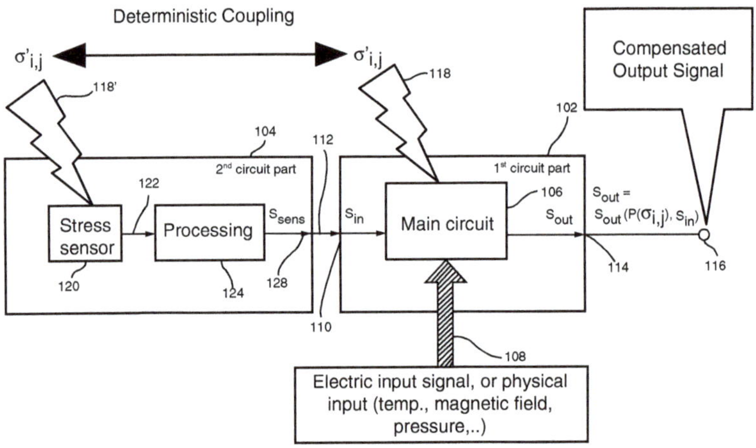

Fig. 16.3 Electronic compensation of drift of main circuit caused by drift of mechanical stress

In the future, more circuits may benefit from electronic compensation of stress related drift: not only do the requirements (accuracy, lifetime stability) increase, but also more inhomogeneous, less predictable, and less stable mechanical stress will be caused by modern package technologies with thinner chips, thinner leadframes, and higher packaging density (Fig. 16.4). Stress compensation circuits also reduce drifts caused by soldering and over-molding.

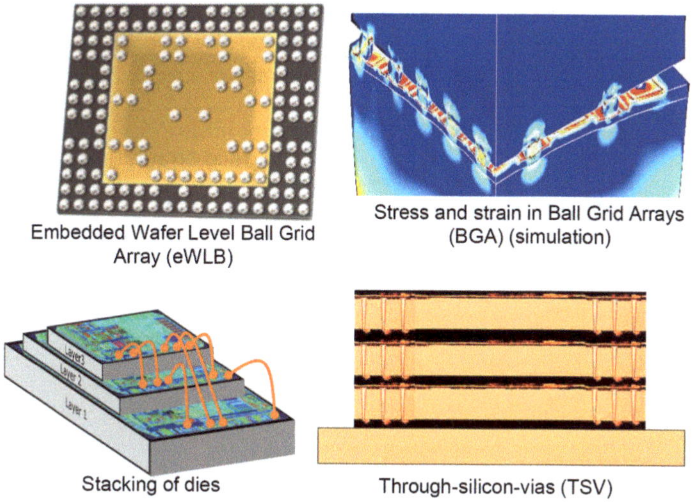

Fig. 16.4 Stress in modern package technologies tends to be higher, less homogeneous, and less predictable with more tensor components

16.2 Some Basics About Mechanical Stress on Micro-Electronic Circuits?

Electronic circuit designers are familiar with temperature drifts and how to compensate them, but usually they are not fluent with mechanical stress and its effects on circuit devices. In spite of some similarities between temperature and stress drift there is one main difference: temperature is a scalar, whereas stress is a tensor quantity. Basically, stress is the force per unit area, which acts inside a rigid body under mechanical load. One can make this force visible, by cutting the rigid body in two parts. Then one has to apply this force on the sectional area of the cuts in order to keep the body under the same load. However, the cut through the body may have various orientations and this also changes the force on the cut area. So the force—which is a vector—depends on the orientation of the cut, too. Thus, this force (= the stress) must have more than three degrees of freedom. Figure 16.5 shows a small block cut out of a rigid body. The block has cut areas perpendicular to x-, y-, z-axes. We can decompose the force on each cut area into components pointing in x-, y-, z-directions. On each cut area we have one component perpendicular to the area: this is the normal stress σ_{xx} with two equal indices specifying the direction of this force. On each cut area we have two mutually perpendicular components parallel to the area. These are the shear stresses σ_{xy}, σ_{xz} with two different indices: the first index specifies the direction of the surface and the second index specifies the direction of the force. In total we have nine components: three cut surfaces, each one with one normal stress and two shear stress components. The forces on the opposite areas (the negative shores with normal vectors in negative x-, y-, z-directions) are equally

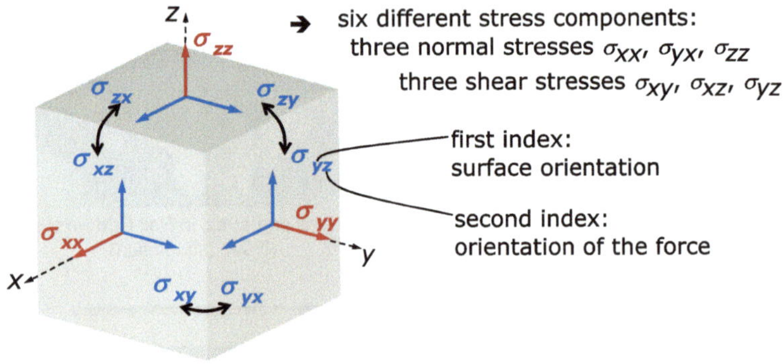

Fig. 16.5 The stress tensor

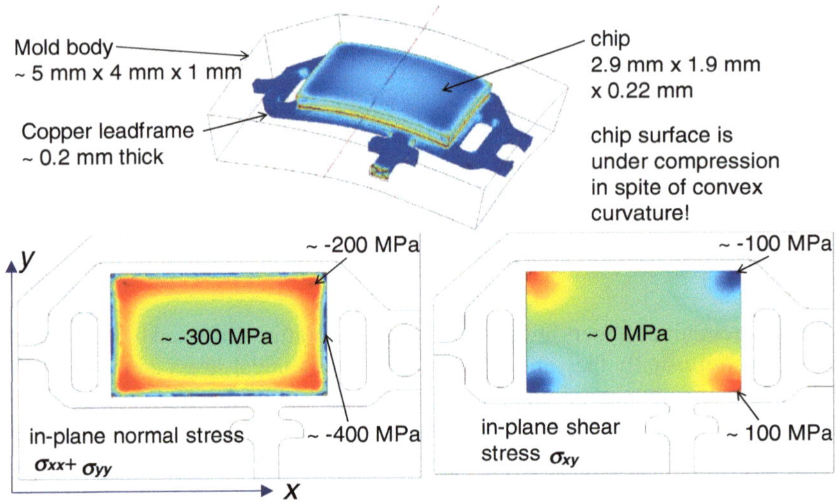

Fig. 16.6 Typical stress pattern in a PG-SSO-4-1 package at 20 °C

large with negative sign. Applying the equilibrium of moments to the block shows that shear stresses are equal when their indices are swapped: $\sigma_{yz} = \sigma_{zy}$, $\sigma_{xz} = \sigma_{zx}$, $\sigma_{xy} = \sigma_{yx}$. So eventually, we end up with six independent components of the stress tensor: three normal stresses σ_{xx}, σ_{yy}, σ_{zz} and three shear stresses σ_{yz}, σ_{xz}, σ_{xy}.

Fortunately, we seldom have to deal with all six stress components, because micro-electronic packages are usually so-called laminates where the lateral (*x*,*y*)-dimensions are much larger than the thickness (*z*-dimension). A typical stress pattern on the chip surface in a package is shown in Fig. 16.6. The PG-SSO-4-1 package serves as an example but other packages of similar technology yet different size show analogous stress patterns. In particular, the amount of stress near the center of the chip surface does not depend on the lateral size of the laminate. Over the main center region of the chip surface (i) the sum of the in-plane normal stress components is homogeneous and negative $\sigma_{xx} + \sigma_{yy} \approx -300$ MPa, and (ii) the in-plane shear stress component vanishes $\sigma_{xy} \approx 0$ MPa. Both values agree with Figs. 5 and 6 in [11]. Negative stress means compressive stress and this is simple to explain, because the mold compound solidifies at elevated temperature and when it cools down, it shrinks more than the silicon chip and the metal leadframe. Interestingly, the curvature of the package is convex: for a homogeneous beam in a bending experiment this shape would mean positive, tensile stress above the neutral fiber, but in the laminate the neutral fiber is shifted (in fact the stress is a zig-zag curve along the thickness [14]). Hence, it is a widespread misconception that the warpage of a package easily defines whether the chip is under compression or tension. Figure 16.6 also shows that near the edges and near the corners all stress components are very inhomogeneous. Thus, we should avoid placing stress sensitive circuitry in an annular region with a width comparable to the chip thickness—this follows from the principle of Saint-Venant [11]. All other stress components (the out-of-plane stress components) vanish on the chip surface, except along the perimeter.

Note that a comparison of stresses in Figs. 16.1 and 16.6 gives $\sigma_{xx} \approx -150$ MPa and $\sigma_{yy} \approx -100$ MPa at positions near the centers of the long edges. In the sequel we will see that some circuit devices respond only to $\sigma_{xx} + \sigma_{yy}$ (which means that they respond equally to σ_{xx} and σ_{yy}) and most other devices can be modified in their layout to have the same stress response. This makes the life for circuit design engineers easy, because one only needs to care about a single combination of tensor components $\sigma_{xx} + \sigma_{yy}$. Fortunately, $\sigma_{xx} + \sigma_{yy}$ is invariant against in-plane rotations so that we can treat this sum of in-plane normal stress components as a scalar just like the temperature in a temperature compensation circuit!

16.3 How Does Mechanical Stress Alter the Parameters of Electronic Devices?

Mechanical stress deforms the crystal lattice, which changes the atomic distances, and this alters the band structure of the semiconductor. The main effect is a change of the curvature of conduction and valence bands, leading to different effective masses of the charge carriers and this affects their mobility.

16.3.1 The Piezo-Resistance Effect

A plain resistor stripe of length L, width W, and thickness t has the resistance

$$R = \rho \, \frac{1}{t} \, \frac{L}{W} \text{ with } \rho = \frac{1}{qn\mu} \tag{16.1}$$

with the specific resistivity ρ, the elemental charge q, the charge density n, and the mobility μ. Mechanical stress alters the resistance according to

$$\delta R = \frac{R - R_0}{R_0} = \frac{\Delta R}{R_0} = \frac{\Delta \rho}{\rho_0} - \frac{\Delta t}{t_0} - \frac{\Delta W}{W_0} + \frac{\Delta L}{L_0} \tag{16.2}$$

where R is the resistance at stress, and all quantities with lower index "0" denote their values at zero stress. The last three terms on the right hand side are the pure geometrical changes (i.e. the strain). The change of specific resistivity is the piezo-resistance effect. In most cases, it is much larger than the strain terms. Resistivity and stress are second rank tensors, which are more conveniently written as six-component vectors

$$\frac{1}{\rho_0}\begin{pmatrix} \rho_{[100],[100]} - \rho_0 \\ \rho_{[010],[010]} - \rho_0 \\ \rho_{[001],[001]} - \rho_0 \\ \rho_{[010],[001]} \\ \rho_{[100],[001]} \\ \rho_{[100],[010]} \end{pmatrix} = \begin{pmatrix} \pi_{11} & \pi_{12} & \pi_{12} & 0 & 0 & 0 \\ \pi_{12} & \pi_{11} & \pi_{12} & 0 & 0 & 0 \\ \pi_{12} & \pi_{12} & \pi_{11} & 0 & 0 & 0 \\ 0 & 0 & 0 & \pi_{44} & 0 & 0 \\ 0 & 0 & 0 & 0 & \pi_{44} & 0 \\ 0 & 0 & 0 & 0 & 0 & \pi_{44} \end{pmatrix} \cdot \begin{pmatrix} \sigma_{[100],[100]} \\ \sigma_{[010],[010]} \\ \sigma_{[001],[001]} \\ \sigma_{[010],[001]} \\ \sigma_{[100],[001]} \\ \sigma_{[100],[010]} \end{pmatrix} \tag{16.3}$$

In (16.3) resistivity and stress are measured in the coordinate system of the single crystal. They are related by the piezo-resistive coefficients $\pi_{i,j}$. For low n- and p-doped ($<10^{18}/\text{cm}^3$) single crystal silicon at room temperature it holds [15]

$$\pi_{11}^{n} = -102.2\ \%/\text{GPa}, \quad \pi_{12}^{n} = 53.4\ \%/\text{GPa}, \quad \pi_{44}^{n} = -13.6\ \%/\text{GPa} \tag{16.4a}$$

$$\pi_{11}^{p} = 6.6\ \%/\text{GPa}, \quad \pi_{12}^{p} = -1.1\ \%/\text{GPa}, \quad \pi_{44}^{p} = 138.1\ \%/\text{GPa} \tag{16.4b}$$

With larger doping the piezo-resistive coefficients become slightly smaller and with larger temperature they decrease roughly by $-0.3\,\%/°\text{C}$ [15]. Transforming (16.3) from the crystal- into the chip-reference frame [16] gives for (100)-silicon

$$\delta R_x \cong \frac{\pi_{11} + \pi_{12} - \pi_{44}}{2}\sigma_{xx} + \frac{\pi_{11} + \pi_{12} + \pi_{44}}{2}\sigma_{yy} + \pi_{12}\sigma_{zz} \tag{16.5a}$$

$$\delta R_y \cong \frac{\pi_{11} + \pi_{12} + \pi_{44}}{2}\sigma_{xx} + \frac{\pi_{11} + \pi_{12} - \pi_{44}}{2}\sigma_{yy} + \pi_{12}\sigma_{zz} \tag{16.5b}$$

$$\delta R_z \cong \pi_{12}\left(\sigma_{xx} + \sigma_{yy}\right) + \pi_{11}\sigma_{zz} \tag{16.5c}$$

We use the "approximately equal"-sign ($\cong$) because we neglect the strain terms against the piezo-resistive terms. The x-direction is equal to the direction [-110] of the single crystal silicon, the y-direction is equal to [-1-10], and the z-direction is equal to [001] (see Fig. 16.7). R_x and R_y are lateral resistances, where the current

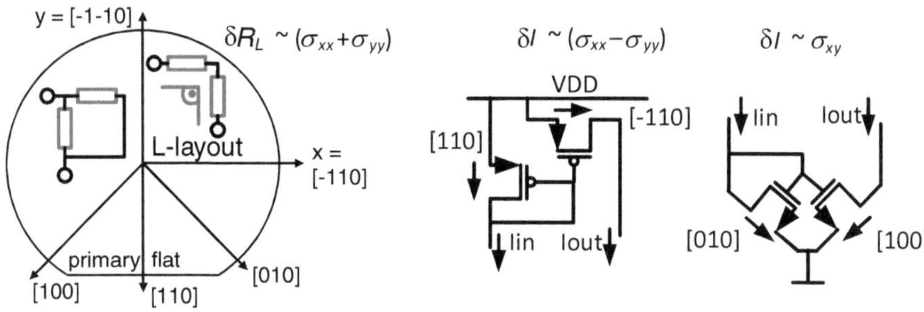

Fig. 16.7 Orthogonal MOS-FET current-mirrors can be used as stress sensors: $\sigma_{xx}-\sigma_{yy}$ (PMOS-FET), σ_{xy} (NMOS-FET). Crystallographic axes in (100)-silicon. Also shown: L-layouts for lateral resistors in parallel and series connection

flows along the x- and y-axes of the chip, respectively, whereas R_z is a vertical resistance with current perpendicular to the chip surface. The vertical resistance has the nice property that it depends equally on σ_{xx} and σ_{yy} (we may neglect σ_{zz} because it vanishes on the main part of the chip surface). The lateral resistances depend differently on σ_{xx} and σ_{yy}, however, we can add both of them in a so-called L-layout (see Fig. 16.7) to obtain the desired stress dependence [17]

$$\delta R_L \cong \frac{\pi_{11}+\pi_{12}}{2}\left(\sigma_{xx}+\sigma_{yy}\right)+\pi_{12}\sigma_{zz} \tag{16.5d}$$

with $R_L = R_x + R_y$ (series connection) or $R_L = R_x\ //\ R_y$ (parallel connection).

Inserting (16.4a) into (16.5a–d) we get for low n-doped resistors

$$\delta R_x^n \cong -17.6\ \%/\mathrm{GPa}\times\sigma_{xx} - 31.2\ \%/\mathrm{GPa}\times\sigma_{yy} + 53.4\ \%/\mathrm{GPa}\times\sigma_{zz} \tag{16.6a}$$

$$\delta R_y^n \cong -31.2\ \%/\mathrm{GPa}\times\sigma_{xx} - 17.6\ \%/\mathrm{GPa}\times\sigma_{yy} + 53.4\ \%/\mathrm{GPa}\times\sigma_{zz} \tag{16.6b}$$

$$\delta R_z^n \cong 53.4\ \%/\mathrm{GPa}\times\left(\sigma_{xx}+\sigma_{yy}\right) - 102.2\ \%/\mathrm{GPa}\times\sigma_{zz} \tag{16.6c}$$

$$\delta R_L^n \cong -24.4\ \%/\mathrm{GPa}\times\left(\sigma_{xx}+\sigma_{yy}\right) + 53.4\ \%/\mathrm{GPa}\times\sigma_{zz} \tag{16.6d}$$

Inserting (16.4b) into (16.5a–d) we get for low p-doped resistors

$$\delta R_x^p \cong -66.3\ \%/\mathrm{GPa}\times\sigma_{xx} + 71.8\ \%/\mathrm{GPa}\times\sigma_{yy} - 1.1\ \%/\mathrm{GPa}\times\sigma_{zz} \tag{16.7a}$$

$$\delta R_y^p \cong 71.8\ \%/\mathrm{GPa}\times\sigma_{xx} - 66.3\ \%/\mathrm{GPa}\times\sigma_{yy} - 1.1\ \%/\mathrm{GPa}\times\sigma_{zz} \tag{16.7b}$$

$$\delta R_z^p \cong -1.1\ \%/\mathrm{GPa}\times\left(\sigma_{xx}+\sigma_{yy}\right) + 6.6\ \%/\mathrm{GPa}\times\sigma_{zz} \tag{16.7c}$$

$$\delta R_L^p \cong 2.8\ \%/\mathrm{GPa} \times \left(\sigma_{xx} + \sigma_{yy}\right) - 1.1\ \%/\mathrm{GPa} \times \sigma_{zz} \tag{16.7d}$$

P-doped vertical resistors and resistor-L have lowest stress dependence [18]. The largest difference in stress dependence between two types of resistors in (100)-Si is for n-doped vertical resistors and lateral resistors in L-layout.

Unsilicided poly-silicon resistors consist of crystallites with a small number of orientations [19], yet there seems to be no preference for x- and y-directions. Therefore, a single poly-silicon resistor has the same isotropic stress-dependence law (16.5d) as single crystal resistors in L-layout. But, the total resistance of poly-silicon is a mixture of inter-grain and intra-grain contributions and this gives piezo-resistive coefficients of equal sign, but smaller magnitude than single crystal resistors in L-layout. For low-ohmic n-poly-Si with 100 Ω/square we measured a $(\sigma_{xx} + \sigma_{yy})$-dependence of −4.9 %/GPa, and for high-ohmic n-poly-Si with 1 kΩ /square we found −8.8 %/GPa.

16.3.2 The Piezo-MOS Effect

Mechanical stress alters the mobility μ of the charge carriers in the channels of MOS-FETs [8], which leads to a piezo-resistive stress dependence. The mobility enters the current gain β but not the threshold voltage V_{th}. The drain current is

$$I_D = \begin{cases} \beta\,(V_{GS} - V_{th} - V_{DS}/2)\,V_{DS} & \text{for} \quad V_{DS} < V_{GS} - V_{th} \\ (\beta/2)\,(V_{GS} - V_{th})^2 & \text{for} \quad V_{DS} \geq V_{GS} - V_{th} \end{cases} \tag{16.8}$$

In (16.8) V_{GS} and V_{DS} are gate-source and drain-source voltages, respectively. The current gain is $\beta = \mu\ (\varepsilon_{ox}\ /t_{ox})\ (W/L)$ with dielectric constant ε_{ox} and thickness t_{ox} of the gate oxide, and with width W and length L of the channel. Thus, NMOS-FETs are comparable with n-doped resistors and PMOS-FETs are similar to p-doped resistors if we replace $\delta\beta = \Delta\beta/\beta_0 \cong \delta\mu = -\delta\rho$ (here again we neglect pure strain effects on the channel geometry). As explained in [8] one can use the piezo-MOS effect to make temperature compensated stress sensors for σ_{xx}–σ_{yy} and σ_{xy} (but unfortunately not for $\sigma_{xx} + \sigma_{yy}$, see Fig. 16.7).

16.3.3 The Piezo-Junction Effect

Mechanical stress alters the mobility μ_B of *minority* carriers in the base of a bipolar transistor [20]. This is in contrast to piezo-resistive and piezo-MOS effects, which act on the *majority* carriers. Stress also acts on the intrinsic carrier density $n_{i,B}$ in the base. Both effects change the saturation current I_S.

$$I_C = I_S \exp\left(V_{BE}/V_T\right) \text{ with } V_T = k_b T/q \text{ and } I_S = k_b T \left(A_E/W_B\right)\left(n_{i,B}^2/N_B\right)\mu_B \tag{16.9}$$

I_C is the collector current, V_{BE} is the base-emitter voltage, V_T is the thermal voltage, k_b Boltzmann's constant, T the absolute temperature, A_E the emitter area, W_B the base width, and N_B the doping of the base. The intrinsic carrier density leads to a notable non-linear dependence of saturation current on mechanical stress, so that we need to add second order terms

$$\delta I_{S,ij} = \Delta I_{S,ij}/I_{S0} = -\zeta_{ijkl}\sigma_{kl} - \zeta_{ijklmn}\sigma_{kl}\sigma_{mn} \tag{16.10}$$

ζ_{ijkl} and ζ_{ijklmn} are 1st and 2nd order piezo-junction coefficients. Not all coefficients are known and the reported values of some are also inconsistent (perhaps due to different doping levels or different intrinsic stress). For vertical bipolars in (100)-silicon we use the following rough numbers

$$\begin{aligned}\delta I_S^{VPNP} \cong &-13.8\ \%/\text{GPa} \times \left(\sigma_{xx} + \sigma_{yy}\right) - 30.8\ \%/\text{GPa} \times \sigma_{zz} \\ &+ 66\ \%/\text{GPa} \times \left(\sigma_{xx} + \sigma_{yy}\right)^2 + 220\ \%/\text{GPa} \times \sigma_{xy}^2\end{aligned} \tag{16.11a}$$

$$\begin{aligned}\delta I_S^{VNPN} \cong &-43.4\ \%/\text{GPa} \times \left(\sigma_{xx} + \sigma_{yy}\right) + 28.4\ \%/\text{GPa} \times \sigma_{zz} \\ &+ 37.5\ \%/\text{GPa} \times \left(\sigma_{xx} + \sigma_{yy}\right)^2 + 366\ \%/\text{GPa} \times \sigma_{xy}^2\end{aligned} \tag{16.11b}$$

Vertical PNPs (VPNPs) are less sensitive to mechanical stress than vertical NPNs (VNPNs) [21]. The situation is more complex for lateral bipolars, because the current flow direction also enters the stress dependence. However, one can again use L-layout to obtain isotropic behavior [21]. Nevertheless, we prefer vertical over lateral bipolars due to practical reasons (less surface effects, size, no L-layout required, availability in standard design packages). For practical use we only need to remember that bipolars should have layouts with 90° symmetry in the (x,y)-plane and they should be placed at regions on the chip where only the in-plane normal stress components are present. Then, their saturation currents have linear and quadratic terms proportional to $\sigma_{xx} + \sigma_{yy}$ and $(\sigma_{xx} + \sigma_{yy})^2$.

16.3.4 The Piezo-Hall Effect

Mechanical stress alters the Hall coefficient, which leads to a change in the current related magnetic sensitivity $S_{i\text{Hall}}$. In (100)-silicon it holds for hall plates [22]

$$\delta S_{i\text{Hall}} = \Delta S_{i\text{Hall}}/S_{i0\text{Hall}} = P_{12}\left(\sigma_{xx} + \sigma_{yy}\right) + P_{11}\sigma_{zz} \tag{16.12}$$

For the piezo-Hall coefficients of low n-doped silicon [22, 23] report $P_{11} =$ –88 %/GPa and $P_{12} = 45$ %/GPa for a doping level of $1.8 \times 10^{14}/\text{cm}^3$ at 20 °C.

At larger doping concentration or larger temperature, the piezo-Hall coefficients get slightly smaller. In our technologies we measured $P_{12} = 42\,\%/\text{GPa}$. (100)-silicon has the largest piezo-Hall effect, whereas (111)-silicon of old bipolar technologies gives the smallest piezo-Hall effect [24].

To sum up, for all common devices we can use layouts which respond equally to normal stress in both in-plane lateral directions (x and y). Other components of the stress tensor can be neglected. The piezo-coefficients have negative temperature coefficients. Thus, at low temperature the devices are most susceptible to mechanical stress. Moreover, thermo-mechanical stress is largest at low temperature, because the package constituents were joined at elevated temperature and stress builds up due to mismatch of thermal expansion between them. Besides, the stiffness of the mold compound is larger at low temperature. Consequently, low temperature is the worst case for stress drifts of circuits.

16.4 How to Make an On-Chip Stress Sensor for $\sigma_{xx} + \sigma_{yy}$?

It is common practice to place stripes of resistors with different direction on a chip in order to measure various components of the stress tensor on the chip surface. However, these are not self-sufficient on-chip stress sensors, because the changes in resistance values are measured by an off-chip Ohmmeter, which is not subject to the stress, and so it can serve as a stable reference. In order to compensate mechanical stress drifts we need a sensor that outputs a signal with large sensitivity to mechanical stress but only small sensitivity to other effects like temperature. Moreover, since it is not possible to apply stress changes during the production and testing of the circuit, the sensitivity of the stress sensor needs to be stable against process spread.

In principle, one can use any two of the devices discussed in the previous Sect. 16.3, and compare their stress-dependent parameters by some circuit.

(1) n/n^+ stress resistor in L-layout:
We may compare low and high n-doped resistors, both of lateral type and in L-layout. They have stress sensitivities according to (16.5d), yet the resistor with the larger doping has a smaller piezo-resistive coefficient than the low-doped one. The maximum differential sensitivity between both resistors is somewhat less than $20\,\%/\text{GPa}$ and for p/p^+ stress resistors in L-layout it is about ten times smaller. The process spread between both types of resistors is large, because the two implantation processes do not correlate with each other: They have different sheet resistance, different temperature coefficient of resistance (TCR), and large spread of both parameters. Yet the piezo-coefficients are less affected by spread of doping concentration. So, one can trim off sheet resistances and TCRs in an end-of-line test. However, the residual errors after trimming are likely to be problematic because of the small stress sensitivity.

(2) *n*/*p* stress resistors in L-layout:
If we compare low *n*- and *p*-doped lateral resistors in L-layout, the situation is slightly better: *n*- and *p*-resistors still have a large mismatch of sheet resistance, but smaller TCR mismatch and their differential stress sensitivity is 30 % higher: 27.2 %/GPa according to (16.6d, 16.7d) [25]. Instead of comparing just two resistors in L-layout one can also use four resistors in L-layout in a Wheatstone bridge circuit [26].

(3) lateral/vertical *n*-doped stress resistors:
A significantly higher stress sensitivity is obtained, if we compare a low *n*-doped lateral resistor in L-layout ($R_x + R_y$) with a low *n*-doped vertical resistor (R_z): 77.8 %/GPa according to (16.6c, d). In BiCMOS technologies with *n*-epitaxial layer and *n*-buried layer we make a vertical resistor where the current flows from a contact to the buried layer, then sideways in the low ohmic buried layer and up again into a second contact ($R^n{}_V$ in Fig. 16.8). Yet the contacts should reach as close as possible to the lateral boundaries of the device to establish straight vertical current flowlines. If there is a spacing between contacts and lateral isolating boundary the current flowlines exhibit some curvature (current crowding into the contact) and this mixes some lateral piezo-resistive portion to the vertical one so that the overall device has less than 77.8 %/GPa stress sensitivity [27].

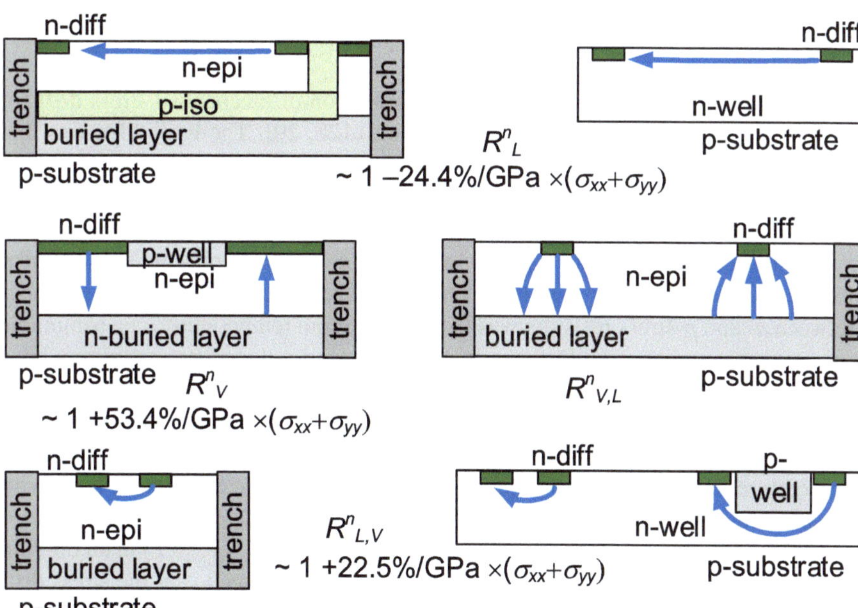

Fig. 16.8 Lateral, vertical and combined stress resistors in technologies with and without buried layer. The *arrows* indicate current streamlines. Only one branch of the L-layouts of resistors is shown

On the other hand, the size of the contact controls the curvature of the current flowlines and so one can achieve nearly all piezo-resistive coefficients between pure lateral and pure vertical current flow ($-24.4 \ldots +53.4$ %/GPa) ($R^n{}_{V,L} \ldots R^n{}_V$ in Fig. 16.8).

We can use the same tub for vertical and lateral resistors, if we split up the large contact into two smaller ones with a spacing in-between ($R^n{}_{L,V}$ in Fig. 16.8). Then the current flow between the two smaller contacts has larger lateral contribution than the vertical current flow. Since the same tub is used, the process spread should have smaller effect on mismatch between the two devices. If the doping profile versus depth is constant, one should be able to realize vertical and lateral resistors with very similar TCR.

In pure CMOS technologies without buried layers one can make lateral resistors having smaller contribution of vertical current flowlines (see $R^n{}_L$ in Fig. 16.8). However the vertical resistors are worse, because they have larger contributions of lateral current flowlines (see $R^n{}_{L,V}$ in Fig. 16.8). Moreover, the piezo-resistive coefficients are smaller because the doping of the n-CMOS well is two orders in magnitude larger than in n-epitaxial layers.

16.5 Circuits for Compensation of Mechanical Stress Drift

16.5.1 Stress Compensation with n/p-Resistors in L-Layout

Historically, we applied the electronic compensation of mechanical stress drift for the first time in a smart Hall sensor (Fig. 16.9) [28, 29]. The basic idea was that the stress sensor detects the very same mechanical stress component, which causes a drift in the Hall sensor. According to (16.12) the relevant stress component is $\sigma_{xx} + \sigma_{yy}$ as long as the Hall-plate is near the center of a (100)-silicon chip. So we used lateral resistors in L-layout with low n- and p-doping as stress sensor and placed them close to the Hall plate. The process spread in sheet resistances between n- and p-stress resistors was trimmed at room temperature via a trimmable ratio of current mirrors. Unfortunately, the low n- and p-doped stress resistors have quite different TCR: they can differ by 0.4 %/°C. An on-chip temperature sensor (T) was added near the Hall-plate and the stress-resistors to remove temperature errors between n- and p-stress-resistors. The signals of stress sensor $S_{\mathrm{raw,sum}}$ and temperature sensor T need only a small bandwidth so that we can digitize them with the same multiplexed ADC. After packaging, the output $S_{\mathrm{raw,sum}}$ of the stress sensor was measured versus temperature and stored as a reference $S_{\mathrm{ref,sum}}(T)$ in an on-chip memory. Later on during operation in the field the system only uses the stress drift dS_{sum}.

$$dS_{\mathrm{sum}} = S_{\mathrm{raw,sum}} - S_{\mathrm{ref,sum}}(T) \tag{16.13}$$

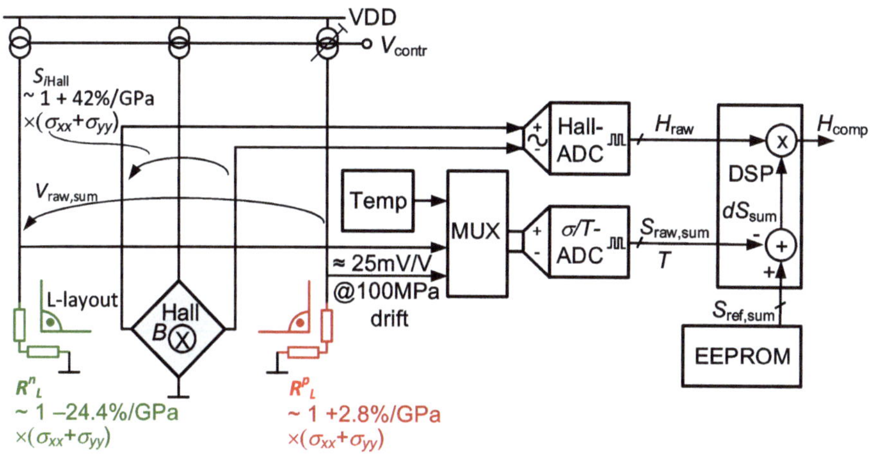

Fig. 16.9 First stress compensation circuit used for long-term stability of magnetic sensitivity of Hall sensor. The stress sensor signal $V_{raw,sum}$ was A/D-converted and a versatile compensation was implemented with digital signal processing

The main signal of the Hall-plate was also digitized with another ADC. In the digital domain the three signals of Hall-plate H_{raw}, temperature sensor T, and stress-sensor dS_{sum} are conveniently combined to remove temperature- and stress-dependencies of the Hall signal.

$$H_{comp} = H_{raw}\left(1 + TCH_1\ T + TCH_2\ T^2 + \cdots\right)$$
$$\left(1 + EPC_1\ dS_{sum} + EPC_2\ dS_{sum}^2 + \cdots\right) \tag{16.14}$$

where H_{comp} is the compensated Hall output signal, H_{raw} is the uncompensated (raw) signal of the Hall plate, TCH_1 and TCH_2 are the first and second order temperature coefficients of the system, and EPC_1 and EPC_2 are the first and second order effective piezo-coefficients of the system. The *effective* piezo-coefficients comprise all effects of stress on all relevant circuit components. In practice we do not need to know which devices and to which extent they are affected by mechanical stress as long as we assure that all delicate devices depend only on $\sigma_{xx} + \sigma_{yy}$ and that the stress values on different devices at different locations are sufficiently correlated. This means that these devices must not be too close to the perimeter of the chip and they must have a layout, which avoids stress dependence different from $\sigma_{xx} + \sigma_{yy}$. The effective piezo-coefficients may again depend slightly on temperature $EPC_1 = EPC_{10} + TCEPC_1 \times T$.

The exact numbers for TCH_1, TCH_2, EPC_1, EPC_2 are conveniently found by empirical investigation, although for the purpose of circuit design one needs rough guesses according to Sects. 16.2, 16.3, and 16.4. To this end we have to subject the chip to different temperatures and different mechanical stress while recording the signals H_{raw}, T, and dS_{sum}. Note that we do not need to know the exact values of

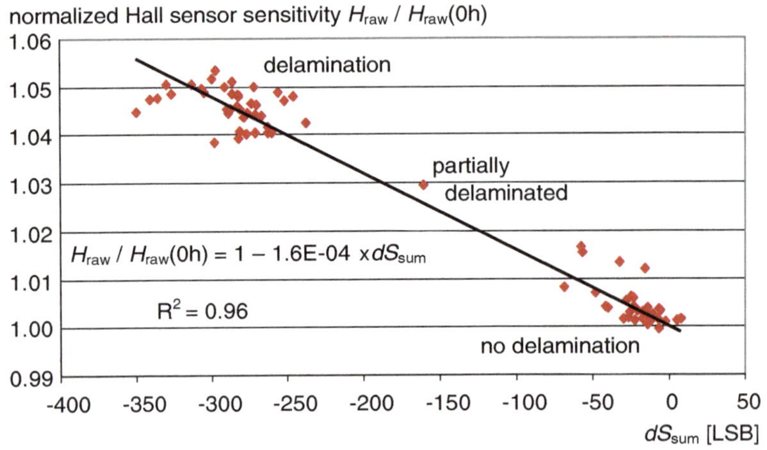

Fig. 16.10 The measurement of normalized magnetic sensitivity of a Hall sensor and the drift of a $\sigma_{xx} + \sigma_{yy}$ stress sensor after a severe lifetime load show good correlation and detectability of delamination in plastic encapsulated packages

mechanical stress (in terms of MPa), because we only need to extract the correlation coefficients between the three signals. The application of different stress can be done by wafer bow experiments [25], by bending of stripes cut out of wafers [30], by comparing pre- and post-package assembling tests, by moisture absorption of plastic encapsulated packages [25], by applying mechanical force to packaged devices [25, 31], or by comparing different packages with different mechanical stress (e.g. with different thicknesses of leadframe and chip or with different mold compounds).

Figure 16.10 shows an example of a strong correlation between drift of normalized magnetic sensitivity of a Hall plate and the stress drift signal dS_{sum}. It was measured in the course of a temperature humidity cycling test which was continued until finally the plastic encapsulated package disintegrated. The delamination at various mechanical interfaces caused a reduction in compressive stress which means a positive stress change on the Hall-plate and this led to an increase in magnetic sensitivity of about $4 \ldots 5\,\%$. With the piezo-Hall coefficient of (16.12) we conclude that the delamination caused a relaxation in stress $\sigma_{xx} + \sigma_{yy}$ by 100 MPa. This is a good example how a stress sensor can detect serious mechanical flaws which might occur during lifetime. In a similar way comparison of stress sensor readouts in pre and post packaging tests can be used to monitor the quality of package assembly process. Finally, stress changes due to soldering the IC-package to a printed circuit board can also be screened.

Figure 16.11 shows the final performance of the stress compensated Hall sensor circuit. Stripes were cut out of the silicon wafer and mechanical stress was applied to them with a four-point bending bridge. The plot compares normalized magnetic sensitivity of the circuit with and without stress compensation at low and high temperature. An improvement of factor 7 can be observed.

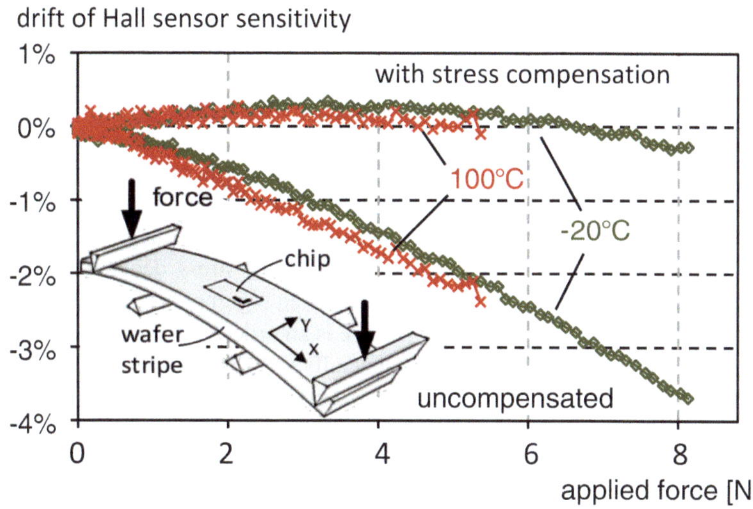

Fig. 16.11 Measured drift of magnetic sensitivity of a Hall sensor with and without stress compensation in a wafer stripe bending experiment

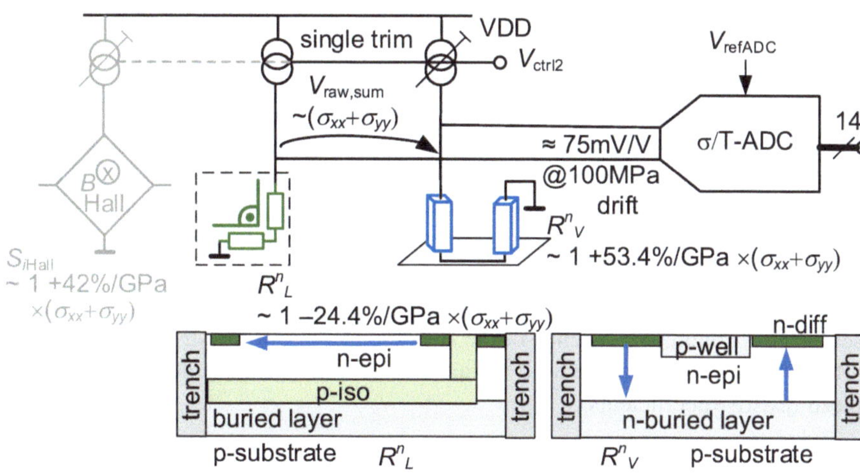

Fig. 16.12 Improved digital stress compensation of a Hall-plate with lateral/vertical n-doped stress resistors having larger stress sensitivity and smaller difference in TCR

16.5.2 Stress Compensation with Lateral and Vertical Resistors

A stress sensor with lateral/vertical *n*-doped stress resistors has three times higher stress sensitivity than the *n*/*p* stress resistors in L-layout (Fig. 16.12). Moreover, the TCRs of lateral/vertical *n*-doped stress resistors differ only in the order <0.1 %/°C, which is four times less than the TCRs of *n*/*p* resistors.

A major problem with all stress sensors for $\sigma_{xx} + \sigma_{yy}$ is the difference in TCR between the two types of stress resistors. A systematic (nominal) TCR difference can be stored as reference $S_{\text{ref,sum}}(T)$, but this requires an on-chip memory and a multi-temperature test after package assembly. For some products this is too expensive. We can reduce the errors caused by nominal TCR difference of the stress resistors by the replica circuit of Fig. 16.13. A bandgap kernel is used to generate a voltage $V_{P1}$of appropriate temperature dependence. If this bandgap kernel also uses devices with mere $\sigma_{xx} + \sigma_{yy}$ stress dependence the voltage shows a small dependence on $\sigma_{xx} + \sigma_{yy}$, which simply adds to the stress dependence of Hall-plate and stress sensors, and it can be compensated together with all other circuits. Besides, the stress dependence of this voltage is an order in magnitude smaller than the stress dependence of the Hall plate, especially if the bandgap kernel uses vertical PNPs and low-ohmic poly-silicon resistors. The input voltage V_{P1} is replicated to the vertical n-doped stress resistor $R^n{}_V$ in a feedback loop comprising a controlled current source and an OTA. With $V_{\text{V}} = V_{\text{P1}}$ the current through $R^n{}_V$ becomes stress and temperature dependent. It is mirrored to the lateral L-shaped n-well resistor $R^n{}_L$. The resulting voltage V_{L} can be adjusted with a trimming at a single temperature to match the voltages $V_{\text{V}} = V_{\text{P1}}$. Yet, even at zero stress the TCR difference between vertical resistor $R^n{}_V$ and lateral resistor $R^n{}_L$ would lead to an unwanted voltage between them. However, since $V_{\text{V}} = V_{\text{P1}}$ and V_{P1} is defined by the bandgap kernel, we can use the same bandgap kernel to construct a further voltage V_{P2}, which has again only tiny stress dependence and whose temperature dependence is chosen such as to guarantee $V_{\text{P2}}(T) = V_{\text{L}}(T)$ at zero stress for all temperatures. In other words, the temperature dependence of V_{P2} is designed to cancel out the systematic TCR differences between $R^n{}_V$ and $R^n{}_L$. Any process spread between V_{P1} and V_{P2} is minimized by the bandgap principle. This circuit technique makes use of the fact that the systematic TCR difference and the stress dependence of the resistors is much more stable than the spread of the sheet resistances (which is about $\pm 20\,\%$). Finally, we can superimpose the same temperature coefficient on V_{P1} and V_{P2} to set the temperature dependence of the stress sensitivity of the stress sensor to any arbitrary value. This can be used to cancel out the temperature dependence of the piezo-resistive coefficients.

16.5.3 Stress Compensation for Different Sensitivity in σ_{xx} and σ_{yy}

The lateral n-doped stress resistor in Fig. 16.13 comprises four resistor stripes connected in a Wheatstone bridge circuit. Two resistors are aligned in x-direction (i.e. [-110] crystal direction) and the other two resistors are aligned in y-direction (i.e. [-1-10]). According to (16.6a, b) the digitized output voltage $S_{\text{raw,diff}}$ of the Wheatstone bridge is proportional to σ_{xx}–σ_{yy} with a sensitivity of 6.8 %/GPa. An even ten times higher stress sensitivity could be obtained if the four resistors were p-doped instead of n-doped. In both cases the output voltage vanishes for $\sigma_{xx} = \sigma_{yy}$

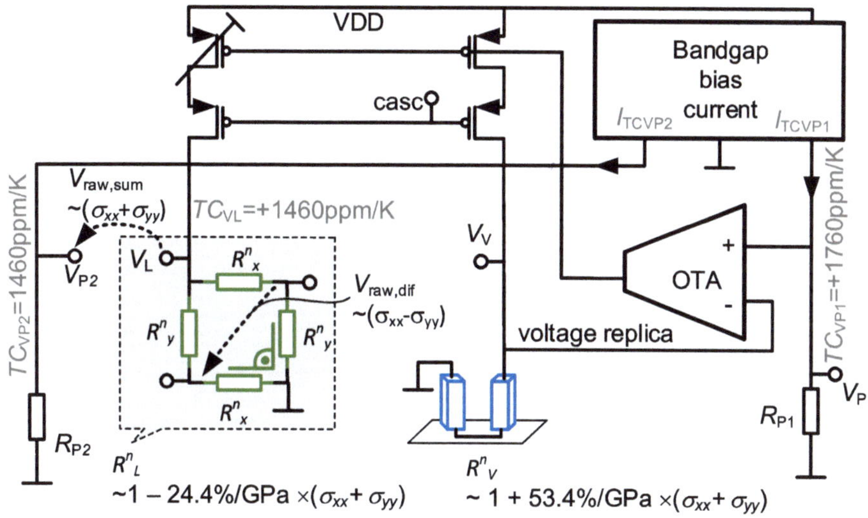

Fig. 16.13 Temperature compensation for measurement of $\sigma_{xx} + \sigma_{yy}$ and $\sigma_{xx} - \sigma_{yy}$

for all temperatures, if we disregard mismatch between the four resistors. So the signal $S_{raw,diff}$ has much less errors caused by process spread and temperature excursion than the signal $S_{raw,sum}$. Therefore it is not necessary to store reference values $S_{ref,diff}(T)$ recorded during a multi-temperature test and compute a subsequent drift $dS_{diff} = S_{raw,diff} - S_{ref,diff}(T)$ as we did in the case of $S_{raw,sum}$ (see above). Instead, we can directly use $S_{raw,diff}$ in the stress compensation algorithm. The accuracy of $S_{raw,diff}$ is quite good and comparable to MOS-stress sensors (Fig. 16.7).

So the system of Fig. 16.13 has in fact two mechanical stress sensors: one for the sum ($S_{raw,sum}$) and one for the difference of in-plane normal stress components ($S_{raw,diff}$). This can be useful if a parameter of a circuit responds differently to mechanical stress in *x*- or *y*-directions. Reasons for such an anisotropic behavior may be a device layout which lacks *x*-*y*-symmetry (because the device is too large to add a second one in L-layout) or silicon technologies, where inhomogeneities in the bulk material (such as deep trenches or the edges of the die) may lead to a cross coupling of different components of the stress tensor. An example of such a case is shown in Fig. 16.14, where the output frequency of an RC-relaxation oscillator increases with 11.7 %/GPa stress in *x*-direction, while it decreases with −2.3 %/GPa stress in *y*-direction. A compensation of the drift of this oscillator frequency caused by mechanical stress can be done like this.

$$f^{osz}_{comp} = f^{osz}_{raw} C_{corr}\left(T, dS_{sum}, S_{raw,diff}\right) \tag{16.15a}$$

$$C_{corr}\left(T, dS_{sum}, S_{raw,diff}\right) = \left(1 + TC_1\ T + TC_2\ T^2 + \cdots\right) \times \left(1 + EPC_1\ dS_{sum} + EPC_2\ dS^2_{sum} + \cdots + EPC'_1\ S_{raw,diff} + EPC'_2\ S^2_{raw,diff} + \cdots\right) \tag{16.15b}$$

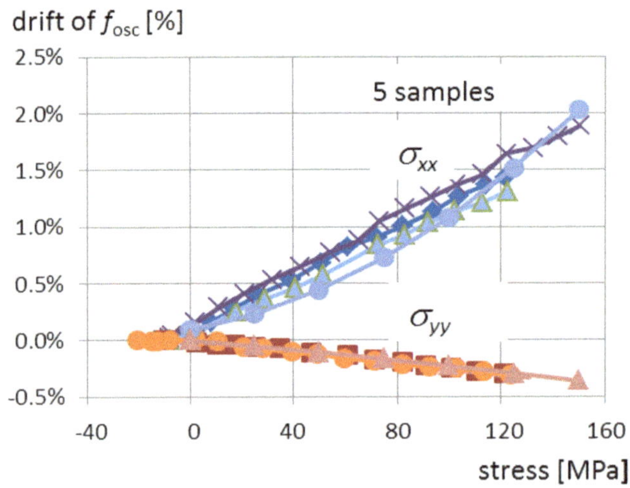

Fig. 16.14 The output frequency f_{osc} of an RC-relaxation oscillator shows different drift when stress is applied in x- and y-directions. Measurement was done in silicon-stripe bending experiments at room temperature

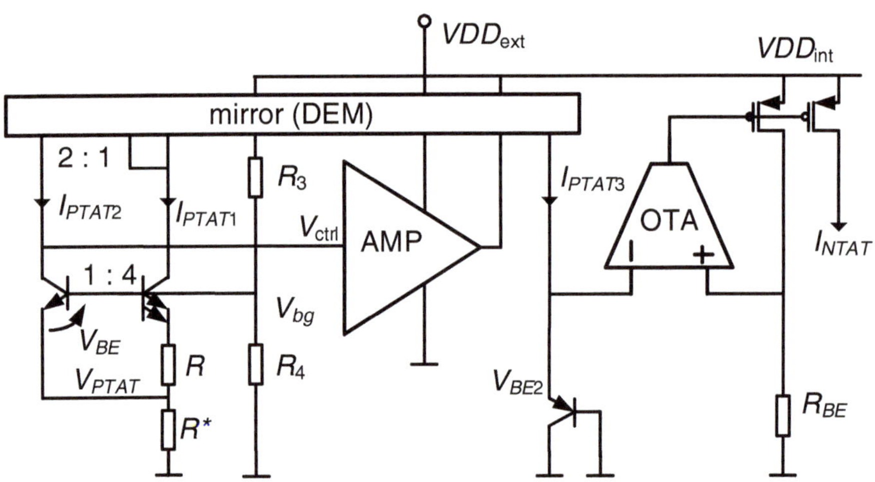

Fig. 16.15 Brokaw-bandgap with VNPNs and NTAT precision bandgap kernel

where EPC'_1, EPC'_2, … are effective piezo-coefficients for σ_{xx}–σ_{yy} stress dependence. In (16.15a) C_{corr} is an analog or digital compensation factor that scales the oscillator frequency. In a practical circuit this can be the integration current or the fractional divider in a feedback loop of a PLL. If the oscillator frequency would also be sensitive to in-plane shear stress σ_{xy} it is straightforward to add a shear stress sensor (e.g. like in Fig. 16.7) in the same way as the σ_{xx}–σ_{yy} sensor and to add according terms in (16.15b). So the compensation principle (16.15) is versatile and can take account of all components of the stress tensor if necessary.

16.5.4 Compensation of Mechanical Stress on Precision Bandgap Circuits

Bandgap circuits in standard plastic encapsulated packages exhibit a notable drift of 1…4 mV due to changes of mechanical stress on the circuit. These drifts are minimized by the use of vertical PNPs [33]. In the past a lot of effort was undertaken to improve temperature stability by chopping, dynamic element matching (DEM), and curvature-correction techniques [32]. Electrical lifetime drift is tackled by cascoding, applying symmetrical voltage stress on matched transistors, and by dynamic element matching DEM. By now the dominant residual error term seems to be mechanical stress drift [35]. The temperature stable voltage in Fig. 16.15 is

$$V_{bg} = \left(R^{*}/R\right) V_{PTAT} + V_T \ln\left(V_{PTAT}\right) - V_T \ln\left(RI_S\right) \tag{16.16}$$

with $V_{PTAT} = \Delta V_{BE} = V_T\ \ln(8)$. The voltage V_{PTAT} is proportional to absolute temperature. It depends only on the ratio of collector current densities of the bipolars, and so it does not depend on mechanical stress. Therefore, it is ideally suited for temperature sensors. The resistors R and R^{*} are p-doped poly—high ohmic resistors with a stress dependence of less than 5 %/GPa. With proper layout their ratio does not depend on mechanical stress. Hence, only the last term in (16.16) is affected by mechanical stress. So the mechanical stress drift of the bandgap voltage is caused by drifts of resistor R and saturation current I_S.

$$V_{bg}\left(\sigma_{ij} + \Delta\sigma_{ij}\right) - V_{bg}\left(\sigma_{ij}\right) = V_T \ln\left(\frac{R\left(\sigma_{ij}\right)}{R\left(\sigma_{ij} + \Delta\sigma_{ij}\right)}\right) + V_T \ln\left(\frac{I_S\left(\sigma_{ij}\right)}{I_S\left(\sigma_{ij} + \Delta\sigma_{ij}\right)}\right) \tag{16.17}$$

In (16.17) we wrote the complete stress tensor σ_{ij} , but for poly-silicon resistors and vertical bipolar transistors not too close to the perimeter of the die only the sum of in-plane normal stress components $\sigma_{xx} + \sigma_{yy}$ is relevant.

The compressive stress on the circuit in a plastic package at dry condition may be $\sigma_{xx} + \sigma_{yy} = -250$ MPa. During operation in the field the stress at the same temperature may relax by 90 MPa to -160 MPa due to moisture uptake. Then (16.11b) predicts a reduction of the saturation current of the VNPN by 4.7 %. With (16.17) this gives a rise of 1.1 mV in the bandgap-voltage. If the resistors R and R^{*} were n-doped vertical resistors with a stress dependence according to (16.6c) they would roughly compensate the drift of the VNPNs and the bandgap-voltage would decrease by 0.15 mV. In fact the stress sensitivity of pure vertical n-resistors is too strong so that it over-compensates the VNPN-drift. Thus, one can use $R^{n}{}_{V,L}$ instead of $R^{n}{}_{V}$ in Fig. 16.8 to adjust its stress dependence. This approach is explained in more detail in Sect. 16.5.6 for Hall sensors, where it is better suited. For bandgap circuits it seems preferable to use resistors with smaller TCR than the low doped stress resistors $R^{n}{}_{V}$ and $R^{n}{}_{V,L}$ because smaller TCR gives less troubles with curvature of bandgap voltage versus temperature. For less stringent requirements of analog

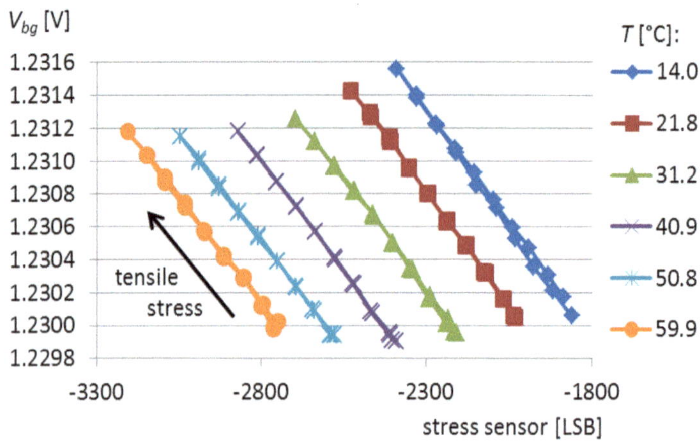

Fig. 16.16 Correlation between bandgap voltage of Fig. 16.15 and output of digital stress sensor for a typical sample. Stress was ramped up and down (0 ... 125 MPa) at various temperatures in a wafer stripe bending experiment

bias currents it is sufficient to mirror out an NTAT current generated by a VPNP, which is less stress dependent than a VNPN (see Fig. 16.15).

We implemented the circuit of Fig. 16.15 in a test chip and added a stress sensor with lateral/vertical n-doped resistors according to section 16.4(3) [34]. Then we measured a strong correlation between bandgap-voltage and stress sensor output during a wafer stripe bending experiment (Fig. 16.16). The bandgap voltage varies linearly with stress sensor output and with a constant slope at all temperatures. In practice the moisture drift in plastic encapsulated packages may reach values up to 90 MPa (for $\sigma_{xx} + \sigma_{yy}$) and in Fig. 16.16 this means a drift of about 1 mV.

A stress compensation of a reference voltage is preferably done digitally on system level instead of trimming it by analog means, which add further random errors. To this end, the system only needs a compensation factor C_{corr} as a function of temperature- and stress-signals similar to the RC-oscillator in (16.15). Figure 16.17 shows the performance of this compensation. There, stress was applied in a wafer stripe bending experiment to five devices at various temperatures. With a single value of effective piezo-coefficient for all five devices, the compensation algorithm reduces the stress drift times seven.

Figure 16.18 shows repeated thermal cycling on a packaged sample in "wet" condition. Without stress compensation, the curve 1 shows a clear hysteresis of 1.4 mV. If the compensation factor C_{corr} uses a constant value for the effective piezo-coefficient, then the hysteresis is only 0.4 mV. However, if C_{corr} uses an effective piezo-coefficient with -0.4 %/°C linear temperature coefficient, the hysteresis gets as small as 0.2 mV in the entire automotive temperature range.

Figure 16.19 shows an example how to combine stress compensation with precision analog circuit techniques like chopping, auto-zeroing, and digital assisted

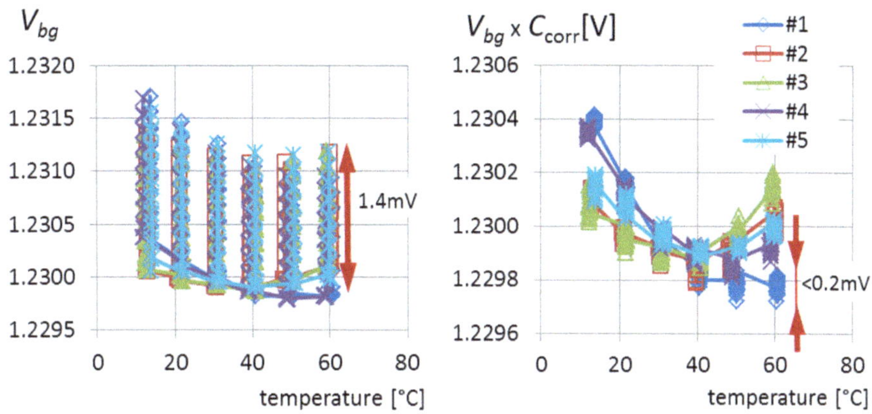

Fig. 16.17 The bandgap voltage of Fig. 16.15 without (LHS) and with (RHS) compensation of mechanical stress. Stress between 0 and 125 MPa was applied on five devices (#1 . . . #5) in a wafer stripe bending experiment

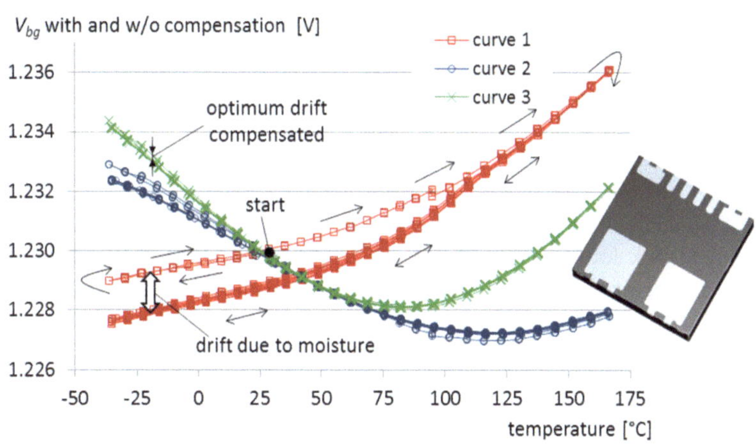

Fig. 16.18 Five thermal cycles applied to the circuit of Fig. 16.15 in PG-TISON-8-1 package measured after moisture soaking for 16 h in water at 85 °C. The effective piezo-coefficient is zero (*curve 1*), a fixed value (*curve 2*), and a temperature dependent value (*curve 3*)

signal processing in an economic and robust way [36]. Offset errors caused by mismatch and mismatch drift are equally large as mechanical stress related drifts, and therefore one has to combine mismatch cancellation techniques with stress compensation. The bandgap kernel is kept as simple as possible to make it predictable and robust against process spread: it only uses VPNPs and resistors with low TCR. Curvature correction, mechanical stress compensation, and single trim are shifted into the digital part. A control loop comprising an OTA pulls the bandgap kernel into its operating point. Offset and offset drift of the OTA (versus temperature, stress, lifetime) are cancelled out by chopping the OTA. This leads to

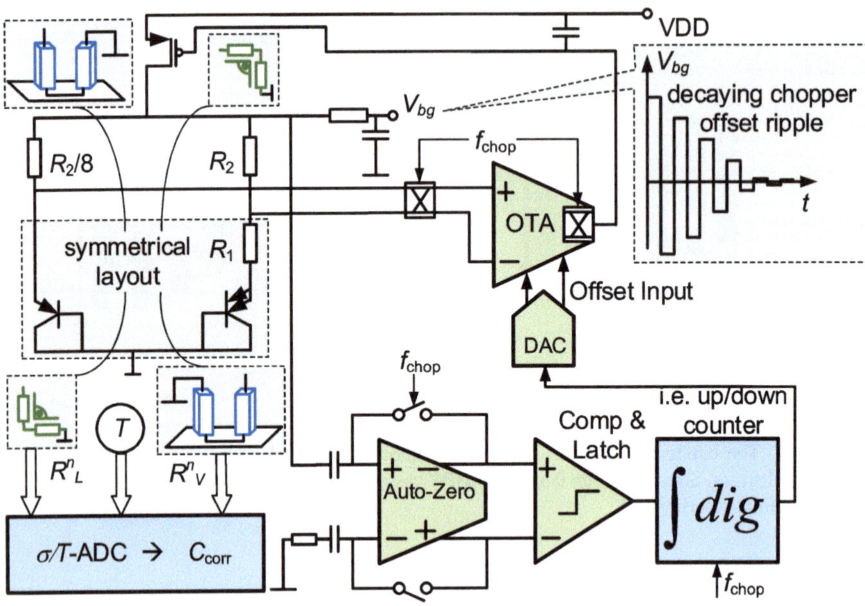

Fig. 16.19 Chopped bandgap reference in 0.13 μm technology with mechanical stress sensors for $\sigma_{xx} + \sigma_{yy}$ and digital assisted chopper ripple compensation

an offset ripple at chopper frequency at the output of the bandgap circuit. A digital assisted feedback loop compensates this offset ripple: The auto-zero comparator shifts the offset ripple from the chopper frequency into the baseband. There the offset is singe-bit quantized, which means that we throw away the amplitude information and only keep the sign. Therefore requirements for gain, speed, and linearity of the auto-zero comparator are relaxed. The sign-signal is integrated digitally by a simple up-down counter to get a multi-bit signal. This signal is fed into a DAC, which compensates the offset in the OTA. The requirements for gain accuracy and linearity of the DAC are also uncritical—it only needs to be monotonic. At transients of temperature or mechanical stress there is a decaying offset ripple at the output. Conversely, at steady state the system may freeze the DAC output, which opens the feedback loop. Then also the chopper can be switched off, to reduce residual glitches at the bandgap output and to save power. The chopper and the feedback loop can be re-activated if notable changes in temperature or stress are detected.

Drifts of mechanical stress and temperature are measured by respective sensors operated at low update rate. These drifts are cancelled out by versatile digital signal processing with a digital compensation factor C_{corr}. The idea is to use the uncompensated bandgap voltage V_{bg} as on-chip reference for ADCs in a system on chip and multiply the ADC output signal by C_{corr}. This digital multiplication is simpler than analog methods and it avoids additional errors.

The bandgap voltage and its residual temperature coefficient can be trimmed simultaneously at a single temperature, because the only errors left after chopping are PTAT errors [37]. At the same instant, the stress sensor output is trimmed to zero. All these trimming values define the digital compensation factor C_{corr}. The temperature dependence of the stress sensor can be designed to be nominally zero (Fig. 16.13) and the piezo-coefficients show little process spread.

16.5.5 Mechanical Stress on an LC-Oscillator

Figure 16.20 shows the drift of an LC-oscillator which was calibrated in a dry plastic package at several temperatures with an accuracy of ±25 ppm [38]. After 48 h storage in wet condition, the oscillator frequency was re-measured versus temperature starting at −40 °C and a larger error of +70/−125 ppm was observed. It was also verified that a water droplet on the top of a naked silicon die does not change the oscillator frequency, which rules out purely dielectric effects. A plausible explanation for the observed drift is that the moisture uptake expands the mold compound, which reduces the compressive stress on the die. Thus, the size of the coil changes and this leads to a drift in the inductance and in the oscillator frequency. Yet, electronic compensation of this drift is more involved, because strain and not stress is at the origin of the drift. In a laminated package, strain and stress are not simply related via Young's modulus of silicon, but parameters of other package constituents also have an influence. The question is, if the correlation between strain and stress is stable enough for a given package so that one may still use a stress sensor. Further investigations are ongoing to examine the possibility for compensation of strain related drift. The target is to replace crystal and MEMs oscillators with on-chip LC-oscillators in certain applications in the future.

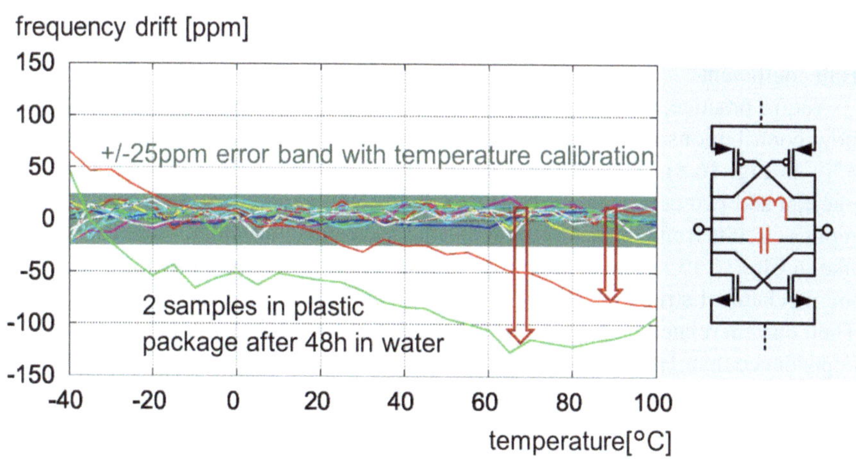

Fig. 16.20 Frequency drift of an LC-oscillator in plastic package: first calibrated vs. temperature in dry condition—then moisture uptake and re-measurement

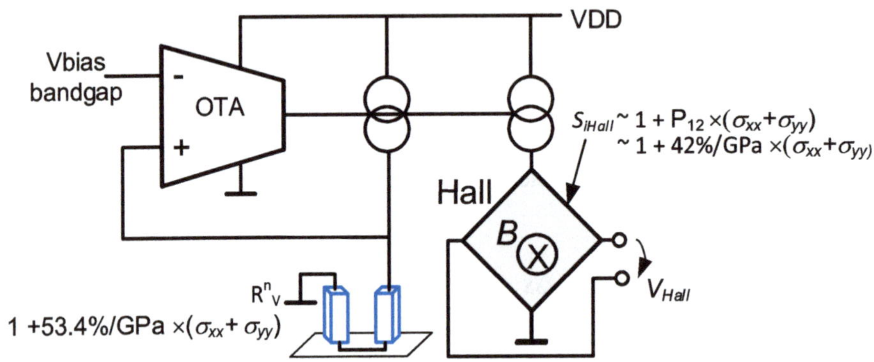

Fig. 16.21 Analog pre-compensation of Hall-plate without explicit stress sensor

16.5.6 Analog Pre-Compensation

In (100)-silicon vertical n-type resistors (incidentally) have only 20 % stronger stress dependence than n-doped Hall-plates (see (16.6c) and (16.12)). This can be used to pre-compensate the magnetic sensitivity against mechanical stress: A voltage of appropriate temperature dependence is generated by a bandgap kernel and copied onto a vertical n-type resistor. The resulting current is used as supply current for the Hall plate (Fig. 16.21). Thus, an increase in mechanical stress by 100 MPa increases the resistance of the vertical n-type resistor by 5.34 % while the bandgap referred voltage remains nearly constant versus stress. Consequently the current through the Hall plate decreases by the same percentage, yet the current related magnetic sensitivity increases by 4.5 %. The total magnetic sensitivity is proportional to the multiplication of Hall input current times current related magnetic sensitivity and therefore it decreases by 5.34–4.5 % = 0.84 %. So the stress dependence of the magnetic sensitivity of the Hall plate is reduced by a factor of 4.5/0.84 = 5.3. In practice the doping of Hall plates is larger than in (16.12), which reduces the piezo-Hall coefficient.

Yet, in practice, vertical n-type resistors always have some small lateral current flow contributions (see the bending of the current streamlines at the contacts of $R^n{}_{V,L}$ in Fig. 16.8). For a better pre-compensation over a large temperature range one can add two currents derived from lateral ($R^n{}_L$) and vertical ($R^n{}_V$) resistors via replica circuits from voltages with dedicated and adjustable temperature dependence like in Fig. 16.13 and adjust their ratio. The same principle can be used to reduce the mechanical stress sensitivity of Vertical Hall-effect devices (Fig. 16.22) [39]. Their current related magnetic sensitivity exhibits a roughly ten times smaller stress dependence than Hall-plates and it is also anisotropic (i.e. they respond differently to normal stress in x- and y-directions). One can adjust the size of the contacts in vertical resistors ($R^n{}_{V,L}$) or the spacing of the contacts in lateral resistors ($R^n{}_{L,V}$) in order to trim the piezo-resistive coefficients of the n-type resistors to perfectly match the piezo-Hall coefficients of the n-type Vertical Hall-effect device.

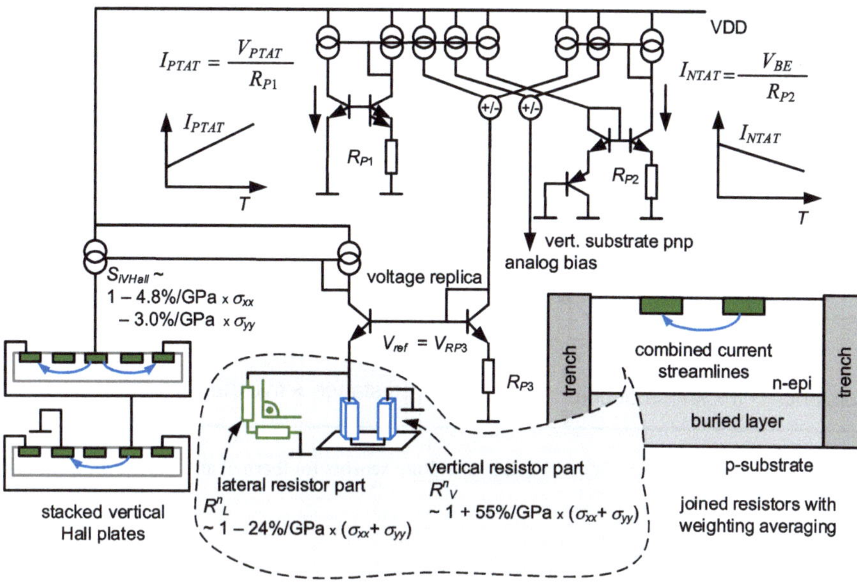

Fig. 16.22 Analog pre-compensation of vertical Hall-effect device

It should be mentioned that for magnetic field sensors it is also possible to use closed loop systems to make the magnetic sensitivity more stable against mechanical stress and temperature (and other) drifts [13]. They use on-chip coils placed near the magnetic sensor element to apply a magnetic reference field. However, the stability of the system is limited by the stability of the coil current. If the current is defined by an off-chip component (resistor), it suffers from tolerance, residual temperature coefficient, parasitics at the additional package terminal and/or EMC hazard and costs. If the current is defined by an on-chip resistor, one still needs a stable one, e.g. p-doped resistors in L-layout or on-chip metal resistors (which have even lower piezo-resistance) and one has to cope with its temperature dependence. Even more problematic is the fact that the system needs significant current (several milli-amps) to produce sufficiently strong field on sensors with moderate magnetic sensitivity like Hall effect devices. Conversely, the stress sensor needs only minute power (some 10 μA), space and extra system resources and it offers the additional advantage to monitor the package quality and detect its end of life.

Figure 16.23 shows layout suggestions how to arrange stress and temperature sensors around circuits which are compensated. These common centroids cancel out errors caused by linear stress gradients and temperature gradients.

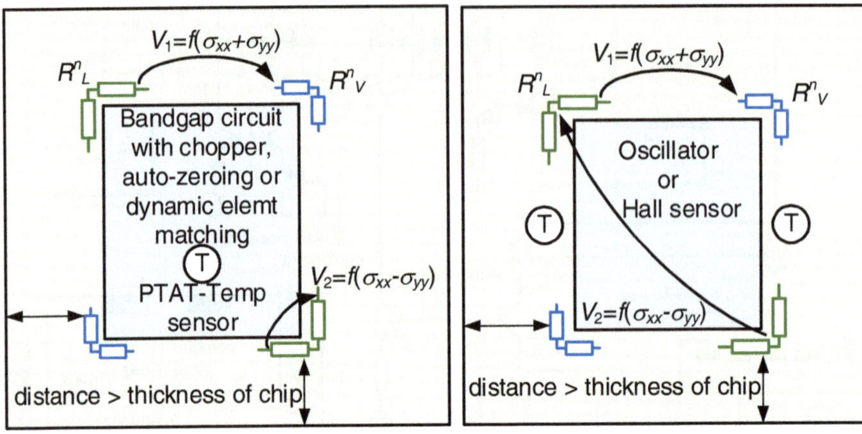

Fig. 16.23 Layouts of stress sensors and temperature sensors for thermal and stress symmetry and linear gradient rejection

16.6 Conclusions

In this paper we proposed electronic compensation circuits, which reduce the drift of integrated precision analog circuits caused by mechanical stress. This method reduces mechanical stress related drifts by a factor of 3 … 10, thereby significantly improving overall accuracy and lifetime stability. This helps to meet the requirements for power efficient Li-ion battery management systems. Practical examples of Hall-plates, vertical Hall effect devices, bandgap voltages, system references, RC- and LC-oscillators were discussed. It was elaborated, that a crucial element for stress compensation circuits is a stress sensor that detects the sum of in-plane normal stress components. A more general mechanical stress compensation method was proposed, which uses additional sensors for other components of the mechanical stress tensor. Significant drifts of mechanical stress in common plastic encapsulated packages were found. They are provoked by moisture absorption, soldering, chemical cure shrink, and package disintegration. Mechanical stress sensors can detect these effects with negligible extra costs in terms of space and power consumption, thereby increasing the reliability of packaged ICs.

Acknowledgements The contributions of Philipp Greiner (TU Graz) for LC oscillator design and measurements, Gerhard Maderbacher (Infineon, Villach), for a bandgap design, Daniel Stoica (Infineon, Bucharest) for design of a vertical Hall sensor and Michael Holliber (Infineon, Villach) for laboratory characterization are greatly acknowledged.

References

1. C.S. Smith, Piezoresistance effect in germanium and silicon. Phys. Rev. **94**(1), 42 (1954)
2. S.M. Hu, Stress-related problems in silicon technology. J. Appl. Phys. **70**(6), R53–R80 (1991)
3. H. Ali, Stress-induced parametric shift in plastic packaged devices. IEEE Trans. Compon. Packag. Manuf. Technol. Part B **20**(4), 458–462 (1997)
4. R.A. Hastings, *The Art of Analog Layout* (Prentice Hall, Upper Saddle River, 2006)
5. J. Bastos, M.S. Steyaert, A. Pergoot, W.M. Sansen, Influence of die attachment on MOS transistor matching. IEEE Trans. Semicond. Manuf. **10**(2), 209–218 (1997)
6. D. Manic, J. Petr, R.S. Popovic, Short and long-term stability problems of Hall plates in plastic packages, in *Reliability Physics Symposium, 2000. Proceeding of 38th Annual 2000 IEEE Intl,* pp. 225–230. IEEE, 2000
7. U. Ausserlechner, M. Motz, M. Holliber, Drift of magnetic sensitivity of smart Hall sensors due to moisture absorbed by the IC-package, in *Sensors, 2004. Proceedings of IEEE,* pp. 455–458. IEEE, 2004
8. R.C. Jaeger, J.C. Suhling, R. Ramani, A.T. Bradley, J. Xu, CMOS stress sensors on [100] silicon. IEEE J. Solid State Circuits **35**(1), 85–95 (2000)
9. H. Kittel, S. Endler, H. Osterwinter, S. Oesterle, F. Schindler-Saefkow, Novel stress measurement system for evaluation of package induced stress, in *2008 2nd European Conference & Exhibition on Integration Issues of Miniaturized Systems-MOMS, MOEMS, ICS and Electronic Components (SSI)*, pp. 1–8. VDE, 2008
10. J.N. Sweet, S.N. Burchett, D.W. Peterson, A.H. Hsia, A. Chen, Piezoresistive measurement and FEM analysis of mechanical stresses in 160L PlasticQuadFlatPacks, in *Proceedings of InterPACK '97*, pp. 1731–1740, 1997
11. S.A. Gee, W.F. Van Den Bogert, V.R. Akylas, Strain-gauge mapping of die surface stresses. IEEE Trans. Compon. Packag. Manuf. Technol. **12**(4), 587–593 (1989)
12. U. Ausserlechner, D. Draxelmayr, Programmable linear magnetic hall sensor for automotive applications, in *Proceedings of 36th International Conference on Microelectronics, Devices and Materials, MIDEM*, Postojna, Slovenia, 28–30 Oct 2000
13. J.M. Cesaretti, W.P. Taylor, G. Monreal, O. Brand, Effect of stress due to plastic package moisture absorption in Hall sensors. IEEE Trans. Magn. **45**(10), 4482–4485 (2009)
14. J. Lau (ed.), *Thermal Stress and Strain in Microelectronics Packaging* (Van Nostrand Reinhold, New York, 1993), chapter 2 by P.M. Hall
15. Y. Kanda, Piezoresistance effect of silicon. Sensors Actuators A Phys. **28**(2), 83–91 (1991)
16. J.F. Nye, *Physical Properties of Crystals*, 1985 edn. (Oxford Science Publications/Clarendon Press, Oxford, 1985), ch. II, sec. 4, p. 45
17. U. Ausserlechner, The piezo-resistive effect in silicon for arbitrary crystal orientation, in *Proceedings of IEEE Sensors*, vol. 3, pp. 1121–1124, 2004
18. S. Komatsu, S. Takahashi, K. Suzuki, M. Wakatsuki, Strain insensitive integrated circuit resistor pair, U. S. Patent 4236832, 29 June 1977
19. P.J. French, A.G.R. Evans, Polycrystalline silicon strain sensors. Sensors Actuators **8**(3), 219–225 (1985)
20. J.F. Creemer, P.J. French, An analytical model of the piezojunction effect for arbitrary stress and current orientations, Transducers' 01 Eurosensors XV. Springer Berlin Heidelberg, 256–259 (2001)
21. F. Fruett, G.C.M. Meijer, A new sensor structure using the piezojunction effect in PNP lateral transistors. Sensors Actuators A Phys. **92**(1), 197–202 (2001)
22. B. Hälg, Piezo-Hall coefficients of n-type silicon. J. Appl. Phys. **64**(1), 276–282 (1988)
23. Y. Kanda, K. Suzuki, The Piezo-Hall effect in n-silicon, in *22nd International Conference on the Physics of Semiconductors*, vol. 1, pp. 89–92, 1995
24. U. Ausserlechner, The piezo-Hall effect in n-silicon for arbitrary crystal orientation, in *Sensors, 2004. Proceedings of IEEE*, pp. 1149–1152. IEEE, 2004

25. U. Ausserlechner, M. Motz, M. Holliber, Compensation of the piezo-Hall effect in integrated Hall sensors on (100)-Si. IEEE Sensors J. **7**(11), 1475–1482 (2007)
26. S. Huber, C. Schott, O. Paul, Package stress monitor to compensate for the piezo-hall effect in CMOS Hall sensors. IEEE Sensors J. **13**(8), 2890–2898 (2013)
27. M. Motz, U. Ausserlechner, M. Bresch, U. Fakesch, B. Schaffer, C. Reidl et al., A miniature digital current sensor with differential Hall probes using enhanced chopping techniques and mechanical stress compensation, in *Sensors, 2012 IEEE*, pp. 1–4. IEEE, 2012
28. M. Motz, U. Ausserlechner, W. Scherr, B. Schaffer, An integrated magnetic sensor with two continuous-time $\Sigma\Delta$-converters and stress compensation capability, in *2006 IEEE International Solid State Circuits Conference-Digest of Technical Papers*, 2006
29. M. Motz, U. Ausserlechner, W. Scherr, E. Katzmaier, An integrated Hall sensor platform design for position, angle and current sensing. In *5th IEEE Conference on Sensors, 2006*, pp. 1008–1011. IEEE, 2006
30. R.E. Beaty, R.C. Jaeger, J.C. Suhling, R.W. Johnson, R.D. Butler, Evaluation of piezoresistive coefficient variation in silicon stress sensors using a four-point bending test fixture. IEEE Trans. Compon. Packag. Manuf. Technol. **15**(5), 904–914 (1992)
31. H. Husstedt, U. Ausserlechner, M. Kaltenbacher, In-situ analysis of deformation and mechanical stress of packaged silicon dies with an array of Hall plates. IEEE Sensors J. **11**(11), 2993–3000 (2011)
32. B.S. Song, P.R. Gray, A precision curvature-compensated CMOS bandgap reference. IEEE J. Solid State Circuits **18**(6), 634–643 (1983)
33. F. Fruett, G. Meijer, A. Bakker, Minimization of the mechanical-stress-induced inaccuracy in bandgap voltage references. IEEE J. Solid State Circuits **38**(7), 1288–1291 (2003)
34. G.C.M. Meijer, Concepts for bandgap references and voltage measure-ment systems, in *Analog Circuit Design*, ed. by J.M. Huising, R.J. van de Plassche, W.M. Sansen (Kluwer, Boston, 1995), pp. 243–269
35. M. Motz, U. Ausserlechner, M. Holliber, Compensation of mechanical stress-induced drift of bandgap references with on-chip stress sensor. IEEE Sensors J. **15**(9), 5115–5121 (2015)
36. G. Maderbacher, S. Marsili, M. Motz, T. Jackum, J. Thielmann, H. Hassander et al., 5.8 A digitally assisted single-point-calibration CMOS bandgap voltage reference with a 3σ inaccuracy of $\pm 0.08\,\%$ for fuel-gauge applications, in *2015 IEEE International Solid-State Circuits Conference-(ISSCC)*, pp. 1–3. IEEE, 2015
37. G.C. Meijer, Thermal sensors based on transistors. Sensors Actuators **10**(1), 103–125 (1986)
38. G.P. Greiner, J. Grosinger, C. Steffan, G. Holweg, W. Bosch, Non-trimmable LC oscillator for all CMOS frequency control, in *ESSCIRC 2015—41st European Solid-State Circuits Conference (ESSCIRC)*, pp. 140–143. IEEE, 2015
39. D. Stoica, M. Motz, A dual vertical Hall latch with direction detection, in *Proceedings of the ESSCIRC (ESSCIRC), 2013*, pp. 213–216. IEEE, 2013

Chapter 17
Power Electronics for LED Based General Illumination

Stefan Dietrich and Stefan Heinen

17.1 Introduction

In 2005, the generation of grid-based electrical light consumed about 19 % of the total global electricity [1]. With an annual growth rate of 2.4 %, it is expected that the generation of light will remain one of the biggest causes of greenhouse gas emissions worldwide [1, 2]. Incandescent lamps are identified by their poor energy efficiency of only 12 lm/W [1]. Modern technological advances in solid-state lighting (SSL) counteracts this trend and leads to higher energy efficiency of electrical light in comparison to traditional electrical light sources [1, 2]. SSL shows the most promising approach to achieve these goals with a reported energy efficiency of more than 220 lm/W [2]. To even speed up the commercial breakthrough of SSL, governments tend to prefer the novel technology with various formalities [3–7].

17.2 Driver Background and Origin

An LED can be operated by controlling the forward voltage of the LED, or by controlling the current through the LED. Due to the diode nature of the device, it is much simpler to control the current through the LED. This parameter has a direct influence on the emitted light of the LED. Controlling the forward voltage leads to defective control, since it differs with the emitted wavelength of the LED and strongly varies with temperature. Various methods to control the LED current are available. Whereas all LED driving topologies offer several advantages and

S. Dietrich (✉) • S. Heinen
Integrated Analog Circuits & RF Systems Laboratory, RWTH Aachen University, Aachen, Germany
e-mail: stefan.dietrich@ias.rwth-aachen.de

© Springer International Publishing Switzerland 2017

A. Baschirotto et al. (eds.), *Wideband Continuous-time ΣΔ ADCs, Automotive Electronics, and Power Management*, DOI 10.1007/978-3-319-41670-0_17

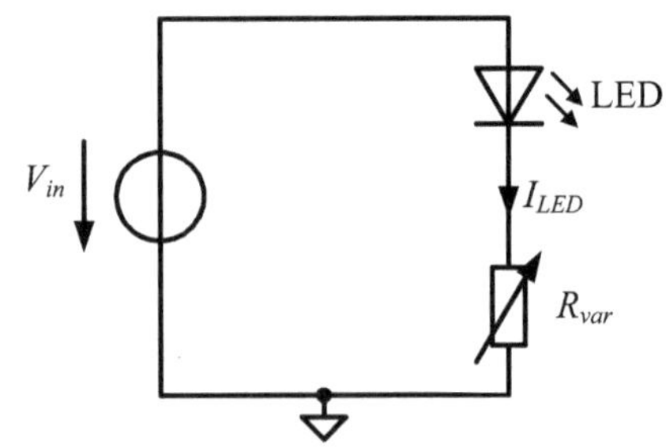

Fig. 17.1 Resistor-based LED driver

disadvantages, it is obvious that this mass product always tries to minimize the costs to compete with low-cost compact fluorescent lamps (CFL). As a matter of fact, higher system costs with integrated circuits always come with an increased functionality. This increase in functionality includes high dimming ranges from 0.1 to 100 %, high efficiency in high dimming ranges, as well as multicolored red-green-blue-white (RGBW) light, and color mixing. Starting from an arbitrary DC input voltage, the easiest way to control the current of an LED is by adding a series resistance to the LED to limit its peak current. The simplest form of an LED driver is given in Fig. 17.1 and is used as a starting point for all discussions.

This technique offers a fixed LED current with a fixed color temperature and already clearly exceeds the power efficiency of light bulbs. When the DC input voltage is close to the LED forward voltage plus the resistive voltage drop, no further DC-DC voltage conversion is necessary. In case of a grid-level AC supply, the DC conversion may be done by passive rectification and a big input capacitor prior to the LED string. To cope with the high DC potential and to provide the LED and resistor from too high forward voltage, the number of series LEDs is increased. In any case, further DC-DC conversion will be required to provide an optimum LED string voltage. This DC-DC conversion can be done by linear regulation, e.g. with LDOs. Or by using switch-mode DC-DC conversion to maintain high system efficiency.

In [8], different switch-mode LED driving methods are introduced. An overview of the topologies is done in Fig. 17.2. Besides single-stage AC-DC LED drivers shown in (a), two-stage AC-DC and DC-DC LED drivers in (b) are used. Single-stage AC-DC LED drivers typically generate light of a fixed color temperature and try to minimize the electrolytic capacitors [9]. To offer higher functionality, two-stage AC-DC and DC-DC LED drivers are favored.

Both topologies provide an LED string voltage V_{out} to supply the LED string and to ensure a reasonable control voltage headroom V_{fb}. Process variations in R limit the application of the resistor. It may lead to variations in the brightness of the LED driver and also leads to higher voltage drop and unnecessary resistive losses in the LED string. Hence, the resistor needs to be substituted in a more reliable solution. To get rid of the increased voltage drop and the lack of current control, a controlled current source is used [10, 11]. Other groups focus on the minimization of the resistive losses in the string and across the resistor itself [12–14].

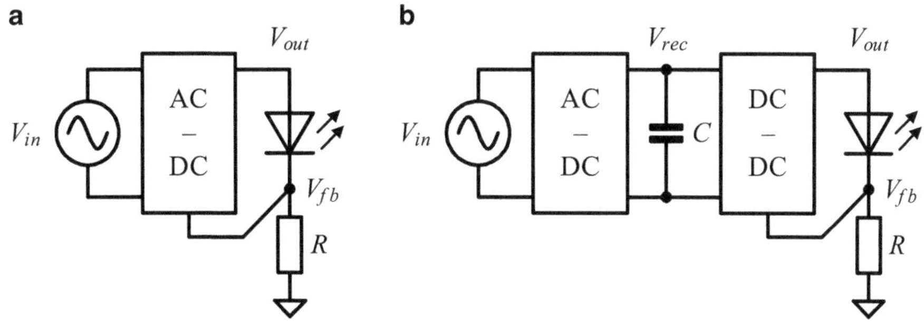

Fig. 17.2 AC connected (**a**) single-stage, (**b**) two-stage LED drivers

17.3 LED Dimming and Control Techniques

LED dimming is done by modulating the LED current. There are two ways to dim an LED. The first way is done by amplitude modulation (AM), whereas the second technique uses pulse-width modulation (PWM) of the LED peak current [15]. Both techniques are followed by several advantages and disadvantages. AM is a resistive control of the LED current and may be done by linear regulation. It is quite simple and achieves high luminous efficacy, but leads to color shifts in the resulting color [15–17]. This effect is based on the LED sensitivity on changing the LED current. PWM is more accurate than AM dimming. Controlling the "on" and "off" times of LEDs is easier than controlling the amplitude of the LED current. In addition, color shifts are reduced with PWM dimming, since the LED is always supplied by the peak current. However, every kind of switch-mode power conversion introduces electromagnetic radiation. Especially LED drivers with their high output power might cause unwanted EMI. One method to reduce the frequency noise is to use a distributed frequency spectrum of the PWM modulated current [18]. When it comes to high dimming levels, e.g. 0.1 % and low PWM frequencies, flickering might occur. This effect is discussed in the IEEE standard PAR1789.

A summary of the standard is given in [19], where a minimum visible frequency of 165 Hz is reported. Hence, the PWM frequency has to stay above the minimum required frequency to avoid flickering. Switch-mode DC-DC converters introduce a continuous conduction mode (CCM) and a discontinuous conduction mode (DCM) [20]. Especially the DCM is used to maintain high system efficiency under light load conditions. In the case of single LED current PWM, these two operation modes can not be used and the LED is operated either with its peak current, or with zero.

To combine the advantages of AM and PWM, hybrid AM/PWM dimming techniques are used [21]. Hybrid AM/PWM increases the color resolution of each dimming method by the resolution of the other technique. It is of higher complexity but also eliminates color shifts of the two techniques. A switch-mode DC-DC LED driver is operated with its optimum operation mode and therefore high system efficiency is realized under every load current. Furthermore, the dimming

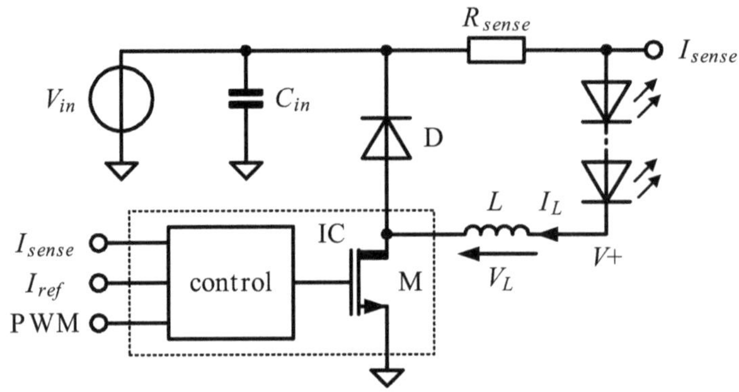

Fig. 17.3 Single-channel DC-DC LED driver

range is increased without flickering. Low LED current then requires an amplitude reduction and remains a high switching frequency to stay above the minimum visible frequency.

17.4 LED Driver Topologies

17.4.1 Single-Inductor Single Output

A single-channel buck type LED driver is presented by Anghel et al. [22] and Park et al. [23]. The topology is shown in Fig. 17.3. The transistor M and diode D are used to charge and discharge the inductor L in series to the LED string [20]. The current control uses a sense resistor R_{sense} to feed back the current information to the converter.

The converter operation is separated into an inductor charge period and an inductor discharge period. To charge the input voltage referred inductor, transistor M is turned "on" and the switch-node equals to zero. The voltage across the inductor equals to the input voltage V_{in} minus the LED string forward voltage $m \cdot V_F$. Due to the positive voltage across L, I_L rises and the LEDs start to emit light.

When the current is above the control threshold of the converter, the voltage drop across the sense resistor is high and the converter turns "off" the transistor M. Due to the remaining inductor current I_L, the high-ohmic switch-node voltage starts to increase. When V_{sw} is a diode voltage above the input V_{in}, the diode is forward biased and discharges the inductor, due to the negative voltage across L. The LEDs still emit light and the converter turns "on" the transistor again, as soon as the voltage drop across the sense resistor is small.

In the topology of Fig. 17.3, the LED string is an active part of the converter operation. The inductor charge and discharge periods are influenced by the forward voltage of the LED string. This is a first step towards a specialized LED driver. In typical DC-DC converters, the load is only passive. The converter introduced by Park et al. [23] neglects the output capacitor. The minimum application integrates the buck converter half-bridge, so that only the external inductor remains. However, even if the form factor reduces to a minimum, the extension of this application towards multichannel drivers to generate tunable white light offers some drawbacks. With the presented single-channel drivers of [22–24] this is only realized by a linear increase in external components with the number of wanted colors.

In particular, the number of inductors increase with the number of wanted colors, e.g. red, green, blue, and white. These colors are typically used to generate so-called tunable white light. External inductors are cost intensive, so that the bottleneck of these systems is the single-channel characteristic. Hence, other topologies have to be used to generate tunable white light.

Converters with a single current control and without storage capacitors for a defined output voltage only apply for PWM dimming. AM dimming and discontinuous operation modes in converters without a defined output voltage are quite challenging. To calculate the inductor charge and discharge periods, inductor volt-second balance or capacitor charge balance are applied [20]. An additional idle period like in DCM demands for both equations to calculate the correct idle time [20]. Due to the absence of capacitors, the set of equations is under-determined. To overcome this issue, typically the output of the feedback error amplifier is used. The current control technique then applies for a constant "on" time and valley current control to detect the zero current.

17.4.2 Single-Inductor Multiple Output

To overcome an increased bill of materials (BOM) for multichannel LED drivers, the single-inductor multiple output (SIMO) concept is presented [25]. It is well-known in battery powered portable electronics and reduces especially the size, weight, and cost of power management devices. Hence, an adaption of this concept in LED background lighting of liquid-crystal displays (LCD)s is identified [26–28]. These topologies only apply for a single inductor and output capacitor. An overview of the boost type converters is given in Fig. 17.4.

The two switches M1 and M2 are used to charge and discharge the inductor L, which supplies the output capacitor C. The string selection transistors S1...Sn multiplex the inductor current and enable color mixing when turned "on". The strings are supplied by the shared output voltage V_{out} and only provide low output current $< 100\,\text{mA}$ per string to drive the RGB colors in the LCD display. Boost type topologies are used, due to the high number of LEDs in a LCD display. In these applications, the number of RGB LEDs in each string typically equals. The required output voltage headroom for each string is in the same range. Hence, only two external components are used, which is the inductor L and output capacitor C.

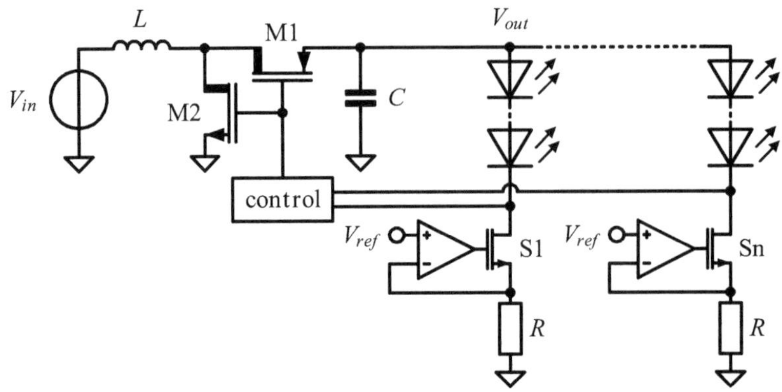

Fig. 17.4 Boost type SIMO LED driver

In LED backlight applications, the number of LEDs in each string is equal. However, the application of general illumination LED drivers for tunable white light differs from backlight applications. In particular, tunable white light demands for LED strings with a much higher number of white LEDs compared to colored LEDs. As a result, the shared output voltage of the parallel LED strings is too high for the strings with colored light. This leads to high string selection transistor voltage and may destroy the devices with a small number of LEDs in series. This drawback has been identified by Zhang et al. [24] and Hsieh and Chen [28] who introduce an output voltage hopping and charge recycle technique. However, in [28] further external components account for the recycle technique. This issue has been identified by Zhang et al. [24] but without the ultimate consequence to completely neglect the output capacitor.

17.4.3 AC-DC LED Drivers

Apart from the two-stage AC-DC and DC-DC LED driver approach reported in [8], single-stage AC-DC LED drivers are numerous. Their compact architecture is the main advantage, since they are also used to retrofit old lamp installations [8, 29]. For stand-alone AC-DC LED drivers, a flyback architecture as shown in Fig. 17.5 is commonly used. The diode-based bridge rectifier generates a non-constant DC potential V_{rec}, which is stored on the input capacitor C_{in}. To reduce the expensive capacitor size C_{in}, the input voltage range of the switch M driver is high and several control strategies are presented. Further EMI filters and circuits for power factor correction (PFC) are optionally implemented prior to the bridge rectifier, or between the input capacitor and the transformer. The transistor M operates the inductor in boundary switch-mode to avoid the transformer from being saturated. The switch current on the secondary side is then rectified by the diode D and capacitor C_{out} to generate the DC output power for the LED string.

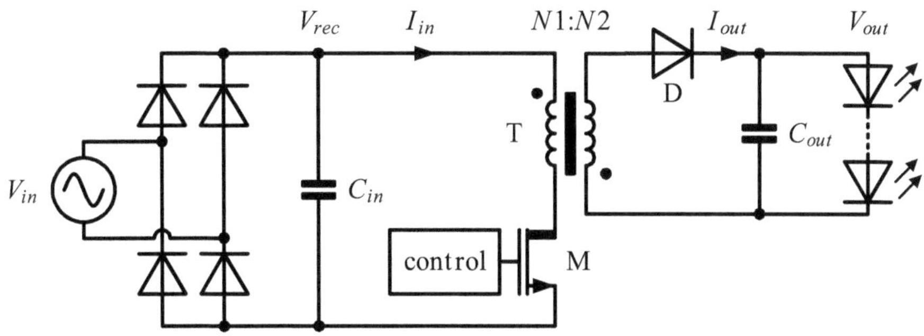

Fig. 17.5 AC-DC flyback LED driver

Flyback AC-DC LED drivers are preferred due to the single high-voltage connected switch M. Furthermore, the galvanic isolation by the transformer in this topology is beneficial. Never the less, especially the bulky capacitor and high-voltage connected driver limit the application. The capacitors are not only expensive, but also shrink the lifetime of the whole LED driver product [9, 30–32]. Therefore, various groups focus on the reduction or negligence of bulky AC-DC input capacitors [9, 30, 32–34]. For high functionality in multicolor LED light, the complex color control is better implemented in the low voltage domain. Hence, for multicolor LED light, the two-stage AC-DC supply and DC-DC LED driver approach is favored.

17.5 Capacitor-Free Single-Inductor Multichannel

Some research efforts are done on floating output single-inductor multichannel devices. The objective of this driver is the minimization of external components to offer the smallest possible form factor. Hence, only a single inductor besides several high current high brightness LEDs is used for general illumination. To generate tunable white light, the number of LEDs in each string differ. Without introducing further components, string isolation and selection is performed by a single double-diffused DMOS transistor per string. An all-digital integration provides the necessary performance for the single current control and color mixing. The topology of the integrated capacitor-free SIMO LED driver is given in Fig. 17.6 [35]. It consists of a high-side PMOS transistor M0 connected to the DC input voltage V_{in}. The inductor L is in parallel to four LED strings for each of the colors white, red, green, and blue. An external sense resistor R_{sense} of nominal 50 mΩ is used for the current control.

Key component of the LED driver is the inductor and a stable steady-state operation as presented in [20] is realized. During a switching period T_s, the inductor is charged by turning "on" the PMOS transistor. The switch-pin voltage $V_{sw} = V_{in}$

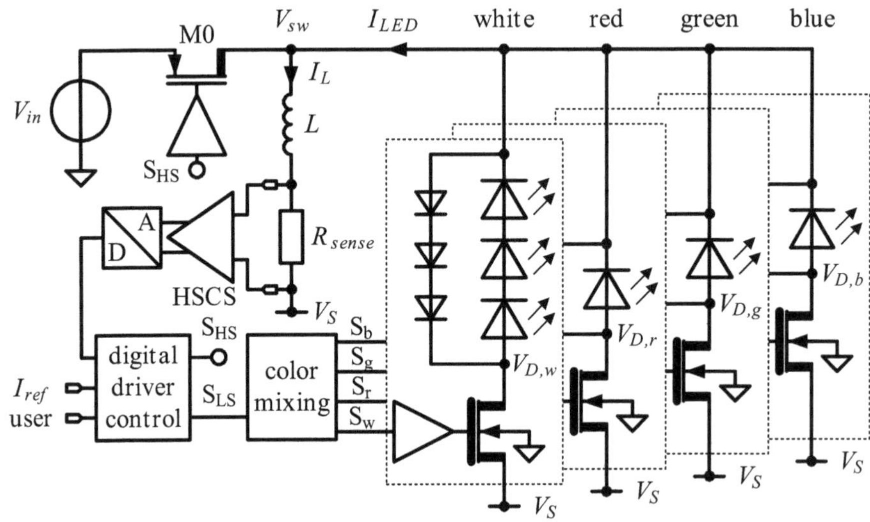

Fig. 17.6 Capacitor-free single-inductor multichannel LED driver

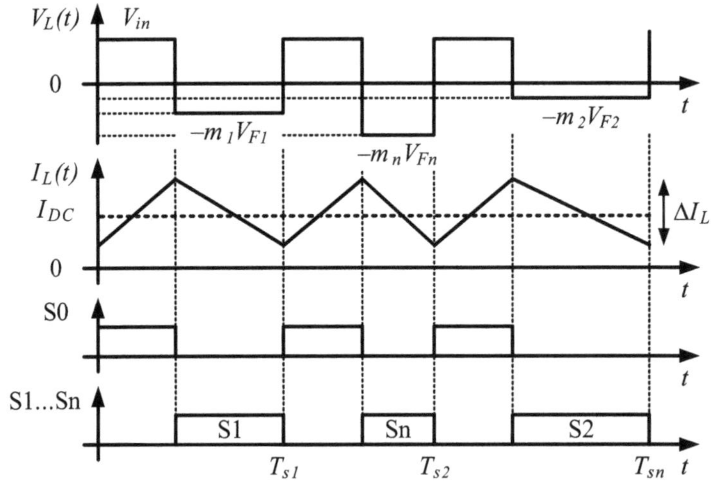

Fig. 17.7 Combined waveforms in CCM

and the inductor is charged. To discharge the inductor, a single LED string is enabled and the free-wheeling inductor current biases the LEDs in forward direction. Hence, the LED current equals to the inductor current and a color is turned "on" during each switching period T_s. The inductor discharge and hence the converter switching periods are a function of the number and forward voltage of each LED string. This results in the characteristic switch-pin waveforms illustrated in Fig. 17.7.

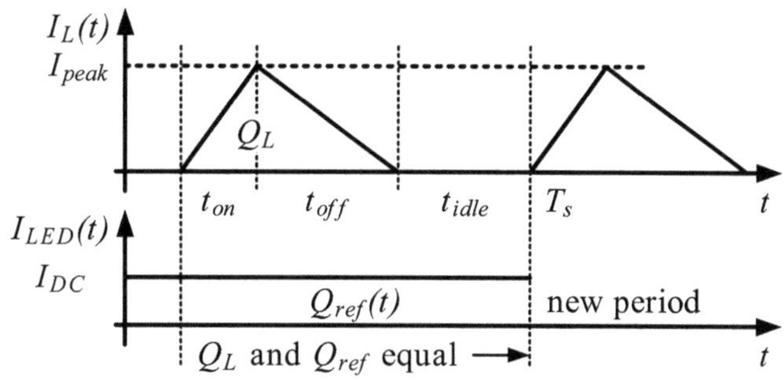

Fig. 17.8 Inductor current integral control in discontinuous conduction mode

The inductor voltage during PMOS "on" is always constant. However, a varying discharge period appears. This is indicated by the variable discharge across the green LED string in period T_{s1}, the white string during T_{s2}, and the red string during T_{s3}. When L is high, the current ripple in the LEDs is low and the LED current is similar to a PWM controlled string in a state-of-the-art converter. The color mixing is therefore a result of a high converter switching frequency and a non-visible integral of several enabled colors in a color sequence [36].

To introduce a safety margin to the minimum visible frequency in [19], the minimum color sequence in the SIMO LED driver is set to 500 Hz. Hence, the resolution of a mixed color during a complete sequence is determined by the minimum visible frequency and the converter switching frequency. With a nominal switching frequency of 500 kHz, 10 bit PWM color resolution is achieved. A further increase in color resolution is the implementation of the 9 bit inductor current AM to realize a hybrid AM/PWM dimming of [21]. This method increases the PWM resolution by the resolution of the current sense and also allows the realization of the DC-DC converter CCM and DCM.

However, a single current control during DCM is complex. To calculate the converter steady-state conversion ratio, capacitor charge balance or inductor volt-second balance is applied [20]. Due to the absence of capacitors, capacitor charge balance cannot be used during DCM to calculate the converter idle time [20]. Hence, a digital inductor charge balance is implemented and the operation is presented in Fig. 17.8 [37].

The inductor current during DCM is identified by the characteristic triangular shape. Whereas the PMOS "on" time is set, the NMOS "on" time is a function of the LEDs in each string. To calculate the idle time, the reference current during a switching period is kept constant and results in a rectangular shape. As soon as the idle time t_{idle} starts, the inductor current triangle is bigger than the reference rectangular. By constantly adding to the reference rectangle, the idle time is calculated to

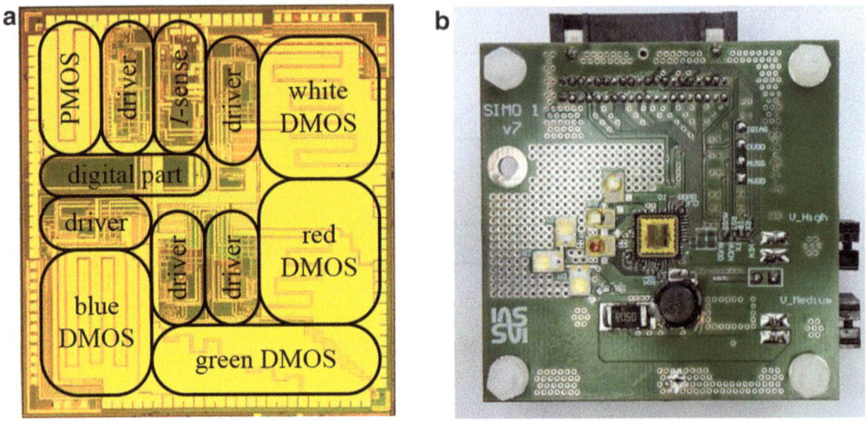

Fig. 17.9 Photograph of the multichannel LED driver (**a**) chip micrograph, (**b**) prototype PCB

$$
\begin{aligned}
Q_{DC} &= Q_L \\
\Rightarrow \quad \frac{1}{2}(D_1 + D_2)\, T_s \cdot I_{peak} &= I_{DC} \cdot (D_1 + D_2 + D_3(t))\, T_s \\
\Rightarrow \quad D_3(t) &= \left(\frac{I_{peak}}{2I_{DC}} - 1\right)(D_1 + D_2)\,.
\end{aligned}
\tag{17.1}
$$

It can be seen from Eq. (17.1) that the idle time is a function of the reference current I_{DC}, the input voltage V_{in}, and the LED specific discharge D_2. As soon as both integrals equal, the idle time is determined and the control decides whether to stay in DCM, or to change to the CCM.

A prototype capacitor-free SIMO LED driver has been fabricated in a 0.18 µm high-voltage (HV) CMOS technology. Figure 17.9 (a) shows a micrograph of the IC and (b) illustrates the prototype PCB with the chip in QFN48 [35].

The die area in (a) is 3.3 mm × 3.4 mm. The main part of the chip consists of the double-diffused low-side DMOS transistors. The size of each transistor is 1.1 mm^2 with its individual driver of 0.39 mm^2. In the upper right corner, the PMOS and its driver is placed. The current sense interface is in next to the PMOS driver. The size of high-side current sense amplifiers is 0.146 mm^2 and the SAR ADC occupies 0.076 mm^2. The digital part to the left occupies 0.4 mm^2. In part (b) of the figure, the prototype PCB is illustrated. It is a two layer FR-4 PCB with an area of 55 mm × 55 mm. The active area of the LED driver is less than 25 mm × 25 mm. At the bottom of the chip, the external inductor can be seen. To its left, the external shunt resistor is mounted. The six LEDs are placed to the left of the IC.

A transient color sequence during steady-state CCM is given in Fig. 17.10. The sequence consists of five converter switching periods. It starts at ≈ 34 µs and represents the colors white-white-red-white-blue. In the upper part of the figure, the inductor current I_L is given. A DC inductor current of 450 mA is achieved and the current ripple is between 250 and 600 mA with a 68 µH inductor.

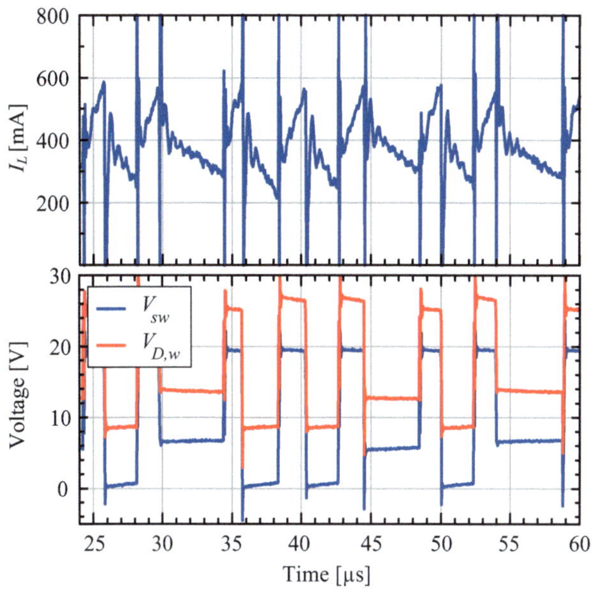

Fig. 17.10 Continuous conduction mode with enabled color sequence

The switching frequency of the converter is $\approx$ 200 kHz, since the five switching periods of the color sequence end at 59 μs. The lower part shows the switch-pin voltage V_{sw} and the drain of the white string DMOS $V_{D,w}$. When the PMOS transistor is turned "on", the switch-pin voltage V_{sw} equals to the input voltage V_{in}. It can be seen from the figure, that the input voltage $V_{in} = 20\,\text{V}$ and $V_S = 9\,\text{V}$ for the transient measurements. The inductor discharge periods across the white LED string are identified by the drain voltage of the white DMOS. When the white LED string discharges the inductor at 36 μs, the drain of the DMOS transistor equals to $V_{D,w} = V_S - |V_{DS}|$.

During the discharge across the colored LED strings, V_{sw} is higher in amplitude. This equals to a discharge across a single colored LED of a parallel string. The discharge times are increased and the switching period is higher than for the white LED string. The white DMOS drain is higher than V_{sw} and higher than V_S. This indicates a disabled DMOS and no white light emission. When the PMOS transistor is turned "on", the white DMOS drain is higher than V_{in}. Due to the slow step response of the diodes compared to the switch-pin voltage V_{sw}, the forward voltage across the LEDs is preserved and the drain is increased. This fact rises the demand for high voltage capability of the low-side double-diffused DMOS transistors above the input voltage V_{in}. Figure 17.11 show the same color sequence during DCM. A sequence is shown between 22 and 78 μs, so that the converter switching frequency reduces to $\approx$ 90 kHz. The explained characteristics of the CCM are also found during DCM.

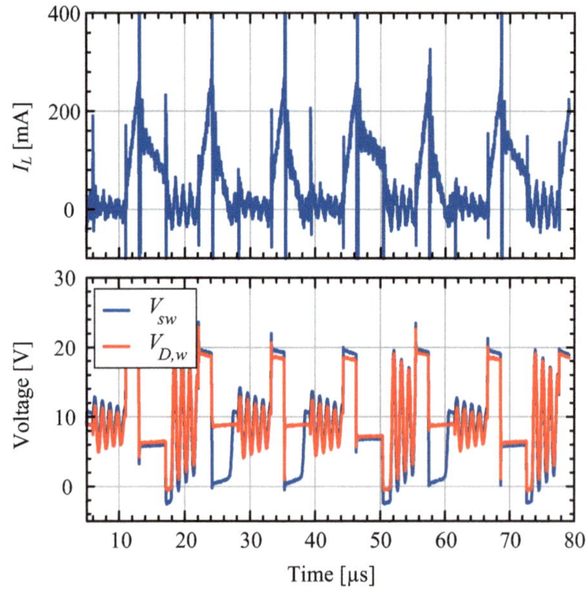

Fig. 17.11 Color sequence during discontinuous conduction mode

Two additional effects arise in the shown color sequence. When the colored LEDs are turned "off" at 50 and 72 μs, parasitic body diode conduction is identified. The explanation is found in the inductor current waveform. The integrated digital DCM control applies for a minimum current before the DMOS transistors are turned "off" to enter the idle period. For the white LED string, the triangle discharge to zero current is achieved. Due to the steeper current slope with the white LED string, the propagation time of the converter control delays the DMOS disable until zero current is reached. However, for a less steep discharge across the colored LEDs, an undefined current at the end of the inductor discharge remains. As soon as the DMOS transistors are "off", the current is dissipated across the body diodes of all parallel LED strings. A current ringing of a rather high amplitude is affected. However, the control is not affected by this effect, since the triangle calculation of the digital control is disabled during the idle period.

17.6 Conclusions

This work presents an overview of state-of-the-art LED driver topologies for general illumination. Starting from the most simple topology to operate LEDs, several control and dimming techniques are introduced. Furthermore, stand-alone AC-DC LED drivers are discussed, as well as DC-DC LED drivers. Throughout this work, a special focus is done on the reduction of external components, such as inductors and capacitors. Since these elements add to the application costs of a LED driver, a minimization has to be taken place to compete with low-cost architectures. As a promising prototype for a minimum application, a capacitor-free single-inductor multiple-output (SIMO) LED driver is presented. The multichannel LED driver accounts for three white high current, high brightness LEDs in series and one LED for red, green, and blue in parallel. The external components have been reduced to a single inductor to offer the smallest possible form factor. Further components are neglected and LED string isolation, as well as color mixing has been provided by double-diffused DMOS switches. A prototype LED driver has been fabricated in a 0.18 μm high-voltage (HV) CMOS technology. The all-digital converter implementation accounts for a single hysteretic inductor current control. With a nominal converter switching frequency of 500 kHz and an integrated 9 bit high-side current sense ADC topology, a modern combined amplitude-modulation and pulse-width modulation (AM/PWM) dimming technique is realized.

References

1. P. Waide, S. Tanishima, *Light's Labour's Lost - Policies for Energy-Efficient Lighting* (International Energy Agency, Paris, 2006)
2. McKinsey & Company Inc., *Lighting the Way: Perspectives on the Global Lighting Market*, 2nd edn. Visual Media Europe (2012). www.mckinsey.com
3. The European Parliament and the Council of the European Union, Directive 2012/874/EC of the European parliament and of the council. Off. J. L **258**, 0001–0020 (2012)
4. The European Parliament and the Council of the European Union, Directive 2009/125/EC of the European parliament and of the council. Off. J. L **285**, 0010–0035 (2009)
5. The European Parliament and the Council of the European Union, Directive 2009/244/EC of the European parliament and of the council. Off. J. L **076**, 0003–0016 (2009)
6. The European Parliament and the Council of the European Union, Directive 2005/32/EC of the European parliament and of the council. Off. J. L **191**, 0029–0058 (2005)
7. The European Parliament and the Council of the European Union, Directive 1998/11/EC of the European parliament and of the council. Off. J. L **191**, 0001–0008 (1998)
8. C. Branas, F. Azcondo, J. Alonso, Solid-state lighting: a system review. IEEE Ind. Electron. Mag. **7**(4), 6–14 (2013)
9. S. Wang, X. Ruan, K. Yao, S.-C. Tan, Y. Yang, Z. Ye, A flicker-free electrolytic capacitor-less AC-DC LED driver. IEEE Trans. Power Electron. **27**(11), 4540–4548 (2012)
10. P. Malcovati, M. Belloni, F. Gozzini, C. Bazzani, A. Baschirotto, A 0.18 μm CMOS 91 %-efficiency 0.1-to-2 A scalable buck-boost DC-DC converter for LED drivers, in *IEEE International Solid-State Circuits Conference on Digest of Technical Papers (ISSCC)* (2012), pp. 280–281
11. P. Malcovati, M. Belloni, F. Gozzini, C. Bazzani, A. Baschirotto, A 0.18 μm CMOS, 91 %-efficiency, 2 A scalable buck-boost DC-DC converter for LED drivers. IEEE Trans. Power Electron. **29**(10), 5392–5398 (2014)
12. L. Lohaus, L. Liao, S. Strache, R. Wunderlich, S. Heinen, Energy efficient current control technique for driving high power LEDs, in *Proceedings of the IEEE 8th Conference on PhD Research in Microelectronics and Electronics (PRIME)* (2012), pp. 75–78
13. L. Lohaus, A. Rossius, S. Dietrich, R. Wunderlich, S. Heinen, A dimmable LED driver with resistive DAC feedback control for adaptive voltage regulation. IEEE Trans. Ind. Appl. **51**(4), 3254–3262 (2015)

14. L. Lohaus, A. Rossius, S. Dietrich, R. Wunderlich, S. Heinen, A dimmable LED driver using resistive DAC feedback control for adaptive voltage regulation, in *Proceedings of the IEEE 5th Annual Energy Conversion Congress and Exposition (ECCE), Denver, USA* (2013), pp. 3126–3133
15. K.H. Loo, W.-K. Lun, S.-C. Tan, Y. Lai, C. Tse, On the driving techniques for high-brightness LEDs, in *Proceeding of the IEEE 1st Annual Energy Conversion Congress and Exposition (ECCE)* (2009), pp. 2059–2064
16. W.-K. Lun, K.H. Loo, S.-C. Tan, Y. Lai, C. Tse, Bilevel current driving technique for LEDs. IEEE Trans. Power Electron. **24**(12), 2920–2932 (2009)
17. Y. Gu, N. Narendran, T. Dong, H. Wu, Spectral and luminous efficacy change of high-power LEDs under different dimming methods, in *Proceeding of the 6th International Conference on Solid State Lighting (SPIE)* (2006)
18. Cypress Perform, AN47372 - PrISM^{TM} technology for LED dimming. Application Note (2011)
19. A. Wilkins, J. Veitch, B. Lehman, LED lighting flicker and potential health concerns: IEEE standard PAR1789 update, in *Proceedings of the IEEE 2nd Annual Energy Conversion Congress and Exposition (ECCE)* (2010), pp. 171–178
20. R.W. Erickson D. Maksimovic, *Fundamentals of Power Electronics*, 2nd ed. (Kluwer Academic Publishers Group, Dordrecht, 2001)
21. S. Beczkowski, S. Munk-Nielsen, LED spectral and power characteristics under hybrid PWM/AM dimming strategy, in *Proceedings of the IEEE 2nd Annual Energy Conversion Congress and Exposition (ECCE)* (2010), pp. 731–735
22. V. Anghel, C. Bartholomeusz, A. Vasilica, G. Pristavu, G. Brezeanu, Variable off-time control loop for current-mode floating buck converters in LED driving applications. IEEE J. Solid-State Circuits **49**(7), 1571–1579 (2014)
23. D. Park, Z. Liu, H. Lee, A 40 V 10 W 93 %-efficiency current-accuracy-enhanced dimmable LED driver with adaptive timing difference compensation for solid-state lighting applications. IEEE J. Solid-State Circuits **49**(8), 1848–1860 (2014)
24. Y. Zhang, H. Chen, D. Ma, A 198 ns/V VO-hopping reconfigurable RGB LED driver with automatic ΔVO detection and quasi-constant-frequency predictive peak current control, in *Proceedings of the IEEE Symposium on VLSI Circuits (VLSIC)* (2012), pp. 184–185
25. R.-L. Lin, C.-R. Pan, K.-H. Liu, Family of single-inductor multi-output DC-DC converters, in *Proceedings of the IEEE 8th International Conference on Power Electronics and Drive Systems (PEDS)* (2009), pp. 1216–1221
26. H. Chen, Y. Zhang, D. Ma, A SIMO parallel-string driver IC for dimmable LED backlighting with local bus voltage optimization and single time-shared regulation loop. IEEE Trans. Power Electron. **27**(1), 452–462 (2012)
27. S.-I. Hong, J.-W. Han, D.-H. Kim, O.-K. Kwon, A double-loop control LED backlight driver IC for medium-sized LCDs, in *IEEE International Solid-State Circuits Conference - Digest of Technical Papers (ISSCC)* (2010), pp. 116–117
28. C.-Y. Hsieh, K.-H. Chen, Boost DC-DC converter with fast reference tracking (FRT) and charge-recycling (CR) techniques for high-efficiency and low-cost LED driver. IEEE J. Solid-State Circuits **44**(9), 2568–2580 (2009)
29. N. Chen, H.-H. Chung, A driving technology for retrofit LED lamp for fluorescent lighting fixtures with electronic ballasts. IEEE Trans. Power Electron. **26**(2), 588–601 (2011)
30. X. Ruan, B. Wang, K. Yao, S. Wang, Optimum injected current harmonics to minimize peak-to-average ratio of LED current for electrolytic capacitor-less AC-DC drivers. IEEE Trans. Power Electron. **26**(7), 1820–1825 (2011)
31. J.T. Hwang, K. Cho, D. Kim, M. Jung, G. Cho, S. Yang, A simple LED lamp driver IC with intelligent power-factor correction, in *IEEE International Solid State Circuits Conference - Digest of Technical Papers (ISSCC)* (2011), pp. 236–237
32. M. Ryu, J. Kim, J. Baek, H.-G. Kim, New multi-channel LEDs driving methods using current transformer in electrolytic capacitor-less AC-DC drivers, in *Proc. IEEE 27th Annual Applied Power Electronics Conference and Exposition (APEC)* (2012), pp. 2361–2367

33. J. Garcia, A. Calleja, E. Corominas, D. Gacio, J. Ribas, Electronic driver without electrolytic capacitor for dimming high brightness LEDs, in *Proceedings of the IEEE 35th Annual Industrial Electronics Conference (IECON)* (2009), pp. 3518–3523
34. L. Gu, X. Ruan, M. Xu, K. Yao, Means of eliminating electrolytic capacitor in AC-DC power supplies for LED lightings. IEEE Trans. Power Electron. **24**(5), 1399–1408 (2009)
35. S. Dietrich, S. Strache, B. Mohr, J. Mueller, L. Rolff, R. Wunderlich, S. Heinen, A 7.5 W-output power 96 %-efficiency capacitor-free single-inductor 4-channel all-digital integrated DC-DC LED driver in a 0.18 μm technology, in *IEEE International Solid State Circuits Conference - Digest of Technical Papers (ISSCC), San Francisco, USA* (2015), pp. 220–221
36. S. Dietrich, S. Strache, R. Wunderlich, S. Heinen, Get the LED out - experimental validation of a capacitor-free single-inductor multiple-output LED driver topology. IEEE Ind. Electron. Mag. **9**(2), 24–35 (2015)
37. S. Dietrich, K. Lu, S. Strache, R. Wunderlich, S. Heinen, All-digital current control for efficiency maximization in novel LED drivers, in *Proceedings of the IEEE International Conference on Renewable Energy Research and Applications (ICRERA), Milwaukee, USA* (2014)

Chapter 18
An Ultra-Low-Power Electrostatic Energy Harvester Interface

Stefano Stanzione, Chris van Liempd, and Chris van Hoof

18.1 Introduction

Solid-state electronics develop enormously every year, thanks not only to circuit design innovation, but also to the development of better technologies. All this allows to constantly reduce power consumption of electronic systems, while keeping the other performance metrics almost unchanged. In this scenario, energy harvesting is becoming more and more interesting as it allows to reduce the cost of an application by avoiding a periodic replacement of the batteries or by reducing their size, or even allowing the implementation of battery-less systems. There are many environmental energy sources that can be exploited for these purposes (e.g. light, temperature gradient, mechanical vibrations, electromagnetic). The choice of the best source of energy clearly depends on the application. Usually the most abundant form of energy is chosen, even if there are exceptions due to system size and shape.

The focus of this chapter is in the field of industrial machines and vehicle monitoring, in which the sensing electronic systems are positioned usually in inaccessible places (the moving parts), where the cabling can be expensive or almost impossible. The most abundant form of energy in the moving parts of machines is clearly mechanical and so vibrational harvesters are chosen in these systems. There are many types of vibrational harvesters: piezoelectric, electrostatic, magnetic. This chapter will focus on the interfacing of electrostatic harvesters. Basically, these devices are charged capacitors with one plate fixed (or just connected to a bigger

S. Stanzione (✉) • C. van Liempd
Holst Centre/imec, Eindhoven, The Netherlands
e-mail: stefano.stanzione@imec-nl.nl

C. van Hoof
Holst Centre/imec, Eindhoven, The Netherlands

KU Leuven, Leuven, Belgium

© Springer International Publishing Switzerland 2017

A. Baschirotto et al. (eds.), *Wideband Continuous-time ΣΔ ADCs, Automotive Electronics, and Power Management*, DOI 10.1007/978-3-319-41670-0_18

inertial mass) and the other moving because of the environmental accelerations. In [1], a circuit capable to both bias the variable capacitor and extract power from its movement is proposed. A drawback of this design is that part of the power extracted is used for the polarization of the capacitor. Additionally, this design strategy limits the capacitor polarization voltages to the maximum voltage ratings of the chosen IC manufacturing technology, which most of the times is lower than the polarization voltage for the harvester itself. A solution for scavenging more efficiently energy from the environmental vibrations consists in building a partially isolated structure, in which the variable capacitor is pre-charged with a certain potential and then isolated from the outside world. In this case, the charged electrode is called electret and the harvesters fabricated in this way are denoted as electret-based electrostatic energy harvesters. Recently, this type of harvesters attracted quite some attention, thanks to their capability to generate a large amount of power, even at low accelerations [2]. Unfortunately, these harvesters have extremely large internal impedances, making challenging the design of interfacing circuits for efficient power extraction. In particular, those circuits need to be at the same time ultra-low-power and resistant to several tens of Volts applied on their inputs. In [3] an interfacing circuit working with an input voltage range between 5 and 60 V has been proposed. The main disadvantage of that circuit was mainly the power consumption, as it was not working under 25 μW of available power.

18.2 Open-Loop Inductive Buck Converter at Low Available Powers

Considering the large voltage ranges required by the applications, the best choice is an inductive converter. As shown in Fig. 18.1a, the basic structure of an inductive DC-DC buck converter is consisting of a switch connected to the input and a diode (implemented with passives or, more often, with active circuits). Given the low available power in energy harvesting applications, the converter should be used in Discontinuous Conduction Mode (DCM), meaning that during the period T there will be some time in which the inductor current will be constant and equal to zero.

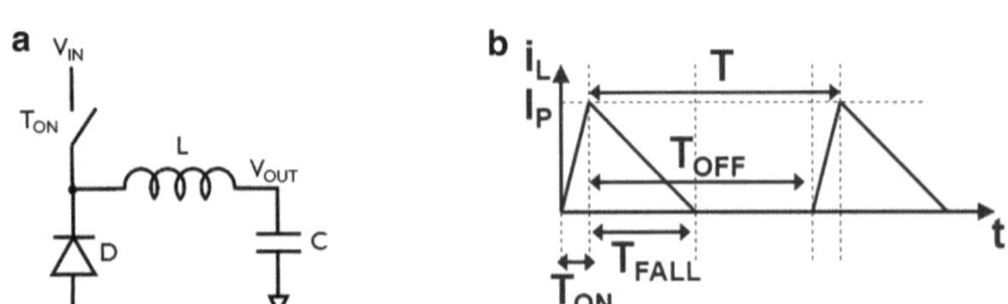

Fig. 18.1 Basic implementation of a DC-DC buck converter (**a**) and inductor current in discontinuous conduction mode (**b**)

This is particularly efficient at low power levels because many control circuits can be turned off when there is no conduction of current to the load. A typical waveform in DCM is shown in Fig. 18.1b. Assuming that the input voltage of the converter is almost constant (the input capacitor has to be large enough), we can easily calculate the input resistance of the converter if operating in open-loop (as charger and not as voltage regulator) by calculating the ratio of the input voltage over the average inductor current flowing during the T_{ON} time (when the switch is ON).

In particular, we get the following expression:

$$R_{IN} = \frac{2LTV_{IN}}{T_{ON}^2 \left(V_{IN} - V_{OUT}\right)} \tag{18.1}$$

Note that, for energy harvesting purposes, the input resistance needs to match the source resistance of the harvester R_S. This means that a Maximum Power Point Tracking (MPPT) algorithm needs to be implemented in order to achieve power matching. So, in a Pulse Frequency Modulation (PFM) scheme, the period T would have to be modified according to the following law:

$$T = \frac{R_S T_{ON}^2 \left(V_{IN} - V_{OUT}\right)}{2LV_{IN}} \tag{18.2}$$

As it can be easily recognized, if the input voltage gets close to the output voltage, the period T needs to be decreased by the MPPT algorithm in order to efficiently extract power from the harvester. Assuming that most of the control circuit consumption happens during the conduction time and considering that the conduction time is $T_{ON}*V_{IN}/V_{OUT}$, the total power consumption is then given by:

$$P_{CTRL} = P_Q + k_1 \frac{T_{ON}}{T} \frac{V_{IN}}{V_{OUT}} + \frac{k_2}{T} \tag{18.3}$$

where P_Q is the quiescent power, k_1 is a constant equal to the power consumed by the blocks turned ON during the conduction phase and k_2 a constant equal to the energy dynamic losses due to the switching operations. By substituting (18.2) in (18.3), we obtain:

$$P_{CTRL} = P_Q + \left(\frac{k_1 T_{ON} V_{IN}}{V_{OUT}} + k_2\right) \frac{2V_{IN}L}{\left(V_{IN} - V_{OUT}\right) T_{ON}^2 R_S} \tag{18.4}$$

Calculating the limit of (18.4) for V_{IN} tending to V_{OUT}, we get that P_{CTRL} tends to infinite. So, when the available power from the harvester decreases, the control power increases, reducing dramatically the converter efficiency. A similar reasoning can be applied to PWM schemes. In this case, the control power results:

$$P_{CTRL} = P_Q + \frac{k_1 V_{IN}}{V_{OUT} T} \sqrt{\frac{2V_{IN} TL}{R_S \left(V_{IN} - V_{OUT}\right)}} + \frac{k_2}{T} \tag{18.5}$$

It can be appreciated that also in PWM schemes the control power tends to go to infinity when the available power tends to zero. Equations (18.4) and (18.5) represent a problem that needs to be solved in order to lower significantly the power consumption of electrostatic harvester interfaces.

18.3 Hybrid Modulation Scheme

Both PFM and PWM modulation schemes have the characteristic of varying only one parameter: T_{ON} or T. However, from equation (18.1) we can see that the period T is proportional to ${T_{ON}}^2$. This means that, while maintaining resistance matching, the period T could be increased of α^2 if T_{ON} is increased of a factor α. Of course, T_{ON} is not a parameter that we can freely choose, because it defines the peak current in the inductor I_P. Since, for a given power train, an optimal peak current for maximizing energy efficiency exists, T_{ON} cannot be increased ad libitum in order to reduce the operating frequency. A good compromise is easily found and consists in fixing the peak current in the inductor to a value close to its optimal value. This means that T_{ON} is chosen at each period in order to keep the peak current constant and then the period T can be calculated:

$$T = \frac{R_S I_P^2 L}{2V_{IN}\left(V_{IN} - V_{OUT}\right)} \tag{18.6}$$

In this case, the period tends to infinity for input voltages close to the output voltage. To verify that this approach solves the problem, we can calculate the power dissipation of the control circuits:

$$P_{CTRL} = P_Q + \frac{2V_{IN}\left(V_{IN} - V_{OUT}\right)}{I_P^2 L R_S}\left[\frac{k_1 V_{IN} L I_P}{V_{OUT}\left(V_{IN} - V_{OUT}\right)} + k_2\right] \tag{18.7}$$

The limit of P_{CTRL} for V_{IN} tending to V_{OUT} is then equal to:

$$\lim_{V_{IN} \to V_{OUT}} P_{CTRL} = P_Q + \frac{2V_{IN}^2 k_1}{I_P R_S V_{OUT}} \tag{18.8}$$

Notice that now the control power tends to P_Q when the available power (${V_{IN}}^2/4R_S$) tends to zero.

18.3.1 *Proposed Implementation*

The design architecture needs then to generate the proper period T at each cycle, following equation (18.6). Given the relative complexity of equation (18.6), it may seem hard to do it in a simple way. In reality, the problem can be solved pretty easily by using the system shown in Fig. 18.2a.

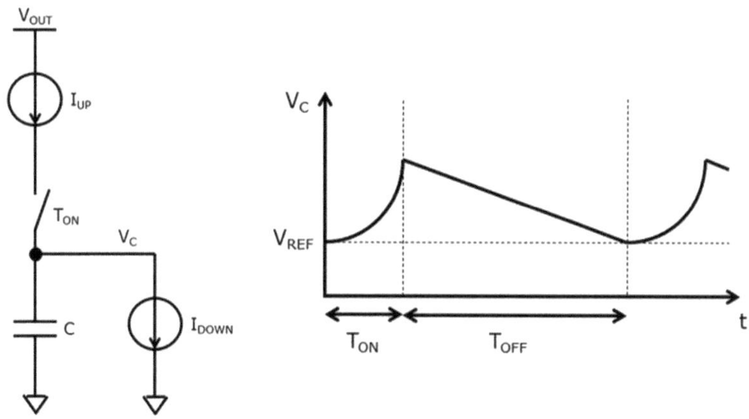

Fig. 18.2 Proposed implementation of the timing control circuit (**a**) and its waveform (**b**)

Assuming now that the conversion period starts when the capacitor voltage V_C becomes lower than a certain reference V_{REF}, we get:

$$T = \frac{\int_0^{T_{ON}} I_{UP}(t)dt}{I_{DOWN}} \tag{18.9}$$

Imposing that (18.9) needs to be equal to (18.2), we get that I_{UP} needs to be a scaled copy of the inductor current and I_{DOWN} a current proportional to the input voltage:

$$\begin{cases} I_{UP} = \frac{i_L}{N} = \frac{(V_{IN}-V_{OUT})}{NL}t \\ I_{DOWN} = k_{MPPT}V_{IN} \end{cases} \tag{18.10}$$

where N is the scaling factor of the inductor current copy and k_{MPPT} is a factor used by the MPPT algorithm in order to change the input resistance of the converter. In particular the input resistance will result equal to $1/(N*k_{MPPT})$. Finally, note that, for fixed values of input and output voltages, I_{UP} increases linearly in time during the interval of time T_{ON} and I_{DOWN} is constant. Therefore, the capacitor voltage V_C will behave like depicted in Fig. 18.2b.

18.3.2 Circuit Implementation

The proposed system architecture is shown in Fig. 18.3, with the integrated blocks highlighted in grey and the die photo on top. Excluding harvester, rectifier and load, there are 4 external components: 2 decoupling capacitors ($C_{IN} = 100$ nF and C_{OUT} larger than 10 μF), 1 inductor (L = 10 mH) and 2 resistors ($R_{DC} = 10$ GΩ and $R_{BIAS} = 100$ MΩ).

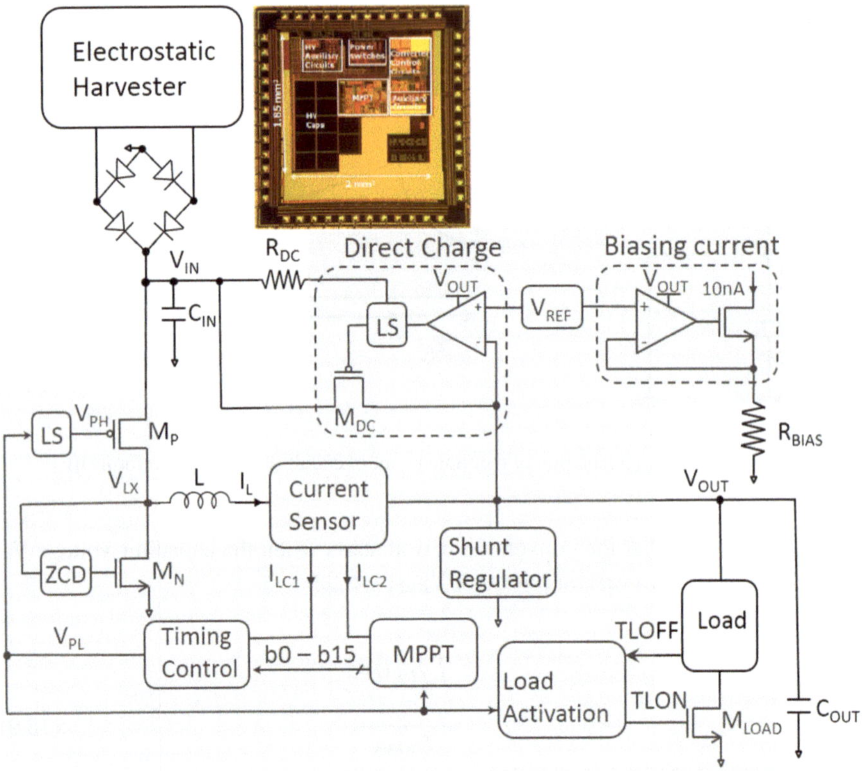

Fig. 18.3 Proposed system architecture and die photo

The resistors R_{BIAS} and R_{DC} are used respectively for defining a bias current of about 10 nA and for providing a basic bias current in start-up conditions. In particular, this start-up bias current is used by a level shifter which shorts the input to the output. All the control mechanisms are based on inductor current sensing. In particular, the sensed copy of the inductor current is used for the timing control, the peak current control and also for the MPPT algorithm, which has to calculate the output power variation. Additionally, in order to allow battery-less operation, the system includes also a shunt regulator and a smart load activation circuit, which connects the load to the output and allows the load to request to be disconnected after it has performed all its operations.

Figure 18.4 shows the circuit implementation of the timing control block. The copy of the inductor current I_{LC1} is mirrored into transistor M_{P2} and integrated by the capacitor C. Simultaneously, the inductor current is also mirrored by M_{P7}, with the aim to perform the peak current control and define the ON-time T_{ON} with the voltage V_{PLN}. As indicated in the previous sub-section, the down current has to be proportional to the input voltage V_{IN}. In order to avoid additional external components, the input voltage is sensed in an indirect way: the inductor current

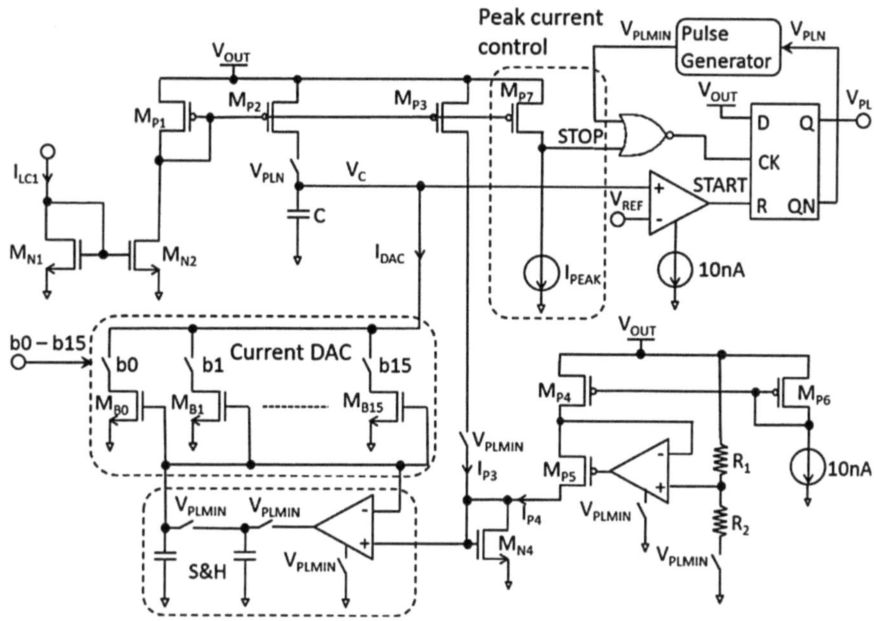

Fig. 18.4 Proposed timing control block

(proportional to V_{IN}-V_{OUT}) is mirrored by M_{P3} and summed to another current I_{P4} proportional just to V_{OUT}. The resulting diode voltage of M_{N4} is then sampled and used to bias a current DAC, which receives as input the MPPT settings b0-b15.

18.3.3 Measurements

The chip has been fabricated in 0.25 μm BCD technology, and in particular using a flavor with 60 V-tolerant MOSFETs. The characterization has been carried out for various values of source resistance, between 500 kΩ and 25 MΩ. The MPPT efficiency (input power over available power) and the end-to-end efficiency (output power over available power) are shown respectively in Figs. 18.5 and 18.6.

Focusing on the left side of the previous figures (around 1 μW available power), the MPPT efficiency is less than 50 %, while the end-to-end efficiency is larger than 10 %. This means that the total power dissipation in this condition is lower than 500 nW, demonstrating the impact of the adopted hybrid modulation scheme. At higher power levels, the MPPT efficiency and end-to-end efficiency reach up to 99 % and 85 %, respectively. As shown in Table 18.1, this work is at state-of-art level even if compared with uncomplete designs (e.g. without MPPT, start-up circuits or load regulation) or not capable to interface high input voltages. In fact, [4] and [5] have no cold-start, only [6] implements an MPPT algorithm but dissipates much more power.

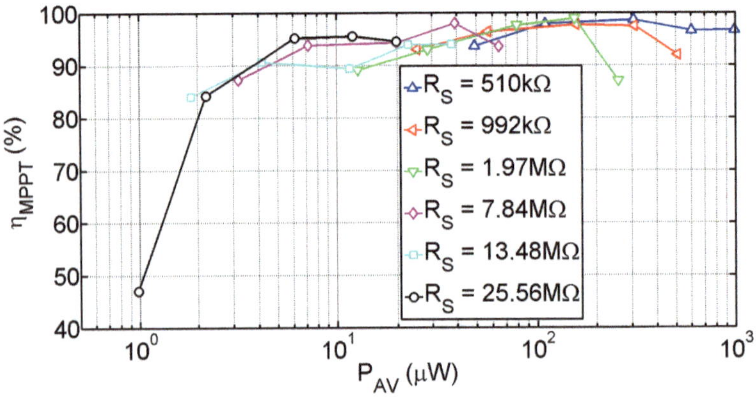

Fig. 18.5 MPPT efficiency as function of available power

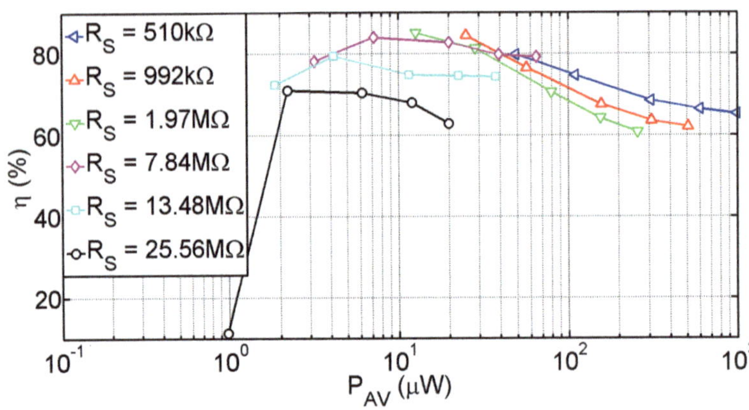

Fig. 18.6 End-to-end efficiency as function of available power

18.4 Conclusions

We have presented an electrostatic harvester interface, based on an ultra-low-power and high-voltage inductive DC-DC converter. Thanks to the proposed modulation scheme, the quiescent current of the converter is lower than 500 nW. The minimum operational available power is 1 μW and the maximum input voltage is 60 V. The chip is fully autonomous and includes start-up, shunt regulation and maximum power point tracking. The converter is capable of tracking source resistance variations between 0.5 MΩ and 25 MΩ.

Table 18.1 Comparison with state-of-art designs

	Hsieh [4]	Huang [5]	Shim [6]	Chen [7]	Danzhu [8]	This work [9]
Process	0.25 µm CMOS	0.25 µm CMOS	0.25 µm BCD	0.18 µm CMOS	0.35 µm	0.25 µm BCD
Max. input voltage (V)	2.5	2.5	7	1	6	60
Output voltage (V)	1	1	1–8	0.35–0.5	2.5	0–5
Input power	1 µW–100 mW	1 µW–50 mW	33 µW–10 mW	50 nW–10 mW	2.5 µW–250 mW	1 µW–1 mW
Function	Output voltage regulation	harvester interface	harvester interface + MPPT	harvester interface	Battery or harvester interface	harvester interface + MPPT
Harvester type	RF	RF	Piezoelectric	Photovoltaic	RF or Vibrational harvester	Electret-based EEH
Architecture	Buck	Buck	Buck-Boost	Buck	Buck	Buck
Control power (nW)	225	217	N.A.	N.A.	100	500
Maximum MPPT efficiency (%)	N.A.	N.A.	99	N.A.	N.A.	99
Maximum end-to-end efficiency (%)	N.A.	N.A.	80	N.A.	N.A.	85
Area (mm2)	0.21	0.39	5.5	1.44	2.88	3

References

1. E.O. Torres, G.A. Rincón-Mora, A 0.7-μm BiCMOS electrostatic energy-harvesting system IC. IEEE J. Solid State Circuits **45**(2), 483–496 (2010)
2. T. Masaki, K. Sakurai, T. Yokoyama, M. Ikuta, M. Doi, T. Seki, M. Oba, Power output enhancement of vibration-driven electret generator using concave electrodes, in *Power MEMES*, pp. 57–60, Dec 2010
3. S. Stanzione et al., A self-biased 5-to-60V input voltage and 25-to-1600μ integrated DC-DC buck converter with fully analog MPPT algorithm reaching up to 88% end-to-end efficiency, in *IEEE ISSCC*, pp. 74–75, Feb 2013
4. C.Y. Hsieh, Y.H. Lee, Y.Y. Yang, T.C. Huang, A battery-free 225 nW buck converter for wireless RF energy harvesting with dynamic on/off time and adaptive phase lead control, in *Symposium on VLSI Circuits*, pp. 242–243. June 2011
5. T.C. Huang, C.Y. Hsieh, Y.T. Yang, T.C. Kang, K.H. Chen, C.C. Huang, Y.H. Lin, M.W. Lee, A battery-free 217 nW static control power buck converter for wireless RF energy harvesting with α-calibrated dynamic on/off time and adaptive phase lead control. IEEE J. Solid State Circuits **47**(4), 852–862 (2012)
6. M. Shim et al., Self-powered 30 μW-to-10 mW piezoelectric energy harvesting system with 9.09 ms/V maximum power point tracking time, in *IEEE ISSCC*, pp. 406-407, Feb 2014
7. P.-H. Chen et al., A 50nW-to-10mW output power tri-mode digital buck converter with self-tracking zero current detection for photovoltaic energy harvesting, in *Prep-Publication IEEE JSSC*, Jan 2016
8. L. Danzhu et al., A 110nA quiescent current buck converter with zero-power supply monitor and near-constant output ripple, in *IEEE CICC*, Sept 2015
9. S. Stanzione et al. A 500nW batteryless integrated electrostatic energy harvester interface based on a DC-DC converter with 60V maximum input voltage and operating from 1μW available power, including MPPT and cold start, in *IEEE ISSCC*, Feb 2015